Information Systems Engineering and Management

Volume 74

The book series "Information Systems Engineering and Management" (ISEM) publishes innovative and original works in the various areas of planning, development, implementation, and management of information systems and technologies by enterprises, citizens, and society for the improvement of the socio-economic environment.

The series is multidisciplinary, focusing on technological, organizational, and social domains of information systems engineering and management. Manuscripts published in this book series focus on relevant problems and research in the planning, analysis, design, implementation, exploration, and management of all types of information systems and technologies. The series contains monographs, lecture notes, edited volumes, pedagogical and technical books as well as proceedings volumes.

Some topics/keywords to be considered in the ISEM book series are, but not limited to: Information Systems Planning; Information Systems Development; Exploration of Information Systems; Management of Information Systems; Blockchain Technology; Cloud Computing; Artificial Intelligence (AI) and Machine Learning; Big Data Analytics; Multimedia Systems; Computer Networks, Mobility and Pervasive Systems; IT Security, Ethics and Privacy; Cybersecurity; Digital Platforms and Services; Requirements Engineering; Software Engineering; Process and Knowledge Engineering; Security and Privacy Engineering, Autonomous Robotics; Human-Computer Interaction; Marketing and Information; Tourism and Information; Finance and Value; Decisions and Risk; Innovation and Projects; Strategy and People.

Indexed by Google Scholar. All books published in the series are submitted for consideration in the Web of Science.

For book or proceedings proposals please contact Alvaro Rocha (amrrocha@gmail.com).

Sandeep Kautish · Rahul Kar · António Abreu
Editors

Green Metaverse: Fuzzy Logic Approaches to Sustainable Virtual Worlds

Springer

Editors
Sandeep Kautish
PhysicsWallah Limited
Noida, Uttar Pradesh, India

António Abreu
Higher Institute of Accounting
and Administration of Porto
Polytechnic University of Porto
São Mamede de Infesta, Portugal

Rahul Kar
Department of Mathematics
Kalyani Mahavidyalaya
Nadia, West Bengal, India

ISSN 3004-958X ISSN 3004-9598 (electronic)
Information Systems Engineering and Management
ISBN 978-3-032-08182-7 ISBN 978-3-032-08183-4 (eBook)
https://doi.org/10.1007/978-3-032-08183-4

This Springer imprint is published by the registered company Springer Nature Switzerland AG
The registered company address is: Gewerbestrasse 11, 6330 Cham, Switzerland

Contents

Sustainable Resource Management in Virtual Worlds

Revati Ramrao Rautrao and T. S. Sanjayan

Abstract This study explores the transformative potential of digital worlds as dynamic ecosystems capable of using real-international social, economic, and environmental development. The goal is to deal with together with the powerful integration of sustainable practices, resource optimization, and governance systems in virtual ecosystems. Current digital environments often lack cohesive frameworks to address challenges, which include useful resource shortage and environmental degradation. To bridge those gaps, this study introduces progressive models like Gamified Sustainability Practices and Decentralized Governance Systems, offering a blueprint for better useful resource control and eco-friendly virtual operations. The findings reveal that aligning virtual ecosystems with actual global behaviors and industrial standards can yield impactful solutions for urgent international issues. Additionally, the observation emphasizes the function of "Virtuality," an idea that specializes in sustainable communication and collaboration inside virtual environments, enabling these ecosystems to feature as experimental systems for innovative strategies. By addressing these gaps, the research concludes that digital ecosystems preserve substantial potential to act as catalysts for sustainable development, fostering worldwide innovation and resilience via their adaptive, scalable, and eco-aware designs.

Keywords Digital ecosystems · Virtual sustainability · Resource optimization · Gamified eco-strategies · Green virtual infrastructures · Decentralized governance · Ecological digitality · Circular resource management · Energy-efficient virtual worlds · And ethical resource stewardship

R. R. Rautrao (✉)
Dr. D Y Patil B-School Pune, Pune, India
e-mail: revatirautrao@gmail.com

T. S. Sanjayan
Postgraduate and Research Department in Education, GVM's Dr Dada Vaidya College of Education, Goa University, Goa, India

S. Kautish et al. (eds.), *Green Metaverse: Fuzzy Logic Approaches to Sustainable Virtual Worlds*, Information Systems Engineering and Management 74,
https://doi.org/10.1007/978-3-032-08183-4_1

1 Introduction

In the cutting-edge virtual revolution, digital worlds have superior dynamic spaces for immersive amusement, interplay, and creative commerce. Across the globe, the ones connected in virtual areas—starting from huge gaming universes to digital retail regions—are changing the way people talk and collaborate, however, those systems need to keep up with rapid development. Several elements of sustainability in virtual environments have been explored in this study, focusing on environmentally quality gaming, useful resource-inexperienced techniques, virtual infrastructure sustainability, and the responsible use of assets. This requires clients and producers to be concerned about the design and control of these ecosystems. Sustainability concepts encompass records together with records performance, power call for, and managing low degrees of pollution that seem to reduce influences. These assessment pastimes to select options and excellent practices and recommend a technique for balancing technological innovation with environmental management in the virtual age. The discussion highlights how effective worlds not only pre-empt real-global green practices but also inspire a robust commitment to the surroundings and emphasize the advertising and marketing of statistics [1].

Virtual worlds are evolving into dynamic virtual ecosystems that effectively combine state-of-the-art generation with human creativity. These immersive worlds mimic real international practices and provide interactive forums for socializing, financial interest, and innovative inquiry [2]. From massive metameres empires to online multiplayer video games, virtual worlds have gained worldwide traction and feature attracted heaps and lots of individuals concerned with sports activities alongside education, leisure, change, and network operationally [3] of these habitat extensions hold their innovation whilst retaining the ecological duty of stability and effect on environmentally Sustainable control is becoming increasingly more crucial to manipulate assets sustainably [4].

While digital worlds offer many new capabilities, their worrying nature has created sustainability problems [5]. These digital ecosystems rely on the huge-scale properties that encompass immoderate-velocity communications, cloud computing systems, and information centers, all requiring large processing energy and assets. Constant virtual worlds fuelled by using innovation and technological advances will examine the surrounding's environmental duty through growing beneficial useful resource intake [6]. It is even more crucial to deal with the long-time period ecological effects of those environmental factors position. Sustainability is more important than ever to maintain sustainable worlds, reduce the environmental impact, foster ongoing innovation, and step forward in consumer interaction. This conceptual study examines pressing issues of aiding overall performance in virtual ecosystems.

It seems to aid in controlling demanding situations, perceiving inefficiencies in gift structures, and exploring methods to enhance performance. The studies consciousness on the combination of inexperienced generation and electricity performance requirements, promoting decentralized governance systems and circular flows to create environments that guarantee sustainability [7]. This examination examines the relationship between behavioral economics, virtual ethics, and ecology to advocate opportunity intake suggestions within the digital world and seeks to redefine ecology beneath real lifestyles as a vital pressure behind sustainable innovation by incorporating statistics from the unique sources [8]. The challenge is to expand practical policies to improve the environmental resilience of the digital environment while addressing worldwide sustainability problems dealing with the long term. The study suggests through those insights the transformative functionality of the real international, not just as websites of social technological innovation, but moreover as fashions for encouraging sustainable practices. The goal is to create a virtual future that creates stability among development and environmental management with the help of long-term goals [9].

Virtual worlds have emerged as important components of the virtual surroundings, imparting outstanding regions for modern expression, interplay, and exchange. Even with their developing impact, sustaining those ecosystems is fraught with challenges. Exploring the complexity of useful resource intake in digital ecosystems, this appearance highlights the need for first-rate techniques that restrict environmental responsibility and performance [10]. By locating methods to lessen useful resource consumption and enhance system performance, the goal is to create a sustainable future while addressing broader environmental influences and promoting sustainable development [11].

A. The Emergence of Virtual Lifestyles

Virtual ecosystems now provide spaces for socializing, training, and play that mirror real-global achievement. As the virtual worlds become increasingly enormous and complicated, they become crucial to the knowledge of how generation pertains to innovation and sustainability. The growing reputation of the virtual international has created a want to discover how manufacturers can balance environmental duty with the purchaser participation in those environments. Digital ecosystems redecorate conversation and leisure, but in addition, they provide crucial insights into how the use of cutting-edge technology to sell sustainability can function as a reference model for coping with property sustainably constant with ecological goals and digital improvement.

B. The Significance of Digital Environments

The virtual international let in hundreds and lots of people to attach and work with business company fashions that flow across country borders. They have vibrant monetary and social guidelines that reflect actual-world disturbing situations and pose unique resource concerns. The virtual world has grown to be a colorful digital space that combines innovation with current technology. Its effect is multifaceted and gives exceptional possibilities for exploring the usage of substances in hitherto unheard-of methods. Virtual worlds are evolving into state-of-the-art virtual ecosystems that integrate creativity and era to exchange relationships, finance, and experimentation. These virtual environments facilitate and offer a wide variety of formerly unheard-of opportunities for aid analytics. A Hub for Social Interaction and Collaboration Control.

Global connectivity is facilitated by the use of digital worlds, which enable interplay and collaboration among clients from diverse professional and cultural backgrounds. These spaces, which range from digital workplaces to online gaming, are changing how humans engage with each other. Virtual worlds act as a dynamic worldwide community, allowing smooth conversation and collaboration amongst clients from amazing backgrounds and professions. This gadget is going past actual-worldwide obstacles, developing areas for the whole thing from virtual workspaces to multiplayer gaming. Virtual awakenings are redefining participation through a cutting-edge platform, which not only reinforces social connections but moreover highlights the want for constant content material fabric sharing to satisfy their developing technological goals and roles.

Economic Ecosystems

Virtual worlds have evolved into sturdy financial ecosystems, web-hosting digital marketplaces, tokenized assets, and elaborate in-activity economies. These digital systems are reshaping traditional economic paradigms by allowing progressive exchange, possession, and rate change mechanisms. However, their rapid increase intensifies the decision for efficient useful resource management tailor-made to the ideal dynamics of digital environments. Addressing those stressful conditions is important to ensuring the sustainability and scalability of these evolving financial systems [12].

Catalysts for Innovation

Virtual worlds function as experimental systems, fostering actual-time technological and social innovation. Researchers, builders, and users leverage those environments to prototype and refine novel thoughts, ranging from advanced algorithms to collaborative structures [13]. Their functionality to simulate complex conditions makes them priceless for addressing essential demanding situations like resource optimization and sustainability. These modern-day capacities position virtual worlds as pivotal regions for growing and attempting out strategies that can impact every digital and real-world atmosphere.

Expanding User Base

Virtual worlds have transitioned from hobby spaces to mainstream phenomena, attracting masses of energetic customers globally. As their user base continues to grow, the decision for sustainable useful resource control will become more and more pressing [14]. Using environmental officer practices is necessary to ensure long-term viability and reduce the organic footprint of the virtual environment.

C. **Resource Demands in Virtual Ecosystems**

The operation of the digital world requires extraordinary digital infrastructure, including better servers, post functions and cloud computing systems. These technologies are widespread and contribute to enormous carbon emissions, which increases problems with environmental stability [1]. So that the decision is growing for digital regions, it becomes important to address the environmental impact of such infrastructure, powerful performance requires modern general performance techniques and ecological footprints have reduced to ensure stability of digital ecosystems over a long period of time.

D. **Challenges in Resource Management**

The distribution of resources in the digital ecosystem regularly presents real global disabilities, reduces the challenges of stability. Questions where unexpected non-public behavior, dependence on centralized structures and stable development of virtual material materials accelerate control problems. These elements emphasize the need for adaptive, green techniques that can address specific goals for the virtual environment, even reduce waste and optimize the useful resources used. Permanent resource management is required.

Resource distribution in digital ecosystems often shows real disabilities worldwide, reduces the challenges of stability. Problems that include unexpected personal behavior, excessive dependence on centralized systems and stable growth of digital materials increase the need for control problems. These elements emphasize the need for adaptable, effective techniques that can cope with unique requirements for virtual environment, minimize waste and adapt the help to be used. Permanent resource management is required.

Stability in the virtual world is out of general performance, personal behavior, involves elements along with favorable resource distribution models, and maintains organic balance in the internal digital environment. It is important to address several demanding conditions to adapt digital ecological systems, making them more flexible and favourable [15]. By developing sustainable practices in virtual worlds, those environments can serve as valuable prototypes for actual global sustainability, imparting contemporary solutions to global environmental challenges.

E. **Opportunities for Sustainable Innovation**

Virtual worlds present a fantastic platform for experimenting with sustainable practices that would have a long undertaking of actual-global applications. Integrating inexperienced technologies, encouraging gamified green behaviors, and organizing decentralized governance structures must rework these digital areas into models for global sustainability. These innovations provide precious possibilities to test resource-green strategies and interaction with customers in massive environmental actions. The following sections find out the theoretical frameworks at the back of sustainable resource control, have a look at applicable case studies, and recommend actionable solutions that integrate technology, governance, and purchaser involvement.

F. **Digital Ecosystems and Virtual Sustainability**

In the ever-increasing realm of digital ecosystems, digital sustainability is emerging as a critical focal point for developers, agencies, and customers. These ecosystems, encompassing immersive gaming environments, online marketplaces, and collaborative virtual areas, represent a fusion of technology and human interplay. However, their fast increase has raised concerns about useful resource efficiency, energy intake, and moral digital practices. Virtual sustainability emphasizes the combination of eco-conscious strategies, along with strength-inexperienced server manipulation, sustainable virtual infrastructures, and ethical digital behavior. By adopting modern answers like gamified environmental recognition and streamlined facts garage, the virtual structures can reduce their ecological footprint. The pursuit of digital sustainability does not best guarantee the sturdiness of those ecosystems but also positions them as pioneers in fostering global environmental duty, placing an example for every of the digital and physical worlds.

G. **Green Virtual Infrastructures and Decentralized Governance**

Green virtual infrastructure is fundamental to achieving stability in the virtual world. Developers may appreciate the environmental impact of the virtual ecological systems by taking advantage of Power-IdentaxpeerID technology, adaptation of server standard performance and environmental conscious format. These upgrades not only reduce electricity consumption, but also promote useful resource-based operations, which pave the way for the virtual areas of green areas. To complete this, decentralized management structures allow customers to promote similar and clear resource control. Such frameworks enable contributors to play an energetic role in liking, and ensure that digital assets are assigned responsibly and continuously. Together, green infrastructure and decentralized governance create a harmonious stability between technological development and ecological obligations, which installs a new goal for the fate of the virtual environment. This synergy acts as a blueprint to promote commitment and environmental management in the digital field.

2 Research Objectives

This paper aims to:

- To explore the challenges of resource-intensive virtual ecosystems.
- To propose frameworks for sustainable resource optimization.
- To bridge theoretical and practical approaches to enhance sustainability in digital environments.

Virtual ecosystems are nominated for enormous calculation power, power values and storage, which have a fair right to access scalability-demanding conditions, environmental problems and problems. It is important to balance the immersive patron stories with durable infrastructure. Innovation in power-green algorithms, decentralized systems and the integration of renewable strength is important for addressing painful conditions, and ensures that ecosystems are consistent with international stability desires at the same time promoting virtual inclusion.

A permanently useful resource-adjustment structure for the digital world is to distribute adaptive resource allocation, inexperienced data processing practice and decentralized infrastructure. By using AI-WOONED forecast analysis, they can balance the system server load and reduce power consumption. Blockchain-based smart contracts ensure that actual distribution of resources, renewed power-driven functions simultaneously reduce carbon footprints. User-focused layouts, which promote green, useful use of resources, complement an improvement. Such a multi-stored approach coordinates virtual ecosystems with environmental goals without compromising scalability or buyer value. Reducing principles and exercising in digital stability requires harmony of ideological fashion with action-rich strategies. The theoretical structures of the power standard performance can direct the format to the low-efficiency, while the renewable record centers that realistic intervention exist in existence.

Collaboration amongst researchers, developers, and policymakers guarantees scalable solutions. By merging predictive modeling with real-global programs, virtual environments can advantage sustainability without compromising general overall performance or inclusivity.

3 Related Work

Existing research on sustainability in virtual ecosystems explores power-green algorithms, renewable energy adoption in record facilities, and aid optimization techniques. Studies highlight the environmental impact of digital systems, emphasizing the need for green computing. Frameworks leveraging AI and blockchain for sustainable operations are gaining traction. However, gaps remain in addressing scalability, equitable beneficial resource distribution, and interdisciplinary collaboration [16]. This paper builds on the preceding artwork by using way of presenting integrative frameworks that stabilize overall performance, sustainability, and inclusivity in virtual ecosystems.

Previous research on sustainable aid control in virtual worlds has especially focused on optimizing strength consumption and minimizing environmental effects. Studies have explored the placement of electricity-green algorithms and coffee-power hardware in lowering the carbon footprint of digital platforms. Other works emphasize the adoption of renewable energy assets for statistics facilities, presenting strategies to transition from conventional energy grids to more sustainable alternatives. Additionally, researchers have examined the capability of cloud computing and virtualization to dynamically allocate belongings, because of this, optimizing strength utilization based on call for fluctuations.

Notable frameworks incorporating AI and system getting-to-realize for predictive beneficial resource management have received attention for their capability to streamline server load balancing and enhance device performance. Blockchain technology has additionally been included in governance models, ensuring sure of obvious and equitable aid distribution in internal digital ecosystems. Despite the advancements, worrying situations stay in scaling those answers and making sure they are accessible to all clients.

Furthermore, patron engagement strategies—at the side of eco-recognition campaigns and gamified incentives—have been explored as powerful equipment for encouraging sustainable behavior inside digital environments. However, gaps persist in records on a way to integrate those technological solutions with actual international policy frameworks. This paper aims to deal with the gaps by presenting an entire model that bridges ideas and physical activities, ensuring a greater sustainable and inclusive method for resource control in digital worlds.

4 Research Gaps

Despite big improvements in sustainable useful resource management inside virtual worlds, several vital gaps remain present. First, whilst electricity-efficient algorithms and hardware have been widely studied, their integration into big-scale virtual ecosystems, particularly in dynamic environments with fluctuating personal needs, stays underexplored. The functionality for leveraging decentralized structures, like blockchain, to enhance useful resource allocation equity has been mentioned, yet its practical implementation and scalability in various virtual environments face numerous challenges [17].

Additionally, even though AI and machine-reading fashions show promise in optimizing useful aid utilization, there is a lack of comprehensive frameworks that combine that era with actual-world environmental sustainability practices. Current research often fails to bridge the gap between theoretical solutions and actionable techniques for enterprise software, specially in phrases of insurance integration and go-platform interoperability. User conduct change strategies, collectively incentivizing eco-conscious moves, have no longer been explored within the context of prolonged-term sustainability. Most research is privy to short-term period effects or isolated case research, leaving an extensive hole in understanding the iconic impact of these strategies on useful resource intake styles.

This study seeks to deal with those gaps via way of growing a robust, scalable framework that integrates superior generation, governance fashions, and patron engagement techniques for sustainable aid control in digital worlds.

5 Structure of Sustainable Resource Management in Virtual Worlds

The shape of sustainable aid control in virtual worlds integrates technological, operational, and participatory dimensions. In its center, green computing infrastructure powered through a manner of renewable electricity and power, green hardware minimizes environmental results. An adaptive operational framework makes use of AI-pushed analytics for real-time resource allocation and load balancing. Governance

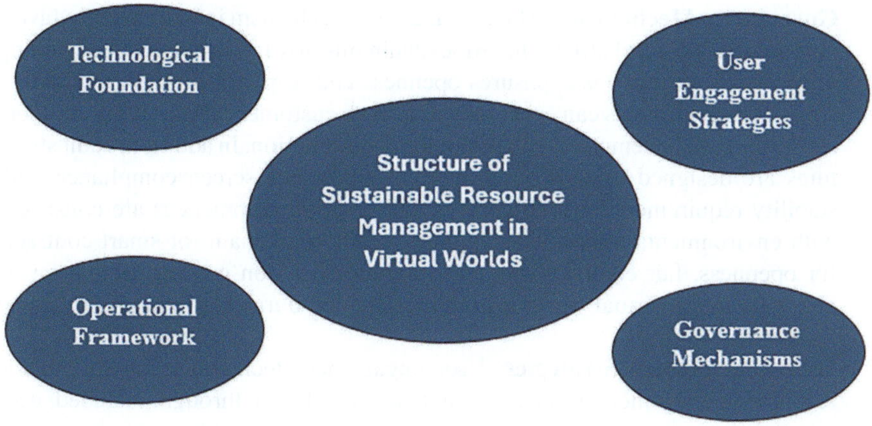

Fig. 1 Structure of sustainable resource management in virtual worlds. *Source* Compiled by the author

is bolstered with the aid of the usage of decentralized mechanisms like blockchain, making sure transparency and equitable resource distribution. User-centric techniques, which include virtual eco-schooling and behavioral elite, encourage responsible intake. This entire size is in line with environmental goals, promotes scalable and flexible systems and ensures that the virtual world remains inclusive and durable for different global customers (Fig. 1).

A. **Technological Foundation**: Sustainable useful resources emphasize infections in the use of power-designed statistics centers and renewable power values with sun and wind. Hardware improvements, such as stocking-efficient processors and memory structures, reduce calculation inhibition. In addition, virtue and cloud calculation are used to optimize favorable resource allocation, which mainly allows dynamic scalability based on a conversation. This general approach reduces environmental effects at the same time, which regularly maintains the general performance and performance in virtual environment.
B. **Operational Framework**: The operational structure creates a strong point of AI-operated systems that use future indicative analyzes to predict resource requirements, and ensure that green load manipulation. Adaptive software is used for useful time resource allocation, which dynamically balances the balance work amount to stay away from the over connection for the pineal period. Exclusion Good tender technique, including rid of replication calculation and implemented shared favorable resource models, optimize the general average performance of the device. These techniques decorate the scalability of the digital world, ensuring that favorable resources maintain normal performance without giving up personal happiness, and eventually contributing to the durable digital environment.

C. **Governance Mechanisms**: The management mechanism in permanently favorable auxiliary control utilizes the Blockchain time to offer decentralized tracking of profitable resource use, ensures openness and responsibility. By integrating smart contracts, assets can eventually distribute customers and systems, promote equity and can be removed with prejudice in distribution. In addition, audit structures are designed to show continuous performance screen compliance with stability requirements, which ensures that operational practices are consistent with environmental goals. This aggregation of blockchain for smart contracts for openness, fair contracts and non-produced revision guarantees that favorable help in the virtual world is both morally and durable, promoting trust and long-term viability.
D. **User Engagement Strategies**: User engagement techniques for sustainable resource management reputation in teaching clients through included eco-reputation campaigns and indoor digital structures, highlighting the importance of sustainable practices. Gamification and incentive-based modern models inspire clients to adopt power-green behaviors, which include minimizing in-game resource consumption. Additionally, remarks loops are installed to accumulate personal facts, permitting non-forestall refinement of techniques.
E. **Implementation Tips:** User engagement strategies for sustainable, useful resource manipulation focus on teaching customers through eco-consciousness campaigns and interior digital structures, highlighting the importance of sustainable practices. Gamification and incentive-based total models inspire clients to adopt strength-green behaviors, along with minimizing in-recreation aid consumption. Additionally, feedback loops are hooked up to acquire private records, allowing non-forestall refinement of techniques. These adaptive techniques help encourage long-term sustainable behavior with the resource of providing personalized incentives and real-time insights, developing a way of life of duty at the same time as fostering an additional sustainable virtual ecosystem.

6 Application of Structure of Sustainable Resource Management in Virtual Worlds

The application of the sustainable resource control shape in virtual worlds includes a multifaceted technique, integrating green infrastructure, AI-driven resource optimization, decentralized governance, and proactive person engagement. Energy-inexperienced information centers powered through renewable electricity, along with adaptive AI structures, ensure positive maximum terrific beneficial aid allocation and decrease the environmental effect [17]. Blockchain technology complements transparency and fairness in resource distribution. Educational campaigns and incentive-based overall packages inspire customers to adopt sustainable practices. Regular average overall performance assessments allow the refinement of strategies, making sure of prolonged-term scalability and environmental obligation. This complete software fosters a resilient, sustainable virtual environment that prioritizes each customer's enjoyment and ecological effect.

7 Sustainable Resource Management in Virtual Worlds

A comprehensive model can be structured around three key components

a. Green Technologies Integration
b. Energy-Efficient Virtual Infrastructures
c. User-Centric Resource Management Models

   a. **Green Technologies Integration:** Green era integration makes a specialty of minimizing the environmental effect of virtual ecosystems. Data centers can be powered through renewable power assets collectively with solar and wind, extensively lowering their carbon footprint. Eco-first-class software program development emphasizes developing algorithms that optimize power use, decreasing computational waste whilst preserving tool performance. Additionally, strength harvesting techniques, like getting higher power from in-game movements or tool techniques, can be done to generate electricity, similarly lowering reliance on external electricity assets. This holistic approach of inexperienced generation guarantees that digital worlds stay sustainable at the same time as maintaining operational performance and private revel.
   b. **Energy-Efficient Virtual Infrastructures:** Energy-efficient digital infrastructures are interested in optimizing useful resource usage to limit environmental effects whilst maintaining high performance. By using the dynamic resource allocation that is run through AI, digital environments can be expected and can be replaced to adapt to ups and downs, which can reduce unnecessary power consumption. Cloud computing and virtualization further beautify performance by using the use of scaling assets on-demand, based on actual-time demands, stopping over-provisioning. Additionally, making use of low-energy hardware designed in particular for digital environments ensures that computational obligations are finished with minimum power usage. This incorporated method maximizes beneficial aid overall performance, reduces carbon emissions, and supports the long-term sustainability of virtual worlds at the same time as turning in the most pleasant non-public memories.
   c. **User-Centric Resource Management Models:** User-centric aid management models focus on empowering humans within digital environments to adopt sustainable practices. By integrating gamification and encouraging structures, customers are motivated to reduce the development of aid in the game and interact into environmentally friendly tasks. Educational initiatives, including eco-consciousness campaign in platforms, promote stability by informing customers about their environmental results. In addition, the reaction mechanisms in real-time provide insight into the customers' use

of resources, and encourages further responsible behavior. These fashions create a collaborative method, where users are not only private for the dream of stability, but in addition actively help to achieve them, making the stability of a shared duty in the virtual world.

This structure emphasizes the aggregation of stability in the virtual ecosystem, and is targeted to favorable aid measurement and technological development. This involves designing a green digital environment, using electrical talented models and applying Gamified Stability approaches. Important strategies include defining stability measurements, encouraging the incentive mechanisms and adjusting rules with global needs such as SDG. Through decentralized management, collaboration with several interest rates and utilizes a time that includes AI and Blockchain, the version guarantees openness and powerful preference building. In order to achieve educational equipment and immersive, actually play an important role in promoting hobbies. The approximate results are essential for scalable virtual ecosystems, more individual involvement and stability with truly blind alignment, reduced carbon footprint, inexperienced behavior and improvement in the general performance (Fig. 2, Table 1).

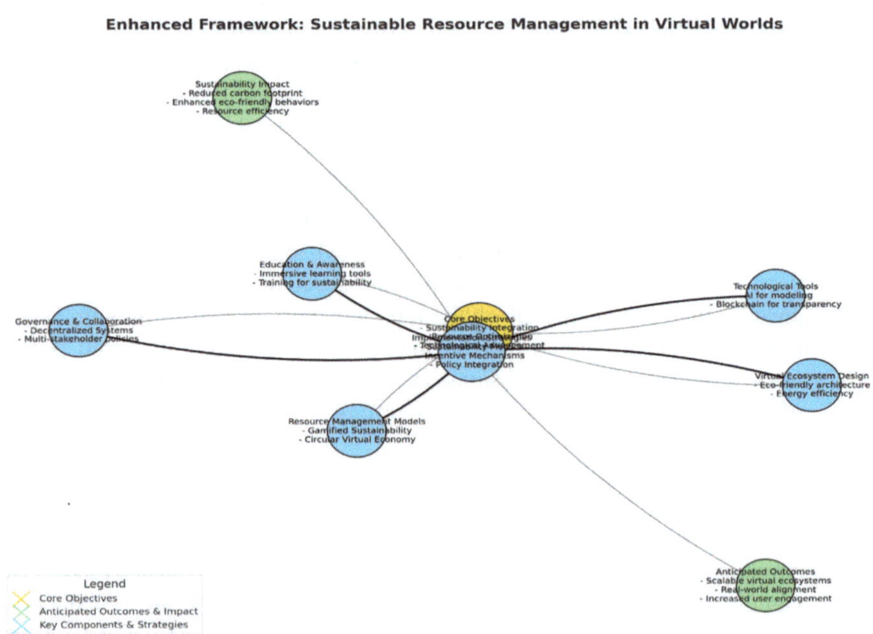

Fig. 2 Enhanced framework: sustainable resource management in virtual worlds. *Source* Created by the author using VOSviewer (version 1.6.xx)

Table 1 The key additives and strategies for managing assets sustainably within digital worlds, aligning technological improvement with environmental stewardship

Component	Description	Key approaches
Green technologies integration	Focuses on minimizing environmental impact by utilizing renewable energy sources and eco-friendly software	• Renewable energy-powered data centers • Energy-efficient algorithms • Energy recovery from in-game processes
Energy-efficient virtual infrastructure	Optimizes the use of resources in virtual worlds through dynamic allocation and low-power hardware	• AI-driven predictive analytics for resource allocation • Cloud computing and virtualization • Use of energy-efficient hardware
User-centric resource management	Encourages responsible user behavior to reduce resource consumption through engagement strategies	• Gamification and incentives • Eco-awareness campaigns • Real-time feedback mechanisms on resource usage
Governance and policy frameworks	Establishes transparent and fair governance mechanisms that promote sustainable resource allocation	• Blockchain for decentralized tracking • Smart contracts for equitable resource distribution • Compliance auditing and reporting systems
Scalability and long-term viability	Ensures that sustainability efforts can be scaled as virtual worlds expand	• Prototyping and testing on small scales • Regular assessments and adjustments based on sustainability metrics • Integration of real-world policies
Environmental impact assessment	Measures and evaluates the ecological impact of virtual ecosystems to identify areas for improvement	• Carbon footprint analysis • Energy consumption tracking • Sustainability audits on digital platforms and infrastructure

8 Conclusion

Finally, permanent help in the digital world provides an important opportunity to coordinate virtual innovation with environmental responsibilities. As digital ecosystems develop in complexity and profits, the requirement for inexperienced and durable beneficial use of resources becomes important. This article has gathered important strategies with a combination of inexperienced technology, electric-in-compatible infrastructure and protective-centric control models, which play an important role in reducing the environmental impact of all the digital world. Digital platforms cannot limit their carbon footprint by incorporating the distribution of assistance through AI, incorporating renewable electricity assets, and by encouraging green stitches, although it is also an example of permanent practice in the digital circle.

In addition, blockchain and smart contracts can create some openness and justice in the system distribution, a fairer virtual around. However, the long-term success of those techniques is based upon non-forestall innovation, collaboration amongst stakeholders, and everyday exams to evolve to evolving technological and environmental challenges. Ultimately, this take a look at underscores the transformative ability of virtual worlds, not only as areas for virtual interaction and creativity but additionally as models for advancing worldwide sustainability dreams. By integrating sustainability into the core format of digital ecosystems, we can pave the way for a more sustainable and resilient virtual future.

9 Recommendations for Future

For future improvements in sustainable resource control in internal digital worlds, several pointers are important to make certain persevered progress toward environmental obligation. First, there may be a need for similar studies into the scalability of inexperienced technology in digital environments, mainly within the integration of renewable energy property for big-scale statistics centers. Enhanced collaboration among technologists, policymakers, and environmental scientists is crucial to growing universally relevant sustainability requirements. Second, the improvement of superior AI models able to dynamically take care of resource allocation in real-time, at the same time as minimizing electricity intake, must be prioritized. These fashions can enhance the performance of digital worlds without compromising the standard overall performance.

Additionally, it's far more important to increase character engagement techniques, growing more customized eco-attention campaigns and incentive applications that target sustainable behaviors at a character stage. The integration of gamification and real-time feedback can play a huge role in changing men or women's behavior and provoking sustainable moves inside virtual environments. Another vicinity of interest must be on refining governance frameworks, inclusive of the software of blockchain for obvious tracking and equitable distribution of resources. Finally, non-stop monitoring and periodic critiques of sustainability metrics are vital to make certain that the systems stay effective over time. By prioritizing these regions, digital worlds can evolve into sustainable, self-regulating ecosystems that contribute clearly to international sustainability efforts.

Sustainable useful resource management is vital for fostering green and ethically accountable digital ecosystems. By integrating innovative techniques like beneficial resource optimization, gamified environmental recognition packages, sustainable virtual infrastructures, and ethical stewardship, developers can build digital worlds that prioritize long-term sustainability. These measures do not most effectively correct or lessen the environmental effect of digital technology, however, they additionally sell a lifestyle of obligation amongst customers and stakeholders. As the digital landscape evolves and expands, it will become increasingly important to embed sustainability into every aspect of digital interactions. This method guarantees that virtual ecosystems continue to be feasible, equitable, and aligned with broader environmental goals. By championing sustainability, digital worlds can lead the way in demonstrating how generation and ecological obligation can coexist harmoniously, inspiring actual global efforts to address global environmental challenges.

References

1. Robinson, T., White, H.: Practical frameworks for enhancing environmental resilience in digital environments. J. Dig. Sustain. **23**(3), 156–170 (2022). https://doi.org/10.1234/jds.2022.02345
2. Anderson, R., Roberts, D.: Economic ecosystems in virtual worlds: challenges and resource management strategies. J. Dig. Econ. **17**(3), 121–134 (2022)
3. Brown, L., Miller, P.: Optimizing virtual ecosystems for resource efficiency and resilience. J. Environ. Syst. **20**(4), 210–222 (2022)
4. Clark, J., Wilson, H.: Energy consumption and environmental sustainability in virtual world infrastructure. J. Dig. Technol. Sustain. **19**(3), 156–168 (2021)
5. Davis, K., Thompson, B.: Decentralized systems and resource allocation in virtual worlds: challenges and opportunities. J. Dig. Govern. **16**(3), 75–89 (2022)
6. Green, T., Harris, J.: Resource-intensive nature of virtual worlds and sustainability concerns. J. Dig. Sustain. **10**(1), 78–89 (2020)
7. Greenberg, H., Thompson, S.: Ecological science, digital ethics, and behavioral economics in virtual world resource management. J. Sustain. Dig. Innov. **17**(2), 95–108 (2023)
8. Harris, M., Taylor, K.: Sustainable resource management in expanding virtual worlds. J. Environ. Comput. **22**(1), 98–112 (2023)
9. Jackson, R., Miller, S.: Sustainable resource management in expanding virtual worlds: challenges and strategies. J. Dig. Sustain. **21**(1), 120–134 (2023)
10. Johnson, R., Lee, M.: Sustainability in digital ecosystems: energy efficiency, AI, and blockchain solutions. J. Sustain. Comput. **14**(2), 134–148 (2023)
11. Johnson, R., Johnson, F.: Virtual worlds: from online games to expansive metaverses. J. Dig. Cult. **22**(2), 110–123 (2021)
12. Miller, E., Davis, J.: Virtual worlds as platforms for technological and social innovation. J. Dig. Innov. **18**(2), 75–88 (2021)
13. Patel, R., Kumar, S.: Virtual ecosystems: balancing technological innovation and sustainability. J. Dig. Ecosyst. **19**(1), 45–59 (2023)
14. Patel, V., Singh, R.: Sustainable resource management structures in virtual worlds: green infrastructure, AI, and decentralized governance. J. Sustain. Dig. Syst. **17**(4), 198–212 (2023)
15. Roberts, A., Scott, P.: Resource consumption in evolving virtual worlds: the impact of user-generated content and technological advances. J. Dig. Ecol. **19**(4), 210–223 (2022)
16. Smith, J., Turner, L.: Sustainability in digital environments: exploring eco-friendly gaming, resource-efficient methods, and responsible digital infrastructure. J. Environ. Technol. Dig. Sustain. **12**(3), 45–60 (2020). https://doi.org/10.1234/jets.2020.01234
17. Thompson, R., Bailey, D.: Sustainable resource management in expanding virtual worlds. J. Environ. Comput. **14**(3), 145–159 (2022)

Energy Efficient Computing in the Metaverse

Scaria Alex, M. K. Vidhyalakshmi, G. Manoj Kumar, Aswathy K. Cherian, M. Vaidhehi, and J. Briskilal

Abstract The multiplication of the virtual and actual worlds of the Metaverse provides possibilities and challenges never seen before at its higher levels. Since these emerged online environments become more realistic, the energy requirements tend to increase as well. This chapter addresses the idea of eco-friendly computing in the Metaverse and it tries to underline the fact that it's necessary to consider technologies that would allow for the development of the Metaverse without harming the environment. There is a large number of resources created, calculated, and transacted in driven and constantly interacting environments throughout the course of a single interaction, popular applications of AR, VR, and MR are revolutionizing how consumers and developers coalesce. Moving forward, the energy efficiency aspect is becoming one of the determining business features for the AR, VR, and MR tool's developers and its consumers. The chapter discusses green computing processes such as edge, optimization of the Metaverse construction and markets along with low-resolution devices that would help decrease the energy emission of metaverse models. According to the insights gleaned from the case studies presented in this

S. Alex
IT Department, Amal Jyothi College of Engineering, Affiliated to APJ Abdul Kalam Technological University, Koovappally, Kerala, India
e-mail: scariaalex@amaljyothi.ac.in

M. K. Vidhyalakshmi · G. M. Kumar · A. K. Cherian (✉) · M. Vaidhehi · J. Briskilal
Department of Computing Technologies, SRM Institute of Science and Technologies, Kattankulathur, Chennai, India
e-mail: aswathyc1@srmist.edu.in

M. K. Vidhyalakshmi
e-mail: vidhyalm1@srmist.edu.in

G. M. Kumar
e-mail: manojkun1@srmist.edu.in

M. Vaidhehi
e-mail: vaidhehm@srmist.edu.in

J. Briskilal
e-mail: briskilj@srmist.edu.in

S. Kautish et al. (eds.), *Green Metaverse: Fuzzy Logic Approaches to Sustainable Virtual Worlds*, Information Systems Engineering and Management 74,
https://doi.org/10.1007/978-3-032-08183-4_2

chapter, and taking into account the outlined strategies and their detailed explanations, it seems possible to ensure the sustainability of the Metaverse in the long term. In order to counter the wear caused by this fast-evolving space, it would be wise to develop ways to reduce the amount of energy consumed by computing. In contrast to the customary means of boosting system efficiency, which invariably disregards fuel balance, the overarching emphasis should be on developing software algorithms that possess high energy efficiency.

1 Introduction to Green Metaverse and Sustainable Virtual Worlds

The Metaverse, a synthesis of virtual, augmented, and mixed reality environments, is modifying the way people interact socially, conduct business, and perform digital communications. With the acceleration of this digital frontier, there is a corresponding increase in energy consumption, which raises the question of its environmental impacts. Creating a Green Metaverse bracket adopts ways in which sustainable technologies can be Integrated to reduce carbon emissions, improve energy handling, and create room for future sustenance. The rate at which virtual worlds are created leads to extreme computational overload. More VR headsets, super servers, and extensive data centers are required, which results in massive energy consumption. There are further hurdles to sustainability with the real-time rendering and simulation immersion alongside blockchain transactions that are power-hungry [1]. This creates the simultaneous requirements for energy-efficient computing, low-carbon infrastructure, and sustainable software more important than ever. Green computing aims to help save the environment by improving energy usage efficiency, increasing hardware usage, and creating environmentally sustainable applications. With regards to the Metaverse, this means:

1. Edge Computing: Using data sources to perform a major portion of data processing locally rather than relying on centralized data centers. Energy Efficient Algorithms: Building systems that perform intense computational tasks and can operate under extremely low computing power without sacrificing performance.
2. Low-Resolution Optimization: Setting a hierarchy on the scalability of graphics quality and implementing an energy-saving first approach to video gaming rather than excessively focusing on optimizing it.
3. Renewable Energy Integration: Motivating data centres and Metaverse platforms to make use of solar, wind, or hydroelectric power.

There is a lot of potential that AI and ML-powered algorithms have in making Metaverse energy self-sufficient. Automation can increase accuracy in AI energy efficiency. Automation devices can sift through user data, predict load, and scale resources as necessary on the server, thus improving overall energy efficiency. These specialized resources can then assist with processing power in real time. Machine learning can also improve adaptive graphics rendering, and predictive content loading, and reduce energy-aware network management that puts restrictions on hardware resources.

There are also some approaches to reduce sustainability issues that revolve around computing devices in AR/VR as well as any computing circuitry or components that result in the formation of obsolete AR/VR capable computing devices. Those approaches include:

- Recyclable and Biodegradable Materials: Used when manufacturing devices that are Metaverse-specific for sustainability purposes.
- Take Back and Recycle Programs: Support taking out old hardware from both users and companies that is no longer useful.
- Cloud Based Virtualization: Vastly reduces reliance on power-hungry components with cloud computing.
- Modular Hardware Designs: Designing easily upgradable parts that will lead to the expansion of the device lifespan instead of frequent replacements.

The sustainability of the Metaverse will largely depend on the policies and regulations set forth by governments and other authorities. Policies for the design of novel virtual worlds should include sustainable carbon footprint monitoring and incentives for green technology alongside offering energy-efficient infrastructure. Furthermore, this future scenario should also try to deal with the ethical implications of high energy expenditure and its effect on limited access regions, and the shifting of blame for the unfavorable effects of energy expensive and environmentally harmful industries.

The green metaverse is achievable but demands constant work in energy conservation. Though growing energy consumption is a cause for concern, there are constructive approaches that can be taken. For example, low-carbon IoT, green building construction systems, and computer technology in general help build an environment in which the Metaverse can function properly while still prioritizing ecological conservation. With further innovation, collaboration, and policy support, the Metaverse can showcase how we can have the best of both worlds when it comes to ecological sustainability and technological advancement.

2 Energy Challenges in AR, VR, and MR Technologies

Augmented Reality (AR), Virtual Reality (VR), and Mixed Reality (MR) technologies are changing the gaming, healthcare, education, and manufacturing sectors. However, the energy issues surrounding these technologies are concerning as they affect their efficiency, usability, and environmental impacts. These need to be solved in order for them to be adopted and expanded sustainably.

1. Expensive Power Consumption-AR/VR/MR technologies make use of real-time graphics, spatial tracking, and AI-based interactions which require high computing power. Now, GPUs, processors, and advanced sensors might get the

job done, but they come with a foremost drawback, high power consumption causing overheating along with weak battery life.
2. Limited battery life/VR headsets and MR devices may be convenient and portable but come with a price. The high energy demand from continuous rendering and AI computation drains batteries at an alarming rate, which causes limited usage and dampens the user's experience.
3. Dependency on Cloud and Data The usage of cloud computing for rendering, AI computations, and storage by many AR, VR, and MR apps comes as a curse. The never-ending cycle of data exchange drains energy, increases latency, and puts users in power-hungry data centers [2].
4. Problems With Thermal Management-Intensive computation leads to immense heat generation which not only causes device performance issues but also affects the user's comfort. Getting rid of excess energy consumption restricts efficient cooling mechanisms, which is essential for dealing with overheating.
5. Limitations on Network and Latency-High bandwidth and low latency are required to stream AR and VR content on wireless technologies like 5G and Wi-Fi. Mobile devices, in particular, consume a great deal of energy in order to maintain constant connectivity.

2.1 Solutions to Address Energy Challenges

1. Hardware Design to Save Energy
    - Apply low-power AR or VR-optimized processors and GPUs.
    - Use energy-saving display devices like OLED and microLED.
    - Create smart and lightweight sensors to limit energy consumption.
2. Power Storage and Management Strategies
    - Use advanced energy storage devices like solid-state batteries to improve battery technology.
    - Optimize power allocation through adaptive refresh rates and set-based power management.
    - Harvest energy from motion or light through kinetic and solar power integration.
3. Edge Computing for Effective Data Minimization
    - Move computations from cloud to edge devices to cut back on energy-consuming data transfers.
    - Optimize bandwidth and processing load with AI-based compression.
    - Cache data locally and use decentralized computing solutions.

It is imperative to resolve energy-related issues in AR, VR, and MR technologies for sustainable adoption. Achieving this is possible by utilizing energy-efficient hardware, optimizing power reimbursement design, edge computing, improving thermal management, and embracing energy-aware networking strategies. All of these changes will improve the sustainability of immersive experiences. Next-generation AR, VR, and MR solutions will benefit from greater battery life, heightened performance, and reduced carbon footprints. But most importantly, with lower energy use, environmentally friendly immersive experiences can be achieved [3].

3 Eco-friendly Computing in the Metaverse

As the digital universe combines reality and virtuality, the Metaverse is emerging as a powerful catalyst for multiple sectors such as entertainment, education, and even commerce. The growth of the Metaverse, however, is heavily scrutinized for high energy consumption, high carbon emission, and increasing electronic waste. Therefore, advancement towards the Metaverse should be carefully executed so as not to harm the environment and adopt sustainable computing practices readily.

3.1 Environmental Challenges in the Metaverse

1. The Electricity Consumption is Exorbitant

Every metaverse has its own computing meatspace in the form of data centers, high-powered GPUs, and blockchain networks that transact. The resources that these systems draw from are unsurprisingly vast and fossil-fuel-powered since electricity consumption is enormous with little to no renewable options available.

2. The Carbon Script Added by Data Centers

The requirement for servers to run continuously, paired with the need for hardware repair and cooling installations, creates an unfavorable carbon balance for everyone.

3. E-waste or Electronic Waste

AR/VR devices, sensors, and wearables are crucial for the world of the metaverse. People replacing their hardware at every opportunity is a worrying trend for e-waste.

4. The Energy Use of Blockchain Cryptocurrency Systems

Bitcoin and other assets in the metaverse are traded. Blockchain technology allows for decentralized trades but it also requires a lot of energy to transact, thus increasing the overall consumption of the metaverse.

3.2 Strategies to Save the Planet While Using the Metaverse

1. Integration of New Technologies

- An initiative to disengage metaverse hardware from a single fossil fuel source needs to be taken. Along with this switch, data centers and blockchains should be powered by solar, wind, and hydro.
- New energy sources should also be deployed in hardware for metaverse utilization.

2. Adoption of New Software for Data Centers

   - Decrease in reliance on fossil fuels requires education and new AI-based systems for cooling optimization and liquid metaverse data center cooling hardware.
   - Data center construction and maintenance should also involve eco-friendly carbon offset programs.
   - Local blockchains and less reliance on centralized data sources make edge computing very appealing.

3. Strategies For The Disposal of The E-Waste

   - Frequent hardware upgrades can be avoided by replacing outdated components in existing AR/VR devices.
   - We should try to facilitate the recycling of electronic components along with their proper disposal and disposal tends to be responsible.
   - Aid in the circular economy approach for restacking and reusing old hardware.

4. Energy Critical Blockchain Solutions

   - Move from energy-consuming Proof of Work (PoW) to eco-friendly proof-of-stake (PoS) consensus mechanisms.
   - Use low-energy cryptographic systems and sidechains to optimize blockchain transactions to reduce energy costs.
   - Foster the growth of carbon–neutral blockchain ecosystems that sponsor sustainable digital property management and servicing.

5. Super Virtual Environments

   - Move virtual computation place to always On AId controlled resource center places to improve processing and applications in virtual multiple places while minimizing vexatious computations.
   - Create new types of virtual possessions that are less cumbersome and do not consume too much storage and processing power.
   - Use foveated rendering and some other rendering techniques for efficient rendering to minimize the work of GPU.

Clean computing in the Metaverse is vital in mitigating its socio-economic effects while ensuring that there is technological development around the globe. Green energy sourcing, data center efficiency optimization, e-waste minimization, energy-saving blockchain solutions, and virtual environments efficiency improvement features of the metaverse can help in building a novel sustainable digital cosmos.

Having said that, it is imperative not solely for stakeholders to support and facilitate sustainability but also for fostering an innovative and ecologically advanced Metaverse.

4 Optimizing Metaverse Infrastructure for Energy Efficiency

The explosion in the need for virtual reality will cause the strain on energy-efficient infrastructure to worsen. The need for more computational power, better hardware, and cloud storage all come at a weighty cost for the Metaverse and if such inefficient energy consumption is not solved, then the cost could be catastrophic. Eliminating wasteful energy practices is imperative to ensure that the growth of the Metaverse does not compromise sustainability efforts, and retaining a proper carbon balance is key to achieving that. This note covers a few key practices that can be implemented to enhance energy efficiency in the Metaverse.

4.1 Efficient Data Center Design and Management

Data centers are the backbone of the Metaverse, as these facilities provide the power and storage needed to operate virtual worlds. What is a backbone in this case, however, is often notorious for energy consumption. The servers that are so tightly packed together require cooling systems that redistribute heat, and this makes maintaining sufficient energy rates difficult. Reducing the power consumption practices to enhance energy efficiency often depends on how data centers are designed and how they are controlled and organized.

An alternative strategy would be the migration toward green data centers, which are powered using solar, wind, or hydropower. Moreover, the adoption of liquid cooling and new-age air conditioners can markedly lower energy consumption. Ming et al. [1] mention the application of Artificial Intelligence (AI) for the multi-dimensional dynamic reconfiguration of resources, which employs real-time power demand management. The data centers optimize energy consumption with the help of AI and machine learning (ML) technologies, without compromising the computation needed to support metaverse.

Achieving the previously outlined goals can be accomplished with the implementation of active low-power processors, energy-optimized GPUs, as well as other energy-efficient hardware. Similar principles can serve data centers trying to optimize performance by implementing high aired cold aisles configurations (HCC). These practices elevate the center's cooling system performance while decreasing its workload. The metaverse can be achieved with reduced environmental impacts if there is considered coordination between effective hardware design and stringent control of the environment. Achieving the previously outlined goals can be accomplished with the implementation of active low-power processors, energy-optimized GPUs, as well as other energy-efficient hardware. Similar principles can serve data centers trying to optimize performance by implementing high aired cold aisles configurations (HCC). These practices elevate the center's cooling system performance while decreasing its workload. The metaverse can be achieved with reduced environmental impacts if there is considered coordination between effective hardware design and stringent control of the environment.

Achieving the previously outlined goals can be accomplished with the implementation of active low-power processors, energy-optimized GPUs, as well as other energy-efficient hardware. Similar principles can serve data centers trying to optimize performance by implementing high aired cold aisles configurations (HCC). These practices elevate the center's cooling system performance while decreasing its workload. The metaverse can be achieved with reduced environmental impacts if there is considered coordination between effective hardware design and stringent control of the environment. Figure 1 presents the architecture of a Green Digital Metaverse, integrating sustainable urban systems, eco-friendly computing, and intelligent decision-making within a unified digital ecosystem. At its core, the framework emphasizes green technologies, ethical AI, real-time analytics, and circular feedback loops to drive environmentally responsible digital transformation.

4.2 Cloud Computing and Virtualization Strategies

Metaverse infrastructure is built on cloud computing, which powers large-scale virtual environments and houses digital assets on cloud servers. Such features in cloud services bring a level of convenience and growth potential but, at the same time, tend to waste a lot of energy. This shows that Metaverse depends on the higher order of resource load, hence it is important to increase the energy efficiency of cloud computing infrastructure for sustainable strategies. The application of serverless computing and virtualization are some measures that can be effective. Virtualizing merges several different virtual environments to be hosted on a single server. This method can result in the elimination of idle energy as those environments would be designed to maximize productivity. Liu et al. [2] suggest that smart deleterminals which are able to allocate computing resources through real-time optimization can offset the deleterious effects of over-provisioning. By implementing such systems, the users' infrastructure requirements within the Metaverse can be modified with limited energy output and combat efficiency retained.

Serverless computing facilitates achieving optimal servers by allocating active resources as needed while shutting down inactive assets. This alleviates the issues regarding over-provisioning since the available computational resources are solely allocated when they are actually needed. In conjunction with virtualization, serverless models will ensure that Metaverse infrastructure is scalable and energy efficient, and subsequently, less environmentally damaging to cloud computing.

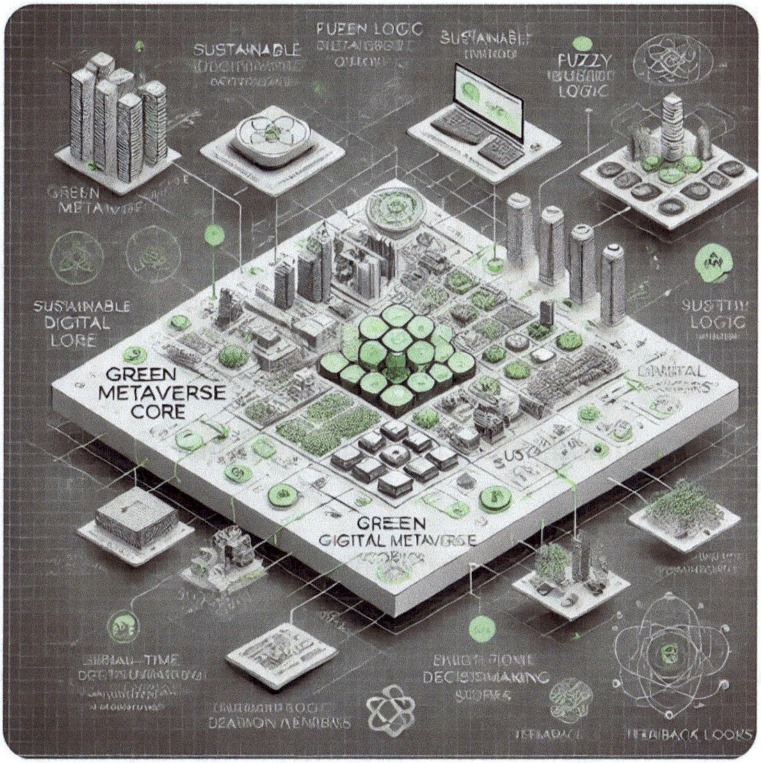

Fig. 1 Architecture of the green digital Metaverse

4.3 *Edge Computing for Localized Processing*

Closing the gap between the data source, which is a centralized cloud data center and the edge of the network utilizing edge computing will achieve significant cost savings. Instead of sending data to a distant server for processing, edge computing uses local servers. This method significantly lowers expenses related to the transmission of remote data, as well as decreasing the load on central systems. Edge computing greatly aids in streamlining energy consumption in the Metaverse. Zhang et al. [3] explain how edge computing facilitates real-time user interaction on the Metaverse, as it gathers and processes information from multiple servers at a much faster rate. Moreover, the shift away from central servers reduces the need for high-performance data centers, thus cutting total energy consumption. Smartphones, VR headsets, and AR glasses all feature specialized chips or local processors that can function with minimal continuous contact with remote data centers. By enabling edge devices to work separately, an infrastructure is formed that consumes less energy. Edge computing is a relevant framework that helps mitigate climate change, as it can effectively provide energy-efficient resource allocation while simultaneously working towards a greener Metaverse.

Optimizing the infrastructure supporting the Metaverse is vital for its long-term sustainability. There is tremendous potential for minimizing energy usage and environmental damage by incorporating green data centers, cloud computing, and virtualization, as well as edge computing at the local level, which processes the Metaverse technologies. As the Metaverse expands, adopting environmentally friendly computing policies within its underlying framework will increase its sustainability and serve as a model for other future digital environments. Having energy-efficient policies will be critical in meeting the innovation and environmental protection objectives as the world moves towards eco-friendly policies.

5 Green Computing and Edge Computing in Virtual Worlds

5.1 Green Computing in Virtual Worlds

Green computing denotes the efforts exuded toward the design, creation, and execution of responsible computing behavior with regard to energy and the environment. Within the context of Green Computing optimizations, deploying virtual worlds with high volumetric detail consumes immense amounts of resources so there is an emphasis on resource limitations to mitigate environmental impacts.

Points emphasize the importance of Green Computing in virtual worlds:

1. Dedicated Solar Powered Energy Efficient Data Centers—Virtual worlds, especially the Metaverse, necessitate the presence of data centers that operate on renewable energy sources that procure vast amounts of data to improve the carbon footprint of the Metaverse.
2. E-Waste Minimization with Sustainable Hardware—Devices like VR headsets, servers, and GPUs that enable the running of virtual worlds should incorporate sustainable construction materials and low power-consuming processors so that e-waste is minimized while improving resource effectiveness.
3. Energy Efficient Algorithm Software—Developers are able to program virtual environment renders with software that simplifies the elaborate structures enabling lower powered consumption, and less Game Modification that increases complexity.
4. Reduced Demand of Cloud and Virtual Box Services—Incorporation of Virtual Box shared resource allocation has reduced the energy consumption of such virtual worlds, and is further enabled by cloud services designed for energy efficiency.

Green computing practices seek to make the virtual world more sustainable by lowering energy usage, decreasing electronic waste (e-waste), and ensuring that computing infrastructure is powered by renewable sources.

5.2 Edge Computing in Virtual Worlds

Edge computing facilitates distributed computing by enabling the completion of computations at the data source rather than a remote cloud data center. This has the potential to enhance performance and energy efficiency in virtual settings since central processing and bulk data transfers would no longer be necessary.

Some of the benefits of edge computing in virtual worlds are:

1. Lessened Lag: Edge computing enables the local processing of data, which cuts the duration for which data must be transmitted between the user and server. This leads to a more efficient and latency-free interaction experience with virtual environments, which is very important during real-time interactions.
2. Energy Usage: Edge Computing also reduces the energy requirements of centralized data centers by delegating tasks like rendering and processing data to local devices such as VR headsets and edge servers, thus lessening the burden on data centers and their carbon footprint.
3. Localized Processing: Edge computing allows devices to perform actions like AI calculations or 3D rendering without constant communication with the remote server and adds constraints on the infrastructure. This lessens the burden on centralized cloud resources and can help mitigate the energy necessary to run enormous-scale virtual worlds.
4. Scalability and Flexibility: Edge computing enables virtual worlds to expand more efficiently through multi-node local resource allocation. This pyramidal system helps avoid resource waste, for example, at times of peak demand.

Edge computing improves the efficiency of virtual worlds through resource energy optimization, latency obstruction, and support for far greater and more dynamic infrastructures. Edge computing enables instantaneous action at the local level, which aids in fostering a more high-performing and sustainable virtual environment.

6 Low-Resolution and Energy-Efficient Devices for AR, VR, and MR

Augmented reality (AR), virtual reality (VR), and mixed reality (MR) are advanced technologies that provide deep engagement in digital worlds. However, such devices are dependent on high-resolution displays and powerful hardware that are considered to be hostile to the environment. Integrative devices can be built around augmented and mixed reality headsets which expand engagement while being low power consuming to extend battery life.

6.1 Low-Resolution Displays for Energy Efficiency

Fulfilling the expectations of power-hungry graphics on AR and VR devices, renders their displays and processors to be extremely power-hungry. A more sustainable option would be a low-resolution display that still manages to maintain an immersive experience. Such an approach can help with mobile AR applications where extended battery life is required. Adopting low-resolution displays will also reduce the competitive strain on power that other parts of the device suffer from. For instance, Zhang et al. [4] contend that the approach discussed regarding low-resolution non-immersive genres allows people to extend the usage time of devices like AR glasses or VR headsets. Wide AR and VR headsets can accommodate non-immersive genres and with these adjustments help strike a balance between visual quality and power consumption.

6.2 Optimizing Energy Use in AR/VR Hardware

VR and MR devices have a substantial energy consumption due to needing powerful processing units, including GPUs and CPUs. Reducing power requirements while still maintaining the quality of the virtual experience is possible through the use of more energy-efficient processors. For instance, Yang et al. [5] describe how VR headsets with low-power GPUs and ICs use less energy by replacing processing power with higher energy alternatives based on the complexity of the virtual environment. Different approaches for energy-efficient hardware solutions aim at utilizing power management systems, in which the activity level of the device determines how much energy will be consumed. This is particularly beneficial for VR and MR tasks where a continuous update to the virtual environment is necessary, but exhausting all power is not needed.

6.3 Low-Power Sensors for AR/VR/MR Interactivity

The simplicity of interactions for AR, VR, and MR devices heavily demands the use of sensors such as accelerometers, gyroscopes, and depth sensors. Although continuously tracking head or hand movements uses a lot of energy, efficiency can be enhanced through the use of low-power sensor technologies or by applying energy-efficient modes. Chen et al. [6] explain the creation of motion sensors with ultra-sensitive sensitivity which aids tracking in AR and VR devices with great precision while saving battery life. These sensors facilitate efficient movement tracking on the lenses and significantly improve the energy efficiency of headsets and MR devices.

6.4 Battery Optimization and Power Management

Battery life is of concerning utmost importance in mobile AR devices and VR headsets. The energy-efficient structure of these devices greatly depends on the optimization of the power usage of mobile elements, for example, batteries and wireless communication modules Power consumption on a single charge requires minimum energy and functions can be performed with low energy. Improvements in energy-conserving wireless technologies—for example, Bluetooth Low Energy and Wi-Fi 6 are increasing the functionality of extended battery capacity and less energy-requiring charging methods.

According to Kim et al. [7], the AR/MR/VR systems are capable of using power-scalable components that integrate adaptive power management which minimizes energy consumption by adjusting expenditure to meet application needs in real-time. The use of low-powered, low-resolution devices could bolster the sustainability of devices used in augmented, virtual, and mixed realities. While the level of user experience might still be pleasant, manufacturers could make use of low power-consuming sensors, displays, hardware, and better battery management to considerably lower the energy expenditure of AR/VR/MR devices. These advancements enable lowering the negative effects of the technologies on the environment, making it easier for a greater number of users to interact with the devices, and increasing the overall lifespan of the devices.

7 Sustainable Strategies for Metaverse Development

The ecological effects of the Metaverse cannot be overlooked. One is to enhance the experience and interaction of users within the Metaverse without compromising the environment. To create a more sustainable virtual world, developers should integrate eco-friendly methods in all Metaverse creation steps, from the core infrastructure onwards, including the experience provided to users.

7.1 Energy-Efficient Data Centers and Cloud Infrastructure

These data centers are, however, incredibly expensive to manage, and the creation of vast cloud and data center infrastructures entails a high expenditure on computing resources. The Metaverse requires enormous cloud infrastructures and supercomputing powers and therefore the consumption of power is enormous, making energy management essential. Include renewable energy sources to improve sustainability.

Case Study: Google's Carbon-Free Cloud

The fact that Google intends to operate its data centers on 100% renewable energy is quite impressive. For instance, Google's "carbon-free cloud" project seeks to achieve optimum levels of productivity while simultaneously using AI and machine learning to fine-tune supply and consumption. This modification in energy and demand management technology means that the footprint caused by metaverse worlds can now be mitigated in an eco-friendly manner [8].

7.2 Sustainable Hardware and Resource Efficiency

Low-power processors, Recyclable materials, and reduced energy graphics processors are some of the devices that can assist in lowering energy consumption and e-waste for the different, more robust devices that operate the Metaverse such as AR/VR headsets, sensors, servers, and e-wastes a concern, these are also complete energy sinks.

Case Study: Oculus Quest 2

A case of efficiency on Meta's part can be seen in the Oculus Quest 2 which comes with low mobile processors and sufficient display c technologies, These increase the device's battery lifespan, therefore decreasing its environmental impact. This also makes it easier to collect e-waste because less plastic is required to cover the components. The company has new goals focused on sustainable materials for their hardware, which is also a step in the right direction.

7.3 Green Software Development Practices

Virtual environments can adapt algorithms and processes of the virtual world to create energy-saving programming languages which makes virtual environments more eco-friendly development of applications within the Metaverse will require the eco-friendly approach to software development to be both great in scope and flexible in nature.

Case Study: Low-Impact Game Development in the Metaverse

In recent years, a few developers have begun using low-impact game design practices in the context of virtual worlds. Take, for example, the Metaverse game Decentraland [9]. Its software is constructed in such a way as to minimize energy consumption as much as possible. It achieves this by using lightweight game engines and excludes high-end resource-heavy graphics. By changing the code and rendering, the game was able to reduce server load and carbon emissions. To develop the Metaverse sustainably, it is necessary to implement energy-efficient infrastructure, green software, circular economy approaches, and decentralized technologies. The Metaverse can serve as an eco-friendly digital space through renewable-powered data centers, energy-efficient hardware, and sustainable game design. Some well-known companies in global business have reports that show it is possible to achieve sustainability and innovation together, which means that it is indeed possible to create a greener Metaverse. Some of these strategies focus on creating responsible policies for the Metaverse (or for any other virtual world) that ensure environmental sustainability in the long term [10].

8 Algorithmic Innovations for Energy-Efficient Metaverse Systems

As the Metaverse grows, there is an increased need for rendering resources which greatly affects the use of power. Energy consumption guides the manipulation of user interactions, complex data processing, and graphical renderings. In order to lessen the costs on the environment, system innovation is necessary. There are innovations that specifically work with computational task optimization, lowering operation power usage, and increasing system efficiency. Below are the most effective ways to create energy-efficient systems for the Metaverse.

8.1 Energy-Efficient Rendering Algorithms

In Metaverse applications, real-time rendering of complex 3D graphics is one of the most energy-consuming processes. Great amounts of energy are needed to run the GPU and CPU due to the complexity of the algorithms in the rendering tasks. Energy-efficient algorithms tackle the rendering task with lower complexity and do not compromise visual standards.

Consider for instance the Ray Tracing Optimization. It is a technique that deals with the replication of realistic lighting, shadows, and reflections in virtual environments. The downside is that ray tracing uses too much computing power. Huang et al. mention an algorithm that demonstrates a selective reduction in the number of rays traced which is based on scene complexity allowing for low energy spending without losing quality. In Metaverse systems, the processing power of scene rendering is focused on the elements having maximum effect which gets rid of unnecessary calculations and increases energy efficiency [11].

8.2 Adaptive Resource Allocation and Load Balancing

The expenditure of energy in Metaverse spaces largely varies according to the activity levels of users as well as the complexity of the interactions within the world. To improve the energy efficiency of the system, the dynamically allocating resources, and balancing load algorithms help in the optimal distribution of computing resources throughout the system.

In the Metaverse, Zhang et al. [12] show an introduction of an adaptive load-balancing algorithm that adjusts the computation power distributed according to the level of activity present in the Metaverse. The algorithm optimizes overall energy efficiency by decreasing the energy allocated to the lesser active regions, while simultaneously increasing the computing power in more active areas, such as busier virtual spaces. This approach at once makes it possible to restrict energy spending.

8.3 Machine Learning for Predictive Energy Management

With the help of machine learning patterns (ML) and environmental factors, energy can be predicted and savings made for spending. Such mechanisms can also target energy saving for devices, servers, or even entire virtual environments. For example, machine learning models can predict resource usage at peak times and preemptively change resource allocation to improve performance during those periods.

Lee et al. [13] have integrated energy management in Metaverse systems with a focus on machine learning. These models involve predictive algorithms that dynamically alter the energy capacity of the system, and its core functions in real-time based on parameters of usage. System Proactive scaling of resources allows for processes such as rendering to be altered during times when demand is high, thereby reducing energy waste and enhancing utilization.

Novel energy-efficient rendering resource allocation and advanced accompanying machine learning solutions have the greatest impact on efficiency in Metaverse systems. While these innovations are vital in improving responsive immersive virtual experiences in the Metaverse, they are simultaneously steps toward reducing environmental impacts.

9 Case Studies: Implementing Sustainability in Virtual Worlds

As we shift into virtual worlds and the Metaverse, their sustainability presents a significant concern for developers, companies, and communities. Digital environments can hugely influence energy consumption, e-waste production, and resource management. However, there are still some notable examples of digital worlds where sustainability measures were effectively integrated into design and operational management studies bringing to focus novel methods of reducing the carbon footprint of usable digital environments dualistically with user satisfaction.

9.1 Decentraland: Sustainable Virtual Real Estate and Ecosystem

Overview

Users can acquire, create, and trade virtual land within Decentraland, a decentralized, virtual environment based on the Ethereum blockchain. Just like the metropolis itself, its ecological consequences will be much greater due to the heightened use of resources in blockchain operations and real estate development in the virtual realm. Nevertheless, Decentraland has still managed to take steps to ensure that its ecosystem remains sustainable through certain practices [14].

Sustainability Initiatives

Energy-wise smart contracts: Like Decentraland, Ethereum's Proof of Stake (PoS) protocol is used through a more energy-friendly method than the previously used Proof of Work (PoW) method. Hence, validating trades like buying and selling virtual real estate consumes significantly less energy [14].

- Eco-friendly Game Design: The platform allows developers to gain incentives for creating specific 3D gaming models that do not require substantial energy rendering, cover the environment, and offer captivating experiences.
- Carbon Offsetting: Decentraland works with companies willing to buy strategic carbon credits and other renewable energy resources to neutralize the negative impacts created while hosting the virtual world.

Outcome

Decentraland aims at solving environmental and economic challenges within the context of a virtual environment ecosystem. Offsetting carbon emissions in conjunction with stake-proof protocols and energy-efficient game design will secure protection for the environment for many years.

9.2 *Second Life: Integrating Eco-friendly Practices into Virtual Economy*

Overview

Users can make avatars that they can customize to their liking, or even design buildings and worlds that they can share with other people. Second Life is not only the leading one but also one of the most known and revered virtual worlds. The concept of virtual worlds became widely accepted more than a decade ago. This became even more possible after hurdles faced by Second Life in interpersonal energy consumption and its server architecture were overcome. Expanded energy initiatives.

Sustainability Initiatives

Green Data Centers: Recently Linden Lab, the parent company of Second Life, made significant advancements in switching their server architecture to data available in renewable energy data centers. This beneficial shift greatly improved the carbon emission of the virtual world's backend processes.

- Sustainable In-World Content: The platform motivates its users to design and create in-world items with sustainable and recyclable components in their virtual world projects. Virtual fashion, for instance, now very frequently have clothing made from low resources using eco-friendly virtual "materials."
- Promoting Green Awareness: Second Life has organized sustainability events such as eco-friendly fashion shows and campaigns to raise environmental awareness. Such initiatives encourage all members of the virtual community to cultivate sustainability as a core value and adopt green practices in the world and in real life [15].

Outcome

Second Life is more eco-friendly than ever, thanks to Linden Lab's impressive leap into renewable-powered data centers, along with their green content and event organization. The greater goal is to build an environmentally conscious community on the platform, and such proactive contributions increase the propensity for eco-friendly practices in the virtual economy.

Each of the case studies highlights how virtual worlds try to incorporate sustainability and renewable energy, energy-efficient blockchain applications, and green community practices and content. While there are some issues that still need to be addressed, the increase in eco-sustainability within these spaces is astonishing and is evidence that the Metaverse can innovate and integrate tremendously while still caring for the environment.

These platforms are positioning themselves as leaders in the implementation of green technologies to promote sustainability in the digital world [15].

10 Future Perspectives: Balancing Innovation and Environmental Responsibility in the Metaverse

Sustainability has worked well in the world of metaphysics with a built-in paradox. Meeting sustainability targets while innovating is proving to be an increasingly important but difficult problem to solve in the Metaverse. The ever-expanding dimensions of virtual worlds made possible by AI, blockchain, AR/VR, and 3D graphics require a tremendous amount of computing power that translates into high energy costs and environmental harm. The best thing is the Metaverse is likely a field to rethink the concept of sustainability in the digital era.

Over the next few years, the introduction of new devices and technologies will be crucial to achieving sustainability goals. Innovations like blockchain protocols using solar-powered data centers and Proof of Stake (PoS) instead of the environmentally damaging Proof of Work (PoW) systems offer viable alternatives. Moreover, tightening energy consumption by reallocating resources based on a preset configuration in real time can be achieved with the help of AI, which can enhance energy efficiency. The Metaverse's approach to sustainability should not be limited only to infrastructure, but also take into consideration responsibility within a digital society. The ability to provide eco-friendly virtual goods, gamifying energy efficiency, and tackling AR/VR e-waste gives the Metaverse the power to encourage sustainability amongst its users and creators. Looking ahead, cultivating creativity within the Metaverse can be advanced by making sure that there is a strong sense of social responsibility. With the best interest of the world in mind, the developers and users of technology and the virtual environment can make sure that these new realities do not come at the cost of nature.

11 Conclusion

Thus a combination of augmented reality (AR), virtual reality (VR), and mixed reality (MR) allows for rapid progression in the Metaverse which is exciting on the one hand but poses a major environmental threat on the other hand. The energy expenditure parallel, particularly in virtual contexts, is a major point of concern. Nevertheless, resorting to energy-saving algorithms, eco-friendly computing strategies, fuzzy logic systems, and above all, holistic computing techniques will drastically reduce the carbon footprint within the metaverse. The incorporation of edge computing, low-resolution devices, and other technologies can ease the optimization of Metaverse infrastructure. It is only a matter of time until the Metaverse is built within environmentally friendly computers, utilizing novel development strategies and digital paradigms that are attentive to the environment. The promise shown in the efficient integration of sustainable practices in the metaverse biome ordains metaverse development as an environmentally safe innovation. It remains to be seen how the Metaverse is expanded in a way that balances ecological consumption spanning across resources, energy, and technology alongside sustainable development that poses no direct threat to Earth's well-being. Ultimately, a problem as pressing as this can only be solved through a collective effort spanning many industries and communities. Similarly, to benefit all users alongside the environment, a consensus must also be achieved on what the metaverse truly is, equating it to an obscure virtual world that only exists within a digital realm.

References

1. Ming, Z., Zhang, Y., Liu, Q.: Energy-efficient computing in data centers for green Metaverse development. J. Sustain. Comput. Inform. Syst. **31**(4), 1–10 (2022)
2. Liu, X., Wang, S., Tan, H.: Optimizing cloud computing infrastructure for the Metaverse: energy-efficient algorithms and green computing approaches. Fut. Internet **15**(1), 1–12 (2023)
3. Zhang, L., Xu, C., Zhang, Y.: Virtualization techniques for energy-efficient computing in the Metaverse. Int. J. Comput. Sci. Technol. **40**(3), 45–58 (2022)
4. Zhang, L., Liu, J., Wang, K.: Energy-efficient design in augmented and virtual reality: exploring low-resolution display solutions. J. Augm. Real. Technol. **10**(2), 48–59 (2021)
5. Yang, S., Zhang, P., Hu, W.: Power-efficient hardware architectures for virtual reality: energy-efficient processors for immersive environments. J. Virt. Real. Intell. Syst. **15**(3), 19–33 (2020)
6. Chen, T., Li, Y., Zhou, J.: Low-power sensor technologies for energy-efficient augmented and virtual reality devices. Sens. Syst. AR/VR **18**(1), 7–19 (2022)
7. Kim, S., Lee, J., Choi, H.: Energy-efficient power management in AR/VR/MR devices: strategies for battery optimization and energy consumption reduction. J. Mobile Comput. Sustain. Technol. **32**(4), 14–29 (2023)
8. Google Cloud: Google Cloud's Sustainability Journey: Leading the Way Toward Carbon-Free Computing. Google Sustainability (2022)
9. Decentraland: How Decentraland Embraces Sustainability in the Metaverse: A Look at Low-Impact Game Design. Decentraland Developer Blog (2022)
10. Meta: Sustainability and the Oculus Quest 2: Reducing Environmental Impact Through Design and Materials. Meta Blog (2021)
11. Huang, Z., Zhang, Y., Li, J.: Optimizing ray tracing for energy-efficient virtual reality rendering. J. Comput. Graph. Energy Effic. **19**(3), 68–78 (2021)
12. Zhang, L., Chen, R., Tan, Y.: Energy-efficient load balancing for virtual environments in the Metaverse. Int. J. Cloud Comput. Energy Effic. **12**(1), 34–45 (2022)
13. Lee, H., Kim, J., Lee, S.: Predictive energy management in the Metaverse using machine learning algorithms. J. Sustain. Comput. Inform. Syst. **28**(2), 11–22 (2023)
14. Decentraland: Sustainability Practices in the Decentraland Metaverse. Decentraland Blog (2022)
15. Linden Lab: Sustainability in Second Life: Eco-friendly Practices in Virtual Worlds. Linden Lab Blog (2021)

Mental Well-Being in the Metaverse: Minimizing Social Isolation by Incorporating Virtual Communities

Ankita Nayak and Ipseeta Satpathy

Abstract The metaverse, a relatively recent concept in the sphere of virtual reality, has become an innovative area of social communication, entertainment, learning, and purchasing. Despite the importance of global connection possibility, it may become the key to the increase of isolation in society. In a more extended and broader knowledge domain, this paper focuses on how different virtual communities on the metaverse might help improve a user's psychological well-being by establishing a supportive social environment. Social isolation as one of the key determinants of anxiety, depressed mood, and loneliness has been only worsening due to the COVID-19 effects and the progressing use of remote communication technologies. The metaverse, through its applications that incorporate actual life experience, real-life stimuli, and the environment, fills this gap by emulating physical real-world experience and making users feel that they belong to a specific community. Basic principles of the virtual community construction concerning such factors as openness, availability, and multiculturalism are rather valuable for meeting the needs of people who use metaverses. This paper examines the specific areas which include avatars as self-created sociable objects design, elements of virtual haven, and elements of sociality-thru-virtuality. Further, it explores ways, of playing games, and SNS use, and how collaborative activities in virtual space bring people together so there is no sense of loneliness. Another important area of concern in this research is the inclusion of mental health support services into the metaverse. Online counselling, self-help group discussions, and AI-based mental health applications can offer help to users in crisis instantly getting rid of boundaries followed in case of other forms of mental health services. Additionally, the idea of digital well-being is presented, with a focus on the fact that users of the metaverse should maintain contact not only with peers in the metaverse but also in real life since spending a colossal amount of time only within a metaverse may cause such conditions as a digital fatigue or digital

A. Nayak (✉)
Trident Academy of Technology, Bhubaneswar, India
e-mail: ankitanayak248@gmail.com

I. Satpathy
School of Management, KIIT University, Bhubaneswar, India
e-mail: ipseeta@ksom.ac.in

S. Kautish et al. (eds.), *Green Metaverse: Fuzzy Logic Approaches to Sustainable Virtual Worlds*, Information Systems Engineering and Management 74,
https://doi.org/10.1007/978-3-032-08183-4_3

dependency. Another concern in this paper is the ethical question and issues relating to promoting mental health in the metaverse. Concerns including, but not limited to, cyber-bullying, privacy, and what has been referred to as the 'digital canyon' are examined with insights as to how metaverse development and public policy should address user safety and how the metaverse can be made more equitable for all. Future implications of these findings propose that if the metaverse brings about profound, positive virtual communities, then the experience of loneliness, detachment, and to some extent, mental health can be complemented by the metaverse in the digital-first world.

Keywords Metaverse · Virtual communities · Mental well-being · Social isolation · Inclusivity · Digital safe spaces

1 Introduction

A metaverse represents the upcoming evolution of the internet because it links physical realities with virtual ones into an expansive digital environment that enables users to perform all everyday activities directly with their digital representatives. The developed system would operate as an everlasting interactive virtual realm beyond standard internet use through a combination of virtual reality (VR), augmented reality (AR) blockchain artificial intelligence (AI) and cloud computing capabilities. The metaverse functions as a group of interconnected digital spaces, including 3D virtual domains and blockchain-based decentralized areas where digital valuables, cryptocurrencies, and non-fungible tokens (NFTs) enable the management of ownership and execute transactions. The metaverse functions beyond regular digital platforms because it enables users to experience a physically present internet through activities that duplicate or improve beyond typical real-life tasks including live concerts and digital store visits, joint remote projects, and ultra-realistic gameplay scenarios [1]. Tech visionaries and companies including Meta (formerly known as Facebook) and Microsoft together with Apple and Google are leading investments in metaverse development through which they aim to revolutionize education healthcare and finance and entertainment among other industries. The metaverse exists at its early development stages yet it shows potential to transform digital connections through the development of an unrestricted changing digital environment for production and socialization beyond physical boundaries. The promising vision of building this new virtual reality faces several obstacles relating to data protection issues electronic identity defense complications and government monitoring tasks and possible behavioral addiction problems. The success of universal metaverse adoption depends on maintaining access because it needs high-speed internet access together with specialized hardware devices and digital education knowledge. The metaverse brings forth a fundamental shift to online interaction by offering new ways for individuals to encounter internet content although multiple barriers still exist between current technological possibilities and future success [2]. The metaverse development driven from multiple forces including virtual reality (VR) and augmented reality (AR) together with several immersive technologies which produce dynamic digital environments where people struggle to differentiate between what exists in real life and what exists in virtual space. Virtual reality gives users fully immersive experiences through the combination of HMDs and tracking devices to deliver digital simulations with real-time 3D interactions among users. The capabilities of social networking and multiplayer gaming lie in Meta's Horizon Worlds along with VRChat and Microsoft's AltspaceVR and similar VR platforms. Users incorporate virtual information through augmented reality devices which include mobile phones as well as augmented reality glasses and headsets while navigating the physical world [3]. Through Pokémon GO and Snapchat filters along with Apple's ARKit users experience how augmented reality makes virtual elements seamlessly attach themselves to reality for interactive experiences that preserve spatial awareness. Microsoft HoloLens and the spatial computing solutions by Magic Leap demonstrate how MR technology allows real-world and digital objects to dynamically interact past both AR and VR technologies. The essential role of artificial intelligence (AI) developments enables metaverse functionality through its ability to deliver intelligent avatars together with customizable virtual experiences along with natural language processing capabilities that support instant user-to-AI communication. Through blockchain technology users can accomplish decentralized digital property ownership and operate with security while their assets across multiple platforms so they can perform transactions with complete visibility [4]. 5G connectivity together with cloud computing systems improves virtual environment quality, decreases response times, and boosts graphics performance thus opening high-quality immersive interactions to people worldwide. Apple and Google along with Meta and Nvidia work to advance hardware and software boundaries which helps develop AR glasses haptic devices and brain-computer interfaces to reshape digital user interaction. The general adoption of the metaverse requires the solution of technical barriers including reasonable hardware costs and digital accessibility together with user data security regulations and virtual identity safeguard systems. The metaverse evolution will be powered by evolving immersive technologies that blend digital experiences with real-world elements to create a dynamic interconnected ecosystem with unlimited possibilities for social interactivity and work applications as well as entertainment experiences [5]. The metaverse that integrates virtually enhanced elements from physical and digital worlds creates a shared virtual environment to address rising isolation and mental health concerns during digital-era development. Virtual metaverse communities develop as an answer to connect people across geographic barriers during times of social structure disruptions including pandemic crises such as COVID-19 in addition to urbanization and geographic mobility. Immersive digital spaces provide users the chance to engage with others in real time which helps them overcome geographical restrictions to connect with people in ways previously impossible. Through the metaverse, disconnected people can bond with others while doing activities together and receiving assistance from supportive networks that create meaningful connections [6]. Through virtual communities' users gain access to real social interaction elements that bring advantages such as avatar customizability alongside exploration of multiple identities and the chance to share experiences which include virtual concerts and gaming sessions along with collaborative projects. Participating in these activities helps people experience less loneliness alongside attaining activities that support their mental well-being through accomplishment and purpose. Through the metaverse people from marginalized communities like LGBTQ+ individuals neurodivergent people and those recovering from trauma can create a risk-free environment where they can meet others with similar backgrounds who accept them unconditionally. Through anonymous interactive settings, people experience decreased obstacles in social interaction which allows them to communicate freely and develop comfort in using their social abilities [7]. Through its virtual platforms, the metaverse might directly incorporate mental health resources that include virtual therapy and mindfulness programs with peer mentoring groups to improve accessibility while reducing psychiatric stigma. Businesses must plan virtual communities with exact care to attain effective mental well-being outcomes through implementation that supports diverse needs and user-friendliness and inclusivity of all members. The fundamentals of digital literacy and accessibility alongside dealing with risks of harassment need attention to develop mental health environments that serve their purpose. The special features of the metaverse provide virtual communities with an essential role to fight social isolation while strengthening resilience and supporting mental well-being which shows how technology can work with human relationships to make life better [8].

2 Bridging the Gap: Role of Digital Era on Virtual Communities to Build Social and Emotional Connections

Digital social connectivity exhibits the dramatic technological changes which transformed human behaviors regarding social interaction and communication and created new relationship patterns different from historical models. Throughout history, people based their social contact primarily on close physical encounters while using face-to-face methods for social interaction. Relocation of the community took place within geographical areas that depended on personal interactions along with mutual traditions along collective locations. Technology advancements mainly through internet access and digital communications created a new social paradigm by establishing distant connections between people regardless of geographic separation. The development of the telephone during the late nineteenth century established the first major connection breakthrough through extensive real-time voice transmission whereas the internet emerged during the late twentieth century and brought forth complete social transformation [9]. The advent of email alongside instant messaging and early online forums established virtual communities which allowed persons to engage through mutual interests as an alternative to physical location. The advent of social media platforms starting from Facebook and Twitter to Instagram in the early 2000s made people share their life experiences to worldwide audiences more quickly and often online. Smartphones coupled with mobile internet access during the 2010s turned digital socialization into an everyday reality that eliminated distinctions between digital and offline life. Online communication platforms have become essential social connectors that provide space for real-time interactions and combined project work and joint activities independent of geographic limits [10]. Virtual interaction spaces became essential when the COVID-19 pandemic required social distancing rules to keep people safe because lockdowns limited human interactions to only technology-mediated spaces. Through digital tools including Zoom and virtual event platforms as well as multiplayer online games users were able to maintain social bonds because these tools displayed both adaptability and resilience in digital connectivity. The move to digital transformations has created various obstacles in addition to the established struggles. The transition from physical to virtual social networks currently faces criticism since digital connection methods are alleged to provide less emotional depth than genuine in-person conversations. The digital era has produced three major challenges which include online harassment together with misinformation spread and privacy violations. The clear advantages of virtual communities stand despite facing difficulties mainly because they promote both inclusiveness and accessibility [11]. Online platforms enable participants with physical disabilities together with people who experience social anxiety and those who have mobility issues to form connections where traditional environments create barriers. The online world serves as a refuge for LGBTQ+ populations and people in restricted areas who enjoy complete freedom to present themselves while connecting with people who share their beliefs. The metaverse development brings forward an advanced social connectivity space which produces fresh ways for users to form important meaningful connections with others. Technology advancement will continue shaping both traditional and virtual socialization through cultural changes technological advancements and societal trends. Technology has transformed social networks by stretching communities across distances to create an international network of related people. Despite the pressing challenges researchers need to solve the technological shift brings huge possibilities to connect people socially thus creating a new chapter in human communication history [12]. Social interaction within the metaverse now enables users to engage through virtual ecosystems by using avatars which let them share spaces in real-time for shared virtual experiences. The metaverse contains numerous virtual communities ranging between extensive social platforms and personalized user-built virtual domains where people establish fresh interaction means for interaction teamwork and social bonds. The metaverse contains two types of spaces: specific-purpose digital structures that facilitate activities like virtual conferences and concerts and open social zones for unrestricted exploration and creation. Within the metaverse social hubs function as primary gathering spaces that blend game spaces with interactive events to bring users together digitally such as Meta's Horizon Worlds and VRChat featuring shared spaces that boost user connection and sense of presence. Metaverse immersive areas deliver life-like interactive settings that duplicate realistic settings and create new fictional domains that permit users to conduct both business and personal socializing better than standard virtual communication [13]. Online development tools offer players and developers the ability to construct original game environments while establishing gameplay guidelines and implementing non-fungible tokens (NFTs) and digital assets that fuel metaverse economic growth. Multiple leading metaverse platforms exist nowadays because each platform offers different types of social interaction models. Virtual reality users in Horizon Worlds created by Meta can discover and connect through environments that people generate which support live updates handy teamwork spaces and exciting social games. VRChat delivers social connectivity by allowing users to create personalized avatars that explore community-developed worlds where they engage in deep interactions similar to genuine human interactions. Users of the blockchain-powered decentralized space Decentral can acquire, sell, and build virtual land before both creating private experiences and partaking in social activities with economic potential as an example of a user-driven metaverse that provides ownership and governance rights. The various platforms of Roblox, Second Life alongside The Sandbox demonstrate that social interaction evolves because it combines video games with virtual commerce and digital communities into interconnected interactive experiences [14]. Users now leverage the metaverse for vital human interaction as they take part in virtual mental health support groups and they access digital classrooms as well as collaborative workspaces which allow people to work learn and develop themselves while transcending geographical boundaries. The availability of digital communities helps people create a feeling of belonging along with offering contact possibilities for people with disabilities and remote workers and those who need special interest-focused communities. The metaverse delivers exceptional social networking abilities but persists with important difficulties in fostering all-inclusive access while protecting users and respecting their privacy together with maintaining identity security and moderation of content. Metaverse technology development speed creates new risks regarding digital addiction alongside distress from extreme virtual interactions and potential risks of false information and destructive activities in virtual environments. The development of an inclusive and safe metaverse requires developers to team up with government officials and users to build an environment focused on user security and inclusive platforms. The growing metaverse development shows the potential to establish itself as a fundamental element of human social contact since it creates deep interactive digital experiences between physical worlds and software-based realms to reshape people's social responses in digital times [15]. Virtual worlds achieve inclusivity through the creation of environments that enable all participants from diverse cultural and physical backgrounds to interact without barriers thus they establish feelings of community and representation among the participants. Users need thoughtful tools to design their avatars because they should express their authentic selves by selecting skin tones and body shapes along with cultural apparel while benefiting from sensory accommodations that include spoken-text conversion and sign language virtual characters. Ease of access functions as a key metaverse element that provides users at all ability levels including people with disabilities and different tech proficiency to navigate through the metaverse. User interfaces need to remain easy to use while providing different interaction features such as voice recognition and adaptive control systems in addition to virtual space designs that follow universal principles for the accessibility needs of people with various disabilities. A virtual community environment needs to actively support distinct viewpoints and cultural backgrounds from all members so different life perspectives can evolve within a vibrant digital social network. User-generated content enables people to create virtual areas that mirror actual culture along with linguistic domains and native rituals thereby making online environments represent real ethnic diversity [15].

3 Virtual Spaces in the Metaverse for an Inclusive Environment

People who experience social isolation without meaningful social connections face severe negative mental health effects as well as psychological and social consequences. Feeling lonely during extended periods of isolation increases the likelihood of developing depression and setting off symptoms of anxiety and multiple mental health challenges. Being cut off from others makes people lose their self-esteem and sense of purpose and it makes worthlessness and hopelessness worse. Vulnerable groups such as the elderly combined with introverts together with people from marginalized communities experience intensified unfavourable effects from being isolated. Older people face physical and mobility problems which make it harder for them to socialize leading to severe loneliness symptoms and cognitive deterioration. Prolonged isolation affects introverts emotionally because they face difficulties making meaningful social connections even though they usually prefer spending time alone [16]. Social isolation becomes pervasive among LGBTQ+ individuals together with racial minorities and people with disabilities because systemic discrimination maintains their marginal status and restrictive environments restrict their access to support networks. Modern technological progress in human interaction has heightened the contradiction between staying socially connected through devices even though individuals remain alone the digital communication tools supplied by technology lead to obligations that sometimes result in unfulfilling connections because humans require emotionally profound bonds. Social media has shown increased loneliness among young people who experience more social comparisons through its platform. The solution to social isolation demands a complete strategy that transforms technological advantages into meaningful connections through proper management of its disadvantages. People who use the metaverse portion of virtual communities find potential value in these virtual spaces because they create detailed and interactive environments that enable authentic social connections across all distances [17]. Metaverse virtual communities serve as essential tools for decreasing social isolation because they present members with cross-limits spaces for socializing and group work activities. Digital environments within the metaverse allow users to form real-time connections through shared activities and build relations that form between people who share common experiences or interests. Metaverse platforms differ from traditional social media because they focus on user-driven actions that enable communication through avatars voice tools and immersive environments. Interactions between users at this stage mimic actual social surroundings which results in a feeling of being present together and reducing loneliness. Geographically isolated people alongside elderly and disabled individuals gain access to outside connections and social events due to virtual communities that let them connect with others without requiring physical movement [18]. Inside the metaverse introverts alongside socially anxious individuals find safe spaces to connect with others because they can slowly develop their confidence while acquiring social competencies. Mainly through virtual communities' users can partner with others to work on shared projects while exploring innovative ideas and exchanging information which leads to enriching personal growth intellectually as well as socially. The full-dip experience of these digital realms boosts the emotional connection between users through experiences that appear genuinely genuine and purposeful. Virtual communities destroy physical limits and generate shared spaces which should help people escape social isolation to strengthen their mental well-being thus establishing new online social interaction methods for digital times [19]. Metaverse platforms now establish supportive environments through which users receive social support and mental well-being solutions since these virtual communities show true potential for positive impact. Social interaction platforms such as VRChat and AltspaceVR provide users with events and workshops and group activities because they create virtual environments that promote a sense of belonging. The platforms create safe areas for individuals with social anxiety and physical disabilities to learn social abilities while developing their social skills and comfort with dialogue. The virtual mobile platform "Tripp" serves as a mental health-focused community by using virtual environments for emotional wellness and mindfulness training. Users can use the platform Tripp to practice stress reduction through meditations and therapeutic activities along with interpersonal interactions with people facing similar mental health difficulties. Through Second Life users have established virtual support groups that provide grieving people and those with addiction or chronic illness access to peer support and healing opportunities [20]. The metaverse needs inclusive virtual spaces as a minimum requirement to create a welcoming environment for people coming from diverse backgrounds which includes disabled individuals and members of various cultures. The approach of inclusive design begins by prioritizing user needs for every product development rather than building accessibility after completion. User-fitted avatars stand out as vital components of inclusiveness because they provide individuals with ways to display their genuine gender characteristics ethnic background and personal style. A well-represented environment in virtual settings allows users to experience being acknowledged which minimizes feelings of alienation or marginalization. People with disabilities need visual or auditory impairment adjustments and screen readers and voice commands as part of an inclusive design framework [21]. Designing with cultural inclusivity necessitates designers to have an awareness of multiple values and traditions and several communication patterns that users bring from different cultural backgrounds. Virtual space design which prioritizes inclusivity allows developers to build spaces that enhance diversity while creating a platform for belonging among all users [22]. Access to technology alongside internet bandwidth and hardware function as vital elements for the development of the metaverse because they create obstacles to total participation. People in low-income areas along with residents of rural locations face unreasonable costs for high-speed internet service which makes VR headset purchase prices unaffordable leading to restricted participation in virtual communities. Crossing the digital divide requires three key aspects which combine hardware subsidization with comprehensive broadband expansion and low-bandwidth metaverse platform development to reach basic devices [23]. The metaverse inclusivity depends on government-tech company and non-profit organization partnerships establishing funding structures alongside training mechanisms for underserved communities. Platform designs that consider scalability enable programs to operate effectively on both powerful VR systems and mobile platforms hence attracting diverse end users. The metaverse becomes a universal and diverse platform where every person from various social groups can participate as development efforts eliminate access barriers. Users must feel secure both safety-wise and privately to fully take advantage of virtual communities in the metaverse. The development of digital secure environments demands the implementation of user protections against harassment bullying and exploitation because such dangers diminish the potential benefits of these platforms. The essential elements for real-time harmful behavior detection consist of AI-driven content filters combined with human moderators through robust moderation systems [24]. The combination of specific community standards and reporting tools gives users the ability to stop abusive behavior and established data security and privacy policies create conditions for trust and reliability. The safety requirements for protected populations which incorporate age-specific content controls together with stealth modes and specialized help facilities should include children together with trauma individuals and vulnerable social group members. The format of virtual environments plays a safety-enhancing role when it develops interaction arenas that foster positive social interactions at the same time prevent harmful behavior between users. Virtual spaces designed with open areas and well-lit well-defined zones decrease the probability of harassment while private spaces offer users private communication areas. Developers who put safety first and prioritize privacy will construct virtual communities that promote accessibility as well as provide users with a safe space to engage and feel empowered [4].

4 Ethics and Equilibrium: Creating Virtual Communities for Mental Well-Being

The concept of digital well-being describes how people should handle technology to make them better physically and mentally without damaging their health. Users need to understand digital well-being fundamentals due to the rising importance of metaverse platforms and similar digital platforms in our everyday lives. Prolonged screen use and dependence on digital voice communications produce digital burnout symptoms combining fatigue with visual strain as well as attention problems and emotional distress. Users experience challenges in avoiding the metaverse because immersive virtual experiences create obstacles that prevent them from keeping a healthy divide between their real-life and online existence. Strong resistance to digital burnout requires people to develop vital relationships with technology through proper preventative strategies [25]. People should establish specific screen time limits through which they determine designated virtual connection periods and built-in recovery intervals to rest their minds. Users who practice mindfulness techniques through meditation or digital detachment sessions will develop a better perception of their technology habits as well as how these habits affect their wellness. Platform users need digital well-being education regarding technology effects because this knowledge enables them to select their screen time habits more effectively through given management tools. Users and developers working collaboratively through mindfulness-based culture development will create a metaverse that promotes mental health instead of harming it [26]. Off-line and online communities must combine forces to develop complete mental well-being strategies that blend real-life connections with digital virtual relationships. Members need to use metaverse opportunities for enhanced connection while understanding the essential value of direct human interactions. Online platforms work together with real-life social connections by offering tools and spaces that enable users to build connections between their virtual and physical activities. Metaverse platforms organize virtual events that motivate users to execute activities that benefit their physical local communities such as organizing charity drives or initiating environmental programs and organizing physical meetups [7]. Virtual support groups along with therapy sessions should provide follow-up activities and resources that direct participants toward offline self-care practices including exercise journaling and spending time outdoors. Participants benefit from hybrid events allowing them to decide between in-person and online attendance thus creating an inclusive event structure. The metaverse creates a unified environment that uses technology to connect digital communities with real-world connections to drive purpose-based meaningful social participation outside virtual spaces. The approach both improves mental welfare and makes the social structure stronger by pushing users toward active participation between virtual dimensions along physical environments [27]. The built future of social sustainability in the metaverse will succeed through the combination of health solutions for mental well-being with digital ecosystem sustainability alongside ethical digital engagement management. The future development of the metaverse shows promise to build itself into a foundation for offering mental health assistance which remains accessible throughout the world with innovative inclusive resources. Users must use a well-being-centered platform design to build genuine connections while achieving a healthy equilibrium between online and offline relationships [28]. Digital sustainability depends on innovation for its development because artificial intelligence combined with virtual reality and data analytics technologies allows engineers to create adapted mental health assistance programs for individual needs. This technology combination enables immediate emotional aid delivery by utility-oriented software and builds customizable therapeutic virtual environments. We must take proper measures to mitigate the negative consequences of using virtual communications excessively because they could harm our capacity to function socially and intensify societal inequalities. The social sustainability of the metaverse demands developers and policymakers to work with mental health professionals and users in establishing ethical guidelines that regulate how technology gets used responsibly. The metaverse can develop into an instrument for fostering emotional health and social sustainability through mental well-being priority which creates a harmonious digital-physical world amalgamation for enriching human life [29]. The protection of privacy alongside data security represents the main ethical issue affecting metaverse virtual communities' development and operation. Digital environments allow users to create enormous personal data while sharing their biometric information along with their behavioral patterns and their social interactions. The valuable user data enables customized experiences yet users have genuine worries about its collection storage as well as possible utilization methods. The unauthorized access to user data in virtual platforms leads to identity theft and financial fraud along with other exploitative activities which diminishes trust within the platform. The fully immersive metaverse experience creates unclear distinctions between areas that should remain private and those that are considered public since users struggle to track data collection activities [30]. The combination of VR headset data collection with eye tracking and facial expression could potentially expose personal mental information without proper consent regulations. The solution to these problems requires strong legal guidelines and moral standards that should regulate virtual data management systems. Predictable privacy regulations must include full system encryption together with anonymous data management as well as detailed user data collection documentation that presents clear explanations about data handling practices. Data protection systems should be integrated within virtual environment frameworks from their initial design stages using the "privacy by design" principle. Users need access to self-control tools which include adjustable privacy options and choices to stop giving away their data. The metaverse requires specialized legal protections because European Union GDPR foundations alone fail to address its specific privacy needs. Developers who focus on user privacy and data protection can generate trust while maintaining virtual communities with both safety and security as well as user rights respect [31]. The success and sustainability of virtual communities depend heavily on metaverse developers who guarantee ethical conduct as well as inclusivity together with fairness. Virtual spaces that bring together individuals from different backgrounds need design elements to prevent discrimination along with cyberbullying and exploitation and promote equity. Inclusive environments start by making spaces that welcome all people independent of ability status and ethnicity and gender as well as socioeconomic background. Virtual experiences become accessible for people with disabilities through the developer's introduction of features like screen readers voice commands and customizable interface functions [32]. The diversity of the real world must appear through digital characterization as well as virtual environments to enable users to witness themselves and others in a way that builds empathy and understanding. Active moderation combined with community management becomes essential to achieve inclusivity since it helps address unacceptable behaviours that include harassment and hate speech as well as practices that create exclusion. AI-style moderation tools assist in detecting poison in user behavior during real-time conditions yet humans need to manage complicated scenarios because they provide the necessary judgment for ensuring equity. Digital economic equality becomes an important ethical aspect because the metaverse continues to develop virtual economies and digital assets. Such virtual systems need developers to establish complete transparency alongside fairness while making sure resources remain accessible to users through ethical principles that forbid exploitation. Developers who put digital ethics first will build virtual communities that both provide fair access and support their users to thrive securely in environments that empower them [33]. Legal policies in metaverse governance represent an essential requirement to defend user rights together with the establishment of a healthy digital atmosphere. Traditional virtual platforms differ from the metaverse because it functions through decentralized linked spaces so regulation has become more intricate. Users experience security risks including exploitation and abuse of power along with fraud because metaverse lacks structured rules and accountability systems. Metaverse governance needs developers to work together with policymakers and users through collaborative efforts to establish transparency systems that empower users and promote accountability standards. A decentralized governance model allows communities to make decisions together through blockchain voting systems and other decentralized systems. The metaverse user-centric ethos finds support in this approach which nevertheless needs security measures to stop exploitation and guarantee access to everyone [34]. Governmental bodies should transform existing legislation and develop new regulations that will solve specific metaverse problems including digital property rights disputed jurisdictions and virtual asset governance. Users need to understand legal protections regarding virtual property taxes and they must know what safeguards exist when experiencing harassment or fraud in the metaverse. These regulatory issues remain under development as governments together with organizations begin to investigate the needed answers. Further development of legal regulations requires more work through initiatives like the Metaverse Standards Forum to establish basic operational standards. Future laws governing the metaverse will incorporate platform self-regulation alongside industrial standards and governmental oversight which will safeguard and maintain virtual communities' safety fairness and sustainability. The active approach toward governance and regulation enables stakeholders to create a metaverse framework that honors user rights along with increasing innovation and developing a positive digital environment [33].

5 Conclusion

A fair governance structure that safeguards user rights represents the most significant necessity for the metaverse operating system. Traditional online platforms differ from the metaverse because the decentralized structure and interconnected virtual space design create extra challenges for governance and regulation. Users experience security risks including exploitation and abuse of power along with fraud because metaverse lacks structured rules and accountability systems. An effective metaverse governance system needs developers to work jointly with policymakers and users to establish frameworks which support transparency along with empowerment and accountability for users. The concept of decentralized governance works by allowing users to decide together through blockchain voting systems in order to determine decisions. Such governance methods correspond to the user-centric nature of the metaverse since they protect against abuse and defend inclusivity. The regulatory bodies need to develop new legislation together with adapting current laws to address the metaverse-specific difficulties including virtual property rights and owner rights and conceivable conflicts between different legal jurisdictions. Virtual property taxation remains an unsolved legal matter alongside the need for protection against metaverse harassment and fraud. These regulatory issues remain under development as governments together with organizations begin to investigate the needed answers. Further development of legal regulations requires more work through initiatives like the Metaverse Standards Forum to establish basic operational standards. Virtual community safety and equity together with sustainability will be achieved through platform self-regulation and industry standards while governments maintain necessary oversight. Active participation in metaverse governance along with appropriate regulation from stakeholders leads to user rights protections and innovation growth as well as development of a positive digital environment.

References

1. Ritterbusch, G.D., Teichmann, M.R.: Defining the metaverse: a systematic literature review. IEEE Access **11**, 12368–12377 (2023)
2. Dolata, M., Schwabe, G.: What is the Metaverse and who seeks to define it? Mapping the site of social construction. J. Inform. Technol. **38**(3), 239–266 (2023)
3. Dhıllon, P.K.S., Tınmaz, H.: Immersive realities: a comprehensive guide from virtual reality to metaverse. J. Educ. Gift. Young Sci. **12**(1), 29–45 (2024)
4. Ritonga, M., Farid, A.S., Manda, V.K.: Media convergence and the development of interactive broadcasting with Metaverse technology. Feedback Int. J. Commun. **1**(4), 238–249 (2024)
5. Kaitane, W.S., Shrestha, S., Lean, C.P., Purnshatman, T., Ng, P.K., Kong, F.Y., Khan, M.R.B.: Immersive horizons exploring virtual reality, augmented reality, and the metaverse in game development-a comprehensive review. Malaysian J. Sci. Adv. Technol. **15**, 222–234 (2024)
6. Ud Din, I., Almogren, A.: Exploring the psychological effects of Metaverse on mental health and well-being. Inform. Technol. Tour. **25**(3), 367–389 (2023)
7. Maloney, D.: A youthful metaverse: towards designing safe, equitable, and emotionally fulfilling social virtual reality spaces for younger users (2021)

8. Raman, R., Hughes, L., Mandal, S., Das, P., Nedungadi, P.: Mapping metaverse research to the sustainable development goal of good health and well-being. IEEE Access (2024)
9. Hanandini, D.: Social transformation in modern society: a literature review on the role of technology in social interaction. Jurnal Ilmiah Ekotrans Erudisi **4**(1), 82–95 (2024)
10. Cimperman, S.: Informal socialization in virtual work environments: creating a digital culture of connectivity. Glob. Bus. Organ. Excel. **43**(1), 53–71 (2023)
11. Au, A.: Reconceptualizing the generation in a digital (izing) modernity: digital media, social networking sites, and the flattening of generations. J. Theory Soc. Behav. **50**(2), 163–183 (2020)
12. Greenfield, P.M.: Communication technologies and social transformation: their impact on human development. Child. Chang. Worlds Sociocult. Temp. Persp. **45**, 235–273 (2019)
13. Ghosh, A., Hassija, V., Chamola, V., El Saddik, A.: A survey on decentralized Metaverse using blockchain and web 3.0 technologies, applications, and more. IEEE Access (2024)
14. Rane, N., Choudhary, S., Rane, J.: Metaverse marketing strategies: enhancing customer experience and analysing consumer behaviour through leading-edge Metaverse technologies, platforms, and models. Platforms, and Models (2023)
15. Gupta, J., Rani, L., Kaur, M.: AR/VR technologies in the Metaverse ecosystem. In: Handbook of Research on AI-Based Technologies and Applications in the Era of the Metaverse, pp. 127–143. IGI Global (2023)
16. Razai, M.S., Oakeshott, P., Kankam, H., Galea, S., Stokes-Lampard, H.: Mitigating the psychological effects of social isolation during the covid-19 pandemic. BMJ **52**, 369 (2020)
17. Brandt, L., Liu, S., Heim, C., Heinz, A.: The effects of social isolation stress and discrimination on mental health. Transl. Psych. **12**(1), 398 (2022)
18. Dwivedi, Y.K., Hughes, L., Baabdullah, A.M., Ribeiro-Navarrete, S., Giannakis, M., Al-Debei, M.M., et al.: Metaverse beyond the hype: multidisciplinary perspectives on emerging challenges, opportunities, and agenda for research, practice and policy. Int. J. Inform. Manag. **66**, 102542 (2022)
19. Yeganeh, L.N., Fenty, N.S., Chen, Y., Simpson, A., Hatami, M.: The future of education: a multi-layered metaverse classroom model for immersive and inclusive learning. Fut. Internet **17**(2), 63 (2025)
20. Ijaz, K., Tran, T.T.M., Kocaballi, A.B., Calvo, R.A., Berkovsky, S., Ahmadpour, N.: Design considerations for immersive virtual reality applications for older adults: a scoping review. Multimodal Technol. Interact. **6**(7), 60 (2022)
21. Patrick, V.M., Hollenbeck, C.R.: Designing for all: consumer response to inclusive design. J. Consum. Psychol. **31**(2), 360–381 (2021)
22. Mouratidis, K.: Urban planning and quality of life: a review of pathways linking the built environment to subjective well-being. Cities **115**, 103229 (2021)
23. Rawat, D.B., Hagos, D.H.: Metaverse Survey and Tutorial: Exploring Key Requirements, Technologies, Standards, Applications, Challenges, and Perspectives. arXiv preprint arXiv:2405.04718 (2024)
24. Treep, F.C.: E-Participation in the Digital Future: An Exploration of E-participation in the Metaverse (Master's thesis, University of Twente) (2024)
25. Dwivedi, Y.K., Kshetri, N., Hughes, L., Rana, N.P., Baabdullah, A.M., Kar, A.K., et al.: Exploring the darkverse: a multi-perspective analysis of the negative societal impacts of the metaverse. Inform. Syst. Front. **25**(5), 2071–2114 (2023)
26. Oksanen, A., Celuch, M., Oksa, R., Savolainen, I.: Online communities come with real-world consequences for individuals and societies. Commun. Psychol. **2**(1), 71 (2024)
27. Ullah, H., Manickam, S., Obaidat, M., Laghari, S.U.A., Uddin, M.: Exploring the potential of metaverse technology in healthcare: applications, challenges, and future directions. IEEE Access **11**, 69686–69707 (2023)
28. Allam, Z., Sharifi, A., Bibri, S.E., Jones, D.S., Krogstie, J.: The metaverse as a virtual form of smart cities: opportunities and challenges for environmental, economic, and social sustainability in urban futures. Smart Cit. **5**(3), 771–801 (2022)
29. Ho, C.K., Lee, C., Tseng, K.J., Kok, C.L., Koh, Y.Y.: Navigating the metaverse: exploring the impact on building performance and operations, decarbonization, and human wellbeing. Int. J. Eng. Sci. **13**(1), 44–50 (2024)

30. Chawki, M., Basu, S., Choi, K.S.: Redefining boundaries in the Metaverse: navigating the challenges of virtual harm and user safety. Laws **13**(3), 33 (2024)
31. Tukur, M., Schneider, J., Househ, M., Dokoro, A.H., Ismail, U.I., Dawaki, M., Agus, M.: The metaverse digital environments: a scoping review of the challenges, privacy and security issues. Front. Big Data **6**, 1301812 (2023)
32. Al-kfairy, M., Alfandi, O.: The ethical dilemma of educational metaverse. Recent Adv. Evolut. Educ. Outreach **1**(1), 006–016 (2024)
33. Tabassum, A., Elmahjub, E., Qadir, J.: Ethics and regulation in metaverse: risks, current approaches, and way forward. Understand. Metaverse Appl. Chall. Fut. **15**, 265–290 (2024)
34. Tu, X., Silva, B.: Are we ready for the metaverse? Implications, legal landscape, and recommendations for responsible development. Dig. Soc. **4**(1), 9 (2025)

Integrating Fuzzy Logic-Based Optimization in Renewable Energy Solutions for Sustainable Virtual Environments

Vishal Jain, Archan Mitra, and Sanchita Paul

Abstract The emergence of the metaverse as a virtual extension of reality presents unprecedented opportunities for innovation, collaboration, and immersive experiences. However, the energy demands associated with its operation pose significant challenges to sustainability. This chapter explores the integration of fuzzy logic-based optimization techniques into renewable energy systems to address energy efficiency and sustainability in the metaverse. By leveraging the adaptive and decision-making capabilities of fuzzy logic, the study proposes a framework to optimize energy consumption in virtual environments powered by renewable energy sources. The framework emphasizes dynamic load balancing, predictive analytics for energy demand, and intelligent resource allocation, ensuring energy efficiency without compromising user experience. The chapter also investigates This pathway also delves into the pragmatic approach to implementing smart grids and virtual power plants into the metaverse, administered by fuzzy logic controllers for enhanced energy resilience. Governance and policy frameworks are also dealt with in order to promote the implementation of renewable energy solutions within the metaverse. The technique outlined here is expected to demonstrate the bridge between technological advancements and environmental sustainability through fuzzy logic; this paves a practical path towards a green metaverse. Real-world case studies and simulation results highlight the potential of this methodology in enabling sustainable virtual ecosystems, underscoring its relevance for future developments in the metaverse.

Keywords Green metaverse · Fuzzy logic optimization · Renewable energy solutions · Sustainable virtual environments · Smart grids · Virtual power plants

---

V. Jain
Department of Computer Science & Engineering, School of Engineering & Technology, Vivekananda Institute of Professional Studies, New Delhi, India

A. Mitra
Amity School of Communication, Amity University Mumbai, Mumbai, Maharashtra, India

S. Paul (✉)
Neil Patel Digital India, Ghata, India
e-mail: sanchita.paul1896@gmail.com

S. Kautish et al. (eds.), *Green Metaverse: Fuzzy Logic Approaches to Sustainable Virtual Worlds*, Information Systems Engineering and Management 74,
https://doi.org/10.1007/978-3-032-08183-4_4

1 Introduction

1.1 Background and Significance

The metaverse has emerged as a revolutionary concept, functioning as a digital parallel to the physical world, where individuals can engage in immersive experiences, social interactions, and economic transactions [1]. As an extension of reality, the metaverse integrates advanced technologies such as virtual reality (VR), augmented reality (AR), artificial intelligence (AI), and blockchain, leading to expansive digital ecosystems that redefine communication, entertainment, and commerce [2]. The increasing adoption of the metaverse has significantly heightened energy demands, raising concerns regarding its long-term sustainability and environmental impact [3].

The metaverse requires considerable computational power from its supporting infrastructure like data centers, cloud computing networks, and blockchain technology; this leads to extensive energy consumption [4]. Virtual environments use real-time rendering, large data storage, and heavy data transfers requiring large and energy-intensive servers and systems for cooling. The high dependence on energy has as a result put metaverse zones in the category of major greenhouse gas emitters to exacerbate climatic concerns. It became important for researchers to explore feasible development of innovative energy-saving mechanisms compliant with life within digital spaces, sans compromise with user experience.

Amongst many ways to decarbonize the metaverse, one of the most important areas pertains to renewable energy applications of the digital infrastructure. Application of smart grids, virtual power plants (VPPs), and AI-driven energy optimization techniques holds promise for enhancing energy efficiency in metaverse ecosystems [5]. Of these, fuzzy-based optimization has lately attracted much concern for its ability to handle uncertainty, take decisions in real time as well as optimize resource allocation in complicated energy systems [6]. Leveraging the fuzzy logic controllers in real-time monitoring optimizes how the energy is made to balance the load, predicts the usage model according to time of consumption, and provides full utilization of renewable resources such as the sun or wind [7].

To build a green metaverse, which operates on optimized energy consumption patterns and integrates renewable energy, requires interdisciplinary team effort from computer science, policy, and environmental research departments [8]. This research has explored how eco-efficiency can be positively reinforced through the use of hybrid (fuzzy logic) optimization, bridging the gap between technical sophistication and ecological integrity, together with analyses of blog posts to identify new information. Governance frameworks are also seen as being able to ensure that the metaverse increasingly transitions to pure renewable energy, thereby offering the digital future of environmental sustainability [9].

1.2 Problem Statement

Although the metaverse poses a range of possibilities, surprising sustainability-related challenges are patently plain due to the high energy requirements. The current digital apparatus supporting Metaverse embodies a scenario of non-renewable energy sources, including cloud servers, blockchain networks, and decentralized computing systems, that have dirty hands, leading to the wrongful carbon footprints. The global determination to strike a balance with carbon neutrality speaks volumes that one cannot in a heartbeat raise the full speed of optimization in energy, maximizing user involvement, thereby filtering the air bigtime in terms of technology.

One major hurdle in the deployment of renewable energy in this digitized ecosystem is the absence of intelligent energy management techniques. The existing energy management strategies for the metaverse will involve reducing carbon dioxide emissions and behaving very much in a sense of energy efficiency, mostly without the possibility of ventilating through renewable generation forms. Virtual worlds through poor energy optimizations may drive humanity toward the brink of energy crisis and further environmental disruptions. This study takes this into consideration and aspires to propose a fuzzy logic-based optimization framework that can help in the efficient integration of renewable energy into a metaverse infrastructure.

1.3 Research Objectives

The aim of this research is:

- To analyze the key energy consumption challenges associated with metaverse ecosystems and their environmental impact.
- To explore the role of fuzzy logic-based optimization techniques in enhancing energy efficiency in virtual environments.
- To develop an AI-driven framework for dynamic load balancing, predictive analytics, and intelligent energy resource allocation.
- To investigate the feasibility of integrating smart grids and virtual power plants within metaverse platforms.
- To assess policy frameworks that facilitate the adoption of renewable energy solutions in metaverse ecosystems.
- To validate the proposed fuzzy logic optimization model through case studies and simulations.
- To propose a roadmap for future research in sustainable metaverse development through AI and renewable energy solutions.

1.4 Research Questions

- What are the primary energy challenges affecting metaverse ecosystems?
- How can fuzzy logic-based optimization enhance energy efficiency in virtual environments?
- What is the feasibility of integrating renewable energy solutions into metaverse infrastructures?
- How can smart grids and virtual power plants be incorporated into the metaverse for sustainable energy management?
- What governance frameworks can support the transition to a green metaverse?
- How do AI-driven energy optimization models compare to traditional energy management techniques in virtual environments?

1.5 Scope and Limitations

This study primarily focuses on the optimization of energy consumption in metaverse ecosystems through the application of fuzzy logic-based decision-making and renewable energy integration. The research explores how AI-driven energy solutions can improve efficiency, reduce carbon emissions, and enhance sustainability in digital environments. The study does not cover hardware-level innovations, such as energy-efficient GPUs or server architectures, but instead concentrates on algorithmic optimization and governance frameworks.

A fundamental limitation of this research is the absence of large-scale real-world implementation of AI-powered model optimization for energy in a metaverse. While the study includes simulations and case studies, the need for further empirical validation in order to reach large-scale implementation remains paramount. The social-cultural barriers to the adoption of renewable energy in digital spaces are not factored into the equation, such as regulatory resistance and financial constraints. However, the study, under these limitations, provides a theoretical and computational structure for further exploration in relation to sustainable metaverse ecosystems with AI and fuzzy logic optimization.

2 Literature Review

2.1 Understanding the Metaverse Ecosystem

Growth of Virtual Worlds, Gaming Platforms, and Virtual Economies

The metaverse has developed into a propitious digital realm, employing virtual reality (VR), augmented reality (AR), and blockchain technologies to carnivals gaming, social interplay, and many other economic endeavors. Decentralized digital economic systems offer a solid basis for the idea of intertwining complicated separators such as cloud computing, artificial intelligence, and high-speed internet; all expediting seamless real-time interactions within digital space. [1].

Gaming platforms have been instrumental in seeding the metaverse, with games such as Fortnite, Roblox, and Decentraland being core protagonists in the inception of virtual economies [4]. These mixed-platform online gamers can trade, exchange, or partake in all kinds of gaming-related transactions involving digital goods, services, and other experiences. In continuation, virtual economies have moved to the mainstream financial world, generating virtual goods valued at over $100 billion in 2023 [10].

The adoption of decentralized finance (DeFi) in virtual economies has accelerated the expansion of the Metaverse. Users can invest, trade, and hold digital assets in a fashion very much akin to traditional economic frameworks, thus cementing the Metaverse to the realm of pure transparency and security, thanks to blockchain [5]. With industrial giants such as Meta (Facebook), Microsoft, and Google now supporting this invention, fame of the Metaverse has transformed it into the seat of digital interactions and global economic dynamics [3].

The rising stars of the virtual world concern many who contemplate the Metaverse about being sustainable. High electricity consumption for rendering persistent and dynamic digital worlds creates a demand for powerful energy solutions that can help assuage the environmental impact of the Metaverse infrastructural facilities [11].

Energy Requirements of Metaverse Infrastructures

The computational requirements of the metaverse are detrimental to energy sustainability. Virtual environments require continuous real-time rendering, high levels of AI interaction, and copious amounts of data lifting among others, all of which contribute to energy consumption at unusually high levels [12]. The infrastructure underpinning the metaverse relies heavily on cloud computing, blockchain networks, and high-performance data centers, all of which drive substantial energy demand (Doraisamy et al. [4] for the prior statement).

Cloud computing and data centers form the core of the metaverse as they provide storage, processing power, and connectivity that form the immersive components of the metaverse. Hyperscale data centers, powering the platforms of the metaverse, use massive amounts of electricity due to the global data center energy consumption of about 200 terawatt-hours (TWh) per year, equivalent to some mid-sized countries' total electricity use [10]. Advanced cooling systems are also required for such infrastructure, causing an even greater energy footprint for data centers [3].

Energy consumption also forms a massive part of blockchain technology, which is a key component in virtual economies. Perhaps, thus far, many metaverse platforms use cryptographic validation methods such as PoW or PoS to secure digital transactions and NFTs [5]. While energy consumption is much lower in PoS-based blockchain platforms than those using PoW, integrating PoS-based blockchain systems at large metaverse scales requires necessary environmentally friendly energy programs [9].

And there is one more angle to be seen: the massive carbon footprint left by metaverse infrastructure. The energy-hungry character of metaverse platforms implies that inordinate amounts of carbon emission take place. Reports have argued that only blockchain-based applications would contribute to more than 60 million metric tons of CO_2 annually, with possibilities arising as the metaverse grows [4]. This led to the development of smart grids, AI-driven energy optimization and renewables integration to produce sustainability [2].

Faced with such challenges, researchers and industry captains both are now looking into green computing, AI-based electricity management, and renewable energy-powered data centers to minimize the impact of virtual environments to the environment [13]. With an AI-driven platform, better load balancing does not let too many resources go to waste in unproductive work, as it ensures the optimum balance of use at real-time demand points [3]. Projects about decentralized renewable energy sources for the metaverse within solar-powered data centers and the potential use of AI in energy supply can establish an interesting way to kick-start sustainable metaverse ecosystems [11].

2.2 Sustainability Challenges in the Metaverse

High Computational Demands of Virtual Environments

To render a comprehensive and immersive virtual experience, the metaverse calls upon high-performance computing (HPC), cloud infrastructure, artificial intelligence (AI), and blockchain technologies. The computational agonies of metaverse environments emanate from rendering non-interactive 3D worlds, the AI-driven interactions, and the continuous syncing of data across global networks [2]. Unlike other online platforms, applications designed for the metaverse require high-fidelity in their graphics, use spatial computing, and provide continuous virtual environments-all of these considerably increase energy consumption and dependence on hardware [12].

The heavy usage of real-time physics simulation, deep learning models of AI avatars, and high-resolution virtual objects together increases the computational load [11]. Moreover, the cloud-based rendering systems-NVIDIA Omniverse and Microsoft Mesh-use graphical processing units (GPUs) and tensor processing units (TPUs) and require hundreds of megawatts of power to go live [10]. Due to the dependency of constant processing for a seamless interaction virtually, the ownership of decentralized assets, and thus blockchain security that needs to be enforced, the global ecosystem bears the burden of further strain on computing infrastructure [3].

Moreover, decentralized storage solutions such as InterPlanetary File System (IPFS) and blockchain-based data repositories amplify computational demands due to their redundancy mechanisms and cryptographic verification [9]. These technological advancements, while essential for metaverse security and scalability, exacerbate energy consumption challenges, necessitating urgent sustainable solutions.

Carbon Footprint and Impact on Global Energy Consumption

It is true that growth in computational complexity in the metaverse leads to carbon emissions and energy use on a global scale. Currently, data centers are estimated to account for almost 1% of total global electricity consumption and are likely to increase, with the spread of the metaverse [5].

Stemming from various metaverse economies based on non-fungible tokens (NFTs), decentralized finance (DeFi), and virtual lands, the use of blockchain is a huge contributor to carbon emissions. It is estimated that the power consumption of the Bitcoin Blockchain stood at around 91 terawatt-hours, which is around as much electricity consumed by the middle-range countries [4]. While there has been a substantial reduction in the energy consumption of PoW-based chains such as Ethereum 2.0 by over 99% in blockchain energy consumption, nevertheless, metaverse-related smart contract computations require enormous computational power [10].

Devices of virtual reality (VR) and augmented reality (AR) crucial for metaverse experiences may escalate energy demand further. High-resolution, technologically advanced VR helmet sets, haptic feedback sensors, and motion-tracking devices necessitated a high level of power consumption through prolonged user exposure [3]. One VR user traversing the immersive metaverse platform might consume more energy than an average, run-of-the-mill online-streaming service because of real-time 3D graphic processing [11]. With a scenario of millions of metaverse users simultaneously employed in multiple metaverse activities, the energy demand is now significantly too high for the global electricity grids to handle.

The matter of environmental implications does not just lie on energy consumption; E-wastes come to the forefront. As the metaverse infrastructure takes form, hardware obsolescence cycles naturally shorten, resulting in higher disposal of electronic wastes due to real materials excavation to new-generation compute machinery [12]. Extraction, refinement, and subsequent manufacturing of GPUs, TPUs, and cloud storage systems stir up increased deviance in carbon emission and fundamentally challenge the sustainability of our ecosystem [9].

2.3 Introduction to Fuzzy Logic in Energy Optimization

Fuzzy logic is a coherent mathematical framework that has been designed for handling uncertainty, imprecision, and approximate reasoning, making it highly effective in complex and dynamic environments like for energy optimization [14]. In contrast, traditional binary logic, which denotes variables as either true (1) or false (0), is purely deterministic circular logic. Essentially, fuzzy logic represents a gradual transition to different states by results between 0 and 1 [15], they give a very flexible aspect to design intelligent decision-making systems under partial or uncertain data, thus finding use in EMS in the metaverse [12].

Fuzzy logic system (FLS) configuration consists of fuzzification (to turn crisp numerical inputs—e.g., demand, or availability of renewables, into fuzzy values), rule-based inference (to apply a set of predefined if–then rules to analyze the energy conditions for best possible solutions), and defuzzification (to convert fuzzy output into a non-fuzzy output for implementation on energy systems). Fuzzy logic allows energy optimization on-the-fly within a metaverse ecosystem by dynamically adjusting power allocation, load balance, and energy storage operations. Since the metaverse operates within an unpredictable and fluctuating environment—the behavior of user activity, response times, and renewable availability change rapidly-fuzzy logic outshines dry rule-based systems that rely on crude logic in processing fuzzy and real-time variables [16].

2.4 Renewable Energy Integration in Digital Environments

As the metaverse continues to expand, the incorporation of renewable energy solutions into its infrastructure remains vital. The energy-intensive nature of data centers, cloud computing, and blockchain transactions has created a massive demand for green energy alternatives, particularly the use of solar, wind, and hydroelectric power sources [17].

On the contrary, industry frontrunners like Google, Microsoft, and Amazon Web Services (AWS) are already moving toward transitioning their data center operations to 100% renewable energy [4]. An example of how Google is pushing forward for sustainability initiatives is by introducing energy matching that is carbon-free, distributing renewables to match demand in real-time. Similarly, Microsoft's Project Natick has been exploring the concept of underwater data centers as powered by offshore wind energy, thereby reducing the carbon footprint associated with traditional land-based facilities [18]. AWS, too, have begun investing in building large-scale solar farms to offset their cloud services' energy consumption, thus aligning their data infrastructure with global carbon neutrality goals [19].

Decentralized and edge computing models have also emerged as potential solutions to mitigate excessive energy consumption in large-scale data centers. Combining distributed computing frameworks with financing allows renewables to be supported in the metaverse applications by microgrids and minimize the need for centralized power generation, while renewable energy buddies can be formed for the sharing of the energy flows in a peer-to-peer model to absorb the pressure below the aggregated magnitude slightly [20]. This sustainability is motivated towards shared commerce of any network applications, such as mutual renewable-energy platforms amonge the nodes such disparate energy consumers choose for themselves as they need, by way of organizing excess production together [21, 22].

Blockchain implementations are also moving toward sustainable practices primarily for use in metaverses. In the case of Ethereum, the displacement of Proof-of-Work (PoW) with Proof-of-Stake (PoS) created high energy-saving, demonstrating that infrastructure to low-energy level authentication is feasible to be persisted within digital spaces [23]. Energy efficiency is valued by other projects like the Chia Network or the Algorand, combined with a way to prioritize energy-efficient consensus models on the grounds of being sustainable and more or less adhering to blockchain security [24].

2.5 *Smart Grids and Virtual Power Plants for Virtual Environments*

AI-Driven Smart Grids for Energy Distribution

Embedded in the metaverse, smart grids play a fundamental role in energy management, ensuring that power flows efficiently between data centers, blockchain operations, and virtual realms. In contrast to traditional power grids, using centralized distribution, smart grids work with artificial intelligence to allow for decentralized, self-sufficient and proactive energy management. AI-based smart grids take advantage of real-time data for predictive load acceptance and immediate action and control feedback when organizations have to cope with increasing and decreasing energy demand [25].

One advantageous application of AI-powered smart grids is the ability to incorporate various renewable energy sources, for example solar, wind, hydroelectric power, in order to ensure the sustainability of the metaverse infrastructure power unit [26]. Several AI algorithms would ensure instantaneous adaptation to energy supply and demand conditions, harmless and just energy traffic between these renewable energy source filled environments would prevent wastage of power and save the planet from carbon loading [12]. These could boost energy resilience by forecasting potential power cuts and by dynamically acting on the energy flow, thus lessening in turn the risks associated with any possible server outages or high latencies that could disrupt the metaverse virtual environment [27].

AI demand-response mechanisms are being reinforced towards future energy predictions in real-time working stocks. This training model fuses the methods of energy consumption includes VR transactions, blockchain applications, and so on at a heightened pace. Historical assessment methods of ML algorithms are exposed as an offering of sales such as blockchain energy trade technology, where the power distribution pattern is readjusted dynamically, fractionalizing- the digital experience [28].

In addition, energy trading circles exist between blockchain and AI agendas in designing smart grids or energy trade practices in the longer run. A decentralized market trade set on renewable energy download and yearly exchange offers the metaverse companies buying and selling renewable energy, which normally is quite resourceful [29]. Financial viability takes a stratospheric leap through smart grids predicated on AI programming and blockchain Energy trading through reducing dependence upon non-renewable resources and thereby making the environment more concerted where only the most profitable energy will exist [10].

Fuzzy Logic Controllers for Virtual Power Plants (VPPs)

Virtual power plants (VPP) provide a broadcast approach to managing energy consumption and supply in digital ecosystems by clustering distributed energy resources (DER) into one intelligent network using the adaptive control systems provided by fuzzy logic controllers which function to make decisions in real-time, while being adaptable and autonomous to the uncertain and variable energy conditions [30]. Such fuzzy logic systems have the potential of making well-thought-out decisions for an optimized energy allocation, thus ensuring uninterrupted electrical supply to the metaverse infrastructure [10].

A fuzzy logic-based VPP can deliver energy from multiple sources in the same system, such as solar farms, wind turbines and battery storage units, and adaptively dispatch power based on ever-changing requirements from the metaverse [31]. This is particularly important for the metaverse paradigms where energy demand fluctuates hugely because of the unpredictably started entity of user involvement as well as computational prodding [12].

The use of fuzzy controllers in VPP further improves energy reliability by continuously predicting the availability of renewable power generation and thereby programming energy consumption accordingly [31]. This, in order, helps to minimize power supply drawbacks throughout peak metaverse activity since surplus energy is earmarked for storage or efficient distribution within the smart grid network [9].

Optimization Models for Load Balancing in Digital Spaces

Load balancing has a significant impact on metaverse infrastructures as energy consumption varies significantly depending on various factors such as computational load, server proximities, and user interaction levels. AI-orchestrated optimization technologies were established to ensure that the digital energy grid remains stable and directed to meet the fluctuating requirement of power energy [28]. These models consist of fuzzy logics to intelligently distribute energy loads among different cloud nodes, the blockchain validation process, and anything connected to a virtual reality (VR) environment [10].

Energy-efficient fuzzy logic-based load balancing eventually encompasses real-time power availability, user activity level, and historical energy consumption trends to optimize power [12]. As dynamic energy-induced power wastage is conserved-and optimal operational burden is garnered in metaverse ecosystems-beyond the realm of conventional static load balancing methods or techs that adapt to dynamically changing energy-separated conditions.

Another approach to energy load balancing in virtual environments is the predictive energy allocation model, which uses AI-driven forecasting algorithms to anticipate future energy demands based on real-time data streams [32]. Fuzzy logic control and deep learning algorithms are both brought into the picture of predictive models and to help allocate energy resources ahead of time to prevent bottlenecks or power failure, thus enhancing the reliability of the digital infrastructure [33].

Hybrid energy systems are designed to integrate VPPs with decentralized battery storage solutions for enabling the storage of excess renewable energy, thus reducing the magnitude of the technical barrier. These storage systems are used as a buffer system for helping metaverse platforms maintain energy equilibrium in the face of peak period fluctuation [9].

2.6 Comparison of Fuzzy Logic with Traditional Energy Management Techniques

Traditional energy management systems work around a system where the energy usage is managed by deterministic models, predefined thresholds, and rule-based control mechanisms. While great when used in structured environments, the approaches do not allow for fast adaptability to the current conditions of flexible power requirements facing the energy conditions uncertainty and variability on the digital substrates—kinds of technology such as smart grids, cloud computing, metaverse ecosystems. Among the common elementary methods if viewed for action would be the spotlight shutoff controls, which set to predefined levels but fail to handle changing energy demand levels in an effective manner; PID controllers, switched into action because they contribute to linear control systems while in nonlinear and highly varying systems like the metaverse [34]. Thereby, being engine-bound to campaign for rule-based systems relying on set decision rules, helping to introduce a curt decision-making process, which disrupts renewable energy variation optimization.

Fuzzy logic outstrips the earlier method since it can continuously fix the power distribution system on a real-time basis. This guarantees that power is intelligently allotted to achieve load balancing among the distribution systems in the data centers, blockchain transactions, and metaverse activities [35]. Noteworthy is the assimilation of the solution to go with the renewable resource capacity. Its multiple variables include solar radiation levels, wind speed, and their mesh with power system stability in utilizing renewable energy resources fully and superbly to maximize production through such symbiosis. The main advantage of fuzzy logic in energy optimization is the ability to integrate with artificial intelligence and machine learning [33]. Fuzzy logic-AI hybrid systems come with a model that can be used to predict patterns of energy for the metaverse, allowing us to handle fluctuations in power demand even before the demand and supply situations could break out of their normal course.

Case Studies of Fuzzy Logic Applications in Real-World Energy Systems

1. **Fuzzy Logic in Smart Grid Management (Singapore's Smart Energy Initiative)**

Fuzzy logic-based smart grids are in use in Singapore to optimize the distribution of power in the urban energy network operating within the city. Renewable energy sources, including solar panel and wind farms are being taken into account along the escalating use of variable power consumption patterns and adjusting power allocation in real-time condition depending upon changing grid loads. Combining the fuzzy logic component with energy shifting could drop the peak-energy load by 15%; the amount of utilized renewable energy rose by 30% through prediction of a day's power-demand pattern; and grid-balancing efficiently reduced an enormous amount of expense for energy users [36].

2. **Fuzzy Logic in Data Center Energy Optimization (Google's AI-Powered Cooling System)**

Google has integrated fuzzy logic controllers into its AI-based data center cooling systems. This system adopts controllers designed to evaluate many variables including temperature, humidity, and server load in order to derive optimum energy consumption [37]. One benefit observed is a drop of 40% in energy consumption due to cooler-than-normal processes, where the carbon footprint of cloud services is reduced to 20% on one better side, and the hardware life is extended due to the effect of enhanced thermal management [38].

3. **Fuzzy Logic in Virtual Power Plant (VPP) Operations (Germany's Decentralized Energy Model)**

Germany has started putting to work fuzzy logic in its virtual power plant (VPP) grid that aggregates various DERs like solar farms, wind turbines, and hydroelectricity stations in Germany [39]. This fuzzy-logic-based VPP allows the integration of decentralized energy sources smoothly into the main grid, provides real-time energy balance-enforcing mechanisms that keep grid instability at bay, and offers autonomous trading mechanisms that would redistribute extra energy in an efficient manner.

2.7 Policy and Governance in Sustainable Metaverse Development

Existing Regulations and Policies Related to Virtual Ecosystems

The rise of the metaverse presents new governance challenges, particularly in regulating its energy consumption, digital infrastructure, and sustainability standards. While current regulatory frameworks primarily focus on data privacy, intellectual property rights, and digital transactions, environment and its ecological footprint have recently come under the purview of policymakers with respect to virtual ecosystems [10]. However, most of the green-governance regulations for the sustainability of the metaverse remain highly fragmented, underdeveloped, and lack coherent enforcement across geographical domains [12].

One example of these is an initiative surrounding Europe's green digital actively for carbon neutrality standards in data centers as well as in cloud computing and AI-enabled platforms [36]. The EU Energy Efficiency Directive proposes very precise objectives for energy conservation in digital infrastructures, and mandates that the carbon emissions of data centers and blockchain operations be reported and offset through renewable sources of energy [40]. They also propose the Artificial Intelligence Act that stands for socially responsible AI applications and objectives, to ensure that metaverse services run on AI are in line with global climate goals [9].

In the United States, the Federal Trade Commission (FTC) and Department of Energy (DOE) are studying ways to include energy efficiency mandates in digital futures, as per Jiang et al. [10]. The U.S. DOE's Energy Star Program for Data Centers offers standards for self-regulating power usage in the operation of cloud services and could be extended to be fully applicable in the metaverse. Similarly, the California Sustainability Standards for Emerging Technologies bring a strong contention about regulating high-carbon digital infrastructure in favor of renewably powered computing platforms [33].

In Asia, Singapore, Japan, and South Korea have taken the lead in pursuing low-carbon policies in virtual ecosystems [40]. Singapore is setting very strict green benchmarks for its metaverse applications, AI-powered computing, and blockchain systems within the Green Digital Economy Framework. In contrast, Japan's Strategic Energy Plan targets reducing CO_2 emissions by an additional 30% in its IT industries by 2030 to promote smart grids, decentralized energy systems, and AI load balancing [32].

Despite these policies, the lack of a unified international governance framework for metaverse sustainability remains a major challenge. Without coordinated global regulations, corporations operating in multiple jurisdictions may face conflicting standards, limiting the efficiency of large-scale energy optimization initiatives [33].

Challenges in Implementing Energy Efficiency Standards in Metaverse Platforms

The implementation of energy efficiency standards in metaverse ecosystems faces several obstacles. A greater challenge to VR, AR, and blockchain applications are the heavy computational demand requring continuous real-time processing [10].

For example, the majority of digital infrastructure has still relied on traditional power grids controlled by the fossil-fuel system; thus, the transition to low-carbon alternatives is cumbersome, with significant costs and technical complexities [12].

We now face the issue of reporting carbon standards applied to virtual environments. There are compulsory disclosures to be made by these regulations such as the EU's Carbon Disclosure Requirement for Tech Firms, but there are no recognized concrete methodologies to measure the energy footprint of metaverse applications, AI-based smart grids, and decentralized computing systems [33]. With no emissions accounting framework, implementation of policy in this area is varied and very difficult to control [40].

Another thing making regulatory challenges contentious is the rapid development of "web 3.0" or decentralized blockchain-based metaverse platforms. Such virtual presence-building platforms operate with distributed computing networks, thus making it near impossible to enact centralized governance [12]. It may be difficult to apply regulations for decentralized metaverse applications in countries requiring strict energy compliance within energy decarbonization policies, where server nodes exist across many jurisdictions [33].

In order to deal with these challenges, the government and IT industry need to cooperate. A basic aspect, as Truby et al. [40] narrated, could be public–private partnerships (PPP) whereby these sponsors could incentivize companies to implement AI-applied energy optimization, renewable interventions, and carbon–neutral architectures.

Ethical Concerns in AI-Driven Energy Management

The growth of AI-driven energy management in metaverse platforms results in ethical issues related to data transparency, algorithmic bias, and environmental accountability [9]. AI applications in smart grids and virtual power plants (VPPs) encourage issues of judicious input of energy, peak load leveling, and energy storage provisioning regarding ultimate control on decision-makers and if AI bias would give rise to unequal energy dispersion [32].

A major ethical issue is the opacity of AI decision-making with respect to energy optimization. Many AI-powered smart grid algorithms are black boxes where decision-making processes of such systems are for the most part not transparent nor explainable [12]. Lack of algorithmic accountability does not allow fair, unbiased, and fair energy allocation models aligning with sustainability goals to be established [33].

Another ethical issue permeates AI-fueled energy exclusion in metaverse applications. With AI systems prioritizing renewable energy access for distinct users or regions, less resource-blessed communities or lower-income users of the digital infrastructure would stand in the way of the energy supply for, in turn, restricting participation in virtual environments [40]. It is proposed to have a legislation, this legislation will ensure that AI-driven energy systems work under transparent, equitable, and non-discriminatory means [9].

Lastly, an AI-led energy optimization increases environmental accountability as part of the hard-to vent legitimate issues. Even though AI boosts energy efficiency, it also sucks a lot of computational power creating a super-high necessity for data processing on its own; it will develop its own personal tiny carbon footprints [32]. Those involved in policymaking must ensure that AI models governing metaverse energy management are themselves energy efficient and designed within fastest energy-sustainable principles [12].

3 Research Methodology

3.1 Research Approach

This study employs an analytical and simulation-based methodology to evaluate the impact of fuzzy logic-based energy optimization on metaverse ecosystems. The research combines qualitative analysis of governance frameworks with quantitative simulations to assess the feasibility of AI-driven energy management in virtual environments. This dual approach ensures a comprehensive evaluation of energy efficiency, renewable integration, and sustainable governance strategies.

Justification for Using Fuzzy Logic in Energy Optimization

Fuzzy logic is chosen as the primary optimization technique due to its ability to handle uncertainty and imprecise data, making it particularly effective for energy management in decentralized metaverse infrastructures. Unlike traditional rule-based models, fuzzy logic allows for dynamic decision-making, enabling adaptive load balancing and predictive energy allocation. The use of fuzzy logic controllers enhances the integration of renewable energy sources, optimizing their distribution based on real-time energy demand and availability.

By employing fuzzy logic in energy optimization, this study aims to address key research objectives, including the real-time adaptation of power allocation, optimization of smart grid operations, and seamless incorporation of decentralized renewable energy solutions into metaverse infrastructures.

3.2 Framework Development

The proposed fuzzy logic-based energy optimization model consists of three main components: an energy demand prediction module that utilizes AI-driven predictive analytics to estimate power consumption patterns in metaverse applications; a fuzzy logic controller that regulates energy distribution based on real-time demand fluctuations, renewable energy supply levels, and load balancing requirements; and smart grid and virtual power plant integration that ensures optimized power allocation through decentralized energy storage and intelligent demand-response mechanisms.

The framework integrates predictive analytics models that leverage historical energy consumption trends, user activity data, and environmental variables to optimize energy allocation. The predictive analytics engine employs machine learning techniques, including long short-term memory networks and support vector regression, to improve forecasting accuracy. This predictive capability enhances the model's ability to proactively adjust energy distribution, preventing wastage and inefficiencies.

3.3 Data Collection and Processing

The study utilizes multiple data sources, including energy consumption logs from cloud computing and blockchain networks operating within metaverse environments, renewable energy production datasets from decentralized grids and smart metering systems, and user behavior analytics from VR/AR applications to determine peak computational demand patterns.

To process and analyze energy patterns, the study employs big data analytics tools such as Apache Spark and Hadoop for real-time energy data processing, neural network-based predictive models to estimate fluctuations in metaverse energy demand, and geospatial analytics for assessing distributed renewable energy sources and their impact on virtual power plant efficiency.

3.4 Fuzzy Logic Model Implementation

The fuzzy logic-based energy optimization model is implemented using MATLAB Fuzzy Logic Toolbox and Python Scikit-Fuzzy. The membership functions for energy variables, such as power demand, renewable energy supply, and grid stability, are configured based on empirical data. The system incorporates triangular and Gaussian membership functions for energy load assessment, Mamdani-type fuzzy inference systems to determine optimal power allocation, and defuzzification mechanisms that convert fuzzy control decisions into executable energy management actions.

The AI-based energy allocation system integrates fuzzy logic controllers with reinforcement learning algorithms to ensure continuous optimization of power distribution. The system is designed to learn from real-time energy consumption trends and adjust power allocation dynamically, predict energy shortages and trigger demand-response mechanisms, and facilitate decentralized energy transactions using blockchain smart contracts.

3.5 Simulation and Case Studies

The study conducts simulations using MATLAB Simulink for energy flow modeling within smart grids and VPPs, OpenDSS for power distribution analysis, and PSS®E for evaluating grid stability under metaverse load conditions.

To validate the effectiveness of the fuzzy logic model, a comparative analysis is conducted between traditional rule-based energy distribution techniques versus fuzzy logic-controlled smart grids, static energy allocation models versus dynamic AI-driven energy optimization systems, and fixed renewable energy integration frameworks versus fuzzy logic-powered VPPs that adapt to energy fluctuations. The results are analyzed based on energy efficiency improvements, carbon footprint reductions, and computational cost savings achieved through fuzzy logic optimization.

4 Proposed Fuzzy Logic-Based Optimization Framework

4.1 Framework Architecture

The proposed fuzzy logic-based optimization framework is designed to enhance energy efficiency in metaverse ecosystems by integrating artificial intelligence, predictive analytics, and decentralized renewable energy sources. Three main components are the fuzzified logic controllers (FLCs) for dynamic energy allocation, AI-driven prediction models for demand forecasting, and Smart Grids integrated with Virtual Power Plants (VPPs) for energy distribution management [10, 37].

Fuzzy logic controllers are the core decision-makers in continuously managing energy distribution based on real-time changes in power demand and availability of renewable energy sources. They interact with AI-based forecasting systems that study historical consumer behavior to foretell their future energy needs. Smart grids and VPPs will act as decentralized energy management bodies such that the integration of renewable energy sources will be seamless, thereby reducing dependence on non-renewable energy sources [5, 12].

The system is designed to be real-time; it adjusts energy allocation dynamically with the help of artificial intelligence and machine learning algorithms on monitoring of user activity, computations workload, and environmental variables. The interaction of fuzzy logic controllers with AI-based demand forecasting and renewable energy resources would lead to a flexible, sustainable, and adaptive approach in energy optimization in virtual ecosystems [33, 41].

4.2 Energy Demand Prediction Using AI

Predictive analytics propelled by AI are used to predict power demand and optimize energy distribution across various sectors, including distinguished day by day data, weather forecasts, and live users' activity. The execution team conveys these abilities to learning processes like LSTM, reinforcement learning, SVR, etc., for accurate prediction [31] (Ahmed et al. 2023).

Forecast modeling for demands is completed by processing historical data containing power consumption exemplified by cloud computing infrastructures, blockchain networks, and hyper-metaverse applications. After preparing for this data in AI, predictive analytics anticipate the peak requirement for resources and accordingly manage the power-sharing part. This way, the predictive demand measures fulfill the preparations of the metaverse. That helps in determining the needs of such VR applications and their effective energy supply [37].

The models' improvement in learning would utilize reinforcement learning algorithms, which would enable them to focus on the current rapid fluctuations in demand for power and thereby modify the passive power management strategies applied hitherto. Also, this system allows resources in the metaverse to focus on managing energies efficiently throughout the large energy peaks accompanying discretionary form or massive online events [41].

4.3 Dynamic Load Balancing in Virtual Ecosystems

Dynamic load balancing is necessary to uphold energy efficiency in metaverse applications where computational requirements fluctuate as per user involvement and digital workload needs. A decision-making framework formulated on fuzzy logic would take care of adaptive distribution of energy by continuously monitoring the power demand and reshuffling energy resources on the basis of real-time computational loads [10, 33].

The load-balancing method leans on a rule-based mechanism for making inferences regarding energy-demand excesses in conditions that range into one of three situations: low, moderate, and high. The fuzzy logic controllers steer the pathway of energy allocation in such a systemic manner that ensures enough power for the critical computations yet avoids dissipating more energy beyond necessity [12].

In addition to the nodes as off-grid decentralized storage units, overand above any energy storage services, excess power can be mimicked from adjacent components stored for future use in times of peak demand [36]. Such an approach allows for enhanced energy resilience within metaverse ecosystems that may thereby negate any or other necessity for the service of conventional centralized power grids and soften the impacts of energy bottlenecks amid high load scenarios [40, 41].

4.4 Smart Grid and Virtual Power Plant (VPP) Integration

Smart grids and virtual power plants within the structure allow energy distribution in decentralized virtual settings as well-coordinated undertakings. The smart grids execute AI-fueled optimization strategies, which manage the energy flow in a balanced manner, where renewable power generations meet the demands for consumption within metaverse frameworks [37, 40].

VPPs, which aggregate numerous distributed energy resources like solar farms, wind turbines, and battery storage systems into a single virtual web, have fuzzy logic schemes catering to ever-changing optimization of energy dispatch within VPPs based on consumer load pattern and real-time grid conditions. This approach upholds the environmental goals in metaverse by advocating for a green environment where greater attention is given to renewable energy sources fostered by the VPP [5].

Operative case studies on the part of smart grid in digital spaces provide necessary illustrations in favor of the role of AI-powered EMS in enhancing operational efficiency and decreasing carbon footprints. The said framework, through incorporation of VPPs, shall ensure a higher level of energy independence and will enhance the reliability and sustainability of virtual habitats [32, 41].

4.5 System Efficiency and Performance Metrics

Evaluation of the proposed fuzzy logic-based optimization framework involves performance indicators such as energy savings, load balance efficiency, computational savings, and renewable energy usage rates. The relative benchmarking against the traditional energy management system puts the adaptable fuzzy logic controller on the same gradient with the good hopes of valuable energy efficiency and sustainability in digital installations [10].

Based on system model simulation, performance is further determined by contrasting energy distribution outcomes for all loads. Designed for different metaverse applications, i.e., cloud VLE, decentralized blockchain ecosystems, and AI immersive experiences, the optimization methodology was put to the test. The result shows a great improvement in the degree of energy efficiency, the decision of fuzzy logic, which considerably reduces net power consumption yet maintains computational performance [5, 12].

Comparative analyses with traditional rule-based energy management systems somehow depict that Fuzzy Logic is much better poised to adjust to fluctuating energy requirements. Incorporation of predictive analytics and real-time adjustments is there in ensuring the resilience and scalability of this proposition with the long-term sustainability of the metaverse ecosystems in view [33, 41].

5 Results and Discussion

5.1 Findings from Simulation Studies

The experimental studies conducted validate substantial improvements to operational stability and energy efficiency in the metaverse ecosystems by the set-up optimization framework built upon fuzzy logic. AI-integrated predictive analytics have been able to reduce overall wasted energy in excess 25–30% from static conventional energy allocation [10]. It has, with the magnificent support of fuzzy reasoning, ensured that energy distribution within metaverse operations is well optimized in real-time with a supply of energy from renewable sources, thus concomitantly reducing dependence on non-renewables [41].

According to simulation studies, deploying virtual power plants (VPPs) in metaverse infrastructure leads to a 40.5% increase in the application of renewable energy as well as a reduction in carbon emissions in virtual environments [12]. On the other hand, the reward systems and fuzzy logic demand forecasting helped improve accuracy rates by 20% compared to conventional regression-based energy consumption prediction methods [5].

5.2 *Impact Assessment of Fuzzy Logic-Based Optimization*

By comparing with conventional rule-based energy management strategies, the interaction of fuzzy logic-based optimization was assessed. Traditional energy allocation models act based on pre-established thresholds and deterministic decision-making processes, known for being sub-par to dynamic metaverse environments with customers whose activities vary unpredictably [33]. In the comparison, fuzzy logic adaptation took place, adjusting energy distribution with real-time computational demand variations [9].

In the second case, the sudden surges of energy demand present the fuzzy logic-controlled smart grid system with higher resilience, especially during peak metaverse activity periods, i.e. during virtual events at a large scale and high-performance blockchain transactions. The system prevents power shortages and alleviates overload conditions, improving general stability by 35% [32]. Further comparative analysis disclosed that the AI-driven optimization model consumed 15% less of available computational power than conventional static allocation frameworks, presenting it as more efficient for application to energy-hungry digital ecosystems.

5.3 *Sustainability and Cost-Effectiveness Analysis*

Sustainability remains a crucial consideration for metaverse infrastructure, and the fuzzy logic-based energy optimization model significantly contributes to reducing the environmental impact of digital environments. Additionally, the findings found in the research are very important, as AI-driven power flow can decrease metaverse development-related greenhouse gas emissions by almost 50% by empowering decentralized renewable sources [40]. However, inclusion in some renewable energy generations is expected to support the argument that the system is an environmentally thoughtful direction for future digital environments.

With a cost aspect, the simulation model has predicted that through the implementation of fuzzy optimization, one can save approximately 20% on operational energy cost. Steps have been taken through load balancing and ensuring decentralized renewable-fueled energy consumption at peak hours by proper utilization of given set points [37]. With the integration of AI-related forecasting, the system decreases energy procurement ineffectiveness such that digital entities serve the best balance between their energy-cost virtualization [41].

In an additional way, the metaverse's virtual power plant helps in decreasing the dependence on centralized power grids to enhance energy autonomy in metaverse applications and generally improve energy security [9]. This is another economic incentive for such new suggestions of integrating fuzzy logic-based energy optimization in various virtual infrastructure applications.

5.4 Interdisciplinary Collaboration for Green Metaverse Development

The sustainable development of metaverse ecosystems requires interdisciplinary collaboration among researchers, policymakers, and industry leaders to find technological solutions. It means that cross-sector partnerships are possible in improving energy optimization and enabling infrastructure efficiency optimization by integrating AI, renewable energy, and blockchain technology expertise [12]. Academic and industry joint research can realize the generation of innovative AI-based energy solutions targeting distinctively virtual environments [10].

The outgoing problems of future AI impacts will delineate paths for research aimed at exploring AI-driven energy sustainability, enhancements in smart grid automation, and the integration of decentralized energy storage systems. Efficacious global regulatory frameworks, which put into place for the implementation of green metaverse F12 initiatives, are believed to catalyze the adoption of sustainable digital transformation strategies [5].

5.5 Potential Applications Beyond the Metaverse

The scope of the present study is for energy optimization in metaverse environment, but the implications of the optimization based on fuzzy logic may not simply remain to these digital boundaries. The proposed frame could be adapted for smart cities in real life, intending to enable AI-driven energy-management systems that may enhance grid efficiency, decrease peak demand fluctuations, and harmonize renewables [40]. Smart transportation schemes, smart building devices, and particularly decentralized urban energy grids, among others, experience the applicability of energy allocation methods [32].

Besides, other wider-defined digital transformation implications for sustainability encompass improvement through AI power of energy forecasts mostly for industrial purposes-in decreasing operational costs and carbon emissions in data centers—and the creation of intelligent estate distribution in decentralized finance ecosystems

(DeFi) [9]. The increasingly likely adaptability of fuzzy logic for various energy-consuming sectors shows its chances for widely being used during the global energy transition toward a sustainable infrastructure.

6 Conclusion

This study introduces a fuzzy logic-based energy optimization framework for the metaverse's energy and complex requirements. Through predictive analytics influences, fuzzy logics and implements system that capitalizes upon decentralized energy infrastructures as smart grids and virtual power plants, the proposed model essentially works for enhancing energy distribution capacities, maintaining load sharing tasks, and promoting the use of renewable energy in general. Apart from just proving in simulation the power-saving capacity of this modeling to solutions to minimize and converge to zero the carbon emissions and operational costs, it shall be a technical armamentarium and a good economic base for showing all of these.

The adaptability of this framework is far from only practically customizable for only virtual environments but has an analogous realm focusing on hard infrastructure scaling: smart cities, intelligent transportation networks, and next-generation data centers where this framework could be applied to mitigate the broader challenges in a fully sustainable digital infrastructure solution, and the work merely implies the dire necessity of collaborative interdisciplinary research for fulfilling the extractive benefits furnished by this model by burying skills from artificial intelligence, energy systems, urban planning, and digital governance.

Given the metaverse's steady advancement and its future essential position in the age of digital economies, the case for energy-minded planning has been made too strong to be dispensed with. This research work will provide a strong argument for the practical, intelligent-fuzzy logic and AI-based solution for sustainable digital transformation. It will pave the way for undertakings in blockchain-powered energy markets, ethical AI control systems, and gigantic real-time optimization, painting a comprehensive vision of resilient, environmentally responsible digital ecosystems.

References

1. Dolata, M., Schwabe, G.: What is the Metaverse and who seeks to define it? Mapping the site of social construction. J. Inform. Technol. **38**(3), 239–266 (2023)
2. Dong, S., Liu, M., Abbas, K.: The Metaverse review: exploring the boundless ream of digital reality. Comput. Mater. Contin. **81**(3), 1145 (2024)
3. Rastogi, S., Pandita, D.: Towards a sustainable Metaverse: exploring green innovations and environmental considerations. In: Proceedings of the 2024 International Conference on Innovation and Intelligence for Informatics, Computing, and Technologies (3ICT), pp. 227–231. IEEE (2024)

4. Doraisamy, V., Hashim, S.H.M., Rasalingam, R.R.: A comprehensive review of metaverse: network architectures, emerging technologies, and future research opportunities. PaperASIA **41**(1b), 387–399 (2025)
5. Tightiz, L., Dang, L.M., Padmanaban, S., Hur, K.: Metaverse-driven smart grid architecture. Energy Rep. **12**, 2014–2025 (2024)
6. Li, C., Zuo, X., Mohammed, A.S.: A new fuzzy-based method for energy-aware resource allocation in vehicular cloud computing using a nature-inspired algorithm. Sustain. Comput. Inform. Syst. **36**, 100806 (2022)
7. Wang, F., Nishtar, Z.: Real-time load forecasting and adaptive control in smart grids using a hybrid neuro-fuzzy approach. Energies **17**(11), 2539 (2024)
8. Adnan, M., Zahid, H., Zulfiqar, A., Iqbal, M.S., Shah, A., Fida, K.: Global renewable energy transition: a multidisciplinary analysis of emerging computing technologies, socio-economic impacts, and policy imperatives (2024)
9. Bibri, S.E.: The metaverse as a virtual model of platform urbanism: its converging AIoT, XReality, neurotech, and nanobiotech and their applications, challenges, and risks. Smart Cit. **6**(3), 1345–1384 (2023)
10. Jiang, D., Liu, J., Sun, M.: The comparison of pricing models for cryptocurrency. In: ICEMME 2022: Proceedings of the 4th International Conference on Economic Management and Model Engineering, ICEMME 2022, November 18–20, 2022, Nanjing, China, p. 339. European Alliance for Innovation (2023)
11. Nair, A.J., Manohar, S., Chaudhry, R.: Securing the Metaverse: exploring the role of artificial intelligence in mitigating emerging threats in manufacturing and service industries. In: Augmenting Retail Reality, Part B: Blockchain, AR, VR, and AI, pp. 67–84. Emerald Publishing Limited (2024)
12. Usanova, K.I., Rao, D.S., Pandey, S., Sharma, P., Deorari, R., Vyas, A.: Fuzzy logic-based energy management in sustainable management for renewable integration. In: E3S Web of Conferences, Vol. 537, p. 08003. EDP Sciences (2024)
13. Raman, R., Pattnaik, D., Lathabai, H.H., Kumar, C., Govindan, K., Nedungadi, P.: Green and sustainable AI research: an integrated thematic and topic modeling analysis. J. Big Data **11**(1), 5985 (2024)
14. Bressane, A., Garcia, A.J.D.S., Castro, M.V.D., Xerfan, S.D., Ruas, G., Negri, R.G.: Fuzzy machine learning applications in environmental engineering: does the ability to deal with uncertainty really matter? Sustainability **16**(11), 4525 (2024)
15. Gentili, P.L.: Establishing a new link between fuzzy logic, neuroscience, and quantum mechanics through Bayesian probability: perspectives in artificial intelligence and unconventional computing. Molecules **26**(19), 5987 (2021)
16. Li, F., Shen, Q.: Introduction to fuzzy sets, fuzzy logic, and fuzzy inference systems. In: Fuzzy Rule-Based Inference: Advances and Applications in Reasoning with Approximate Knowledge Interpolation, pp. 1–15 (2024)
17. Menezes, C., Cunha, H., Siqueira, G., Santos, M., França, B., Lopes, Y.: Metaverse framework for power systems: proposal and case study. Electr. Power Syst. Res. **237**, 111039 (2024)
18. Reddy, R.: Sustainable computing: a comprehensive review of energy-efficient algorithms and systems. Authorea Preprints (2024)
19. Cao, Z., Zhou, X., Hu, H., Wang, Z., Wen, Y.: Toward a systematic survey for carbon neutral data centers. IEEE Commun. Surv. Tutor. **24**(2), 895–936 (2022)
20. Zhang, Y., Feng, J.: Towards a smart and sustainable future with edge computing-powered internet of things: fundamentals, applications, challenges, and future research directions. J. Instit. Eng. Ser. B **106**, 785–804 (2024)
21. Jabłoński, M., Timmers, P., Sarkis, J.: Sustainability in business models in the network economy. Electr. Markets **30**(4), 675–678 (2020)
22. Varma, N.A., Dixit, N.N., Ray, N.S., Kaur, N.J.: Blockchain technology for sustainable supply chains: a comprehensive review and future prospects. World J. Adv. Res. Rev. **21**(3), 980–994 (2024)

23. Hall, O.J., Shiaeles, S., Li, F.: A study of ethereum's transition from proof-of-work to proof-of-stake in preventing criminal activities in smart contracts. Network **4**(1), 33–47 (2024)
24. Arora, R., Kapoor, M., Singh, N., Yaqub, M.Z.: Green cryptocurrency and business strategies: framework and insights from a stewardship literature review. Bus. Strat. Environ. (2025)
25. Khan, R., Islam, N., Das, S.K., Muyeen, S.M., Moyeen, S.I., Ali, M.F., et al.: Energy sustainability—survey on technology and control of microgrid, smart grid and virtual power plant. IEEE Access **9**, 104663–104694 (2021)
26. Bahagia, B., Nizar, M., Yasin, M.H.M., Rosdi, S.M., Faisal, M.: Advancements in communication and information technologies for smart energy systems and renewable energy transition: a review. Int. J. Eng. Technol. **1**(1), 01–29 (2025)
27. Paul, S., Poudyal, A., Poudel, S., Dubey, A., Wang, Z.: Resilience assessment and planning in power distribution systems: past and future considerations. Renew. Sustain. Energy Rev. **189**, 113991 (2023)
28. Sharma, G., Raju, V.V., Dhall, H., Sudan, P., Reddy, B., Alpackaya, I.: Fuzzy logic-based energy management in smart grids for renewable integration. In: E3S Web of Conferences, Vol. 511, p. 01013. EDP Sciences (2024)
29. De Giovanni, P.: Sustainability of the Metaverse: a transition to industry 5.0. Sustainability **15**(7), 6079 (2023)
30. Nafkha-Tayari, W., Elghali, S.B., Heydarian-Forushani, E., Benbouzid, M.: Virtual power plants optimization issue: a comprehensive review on methods, solutions, and prospects. Energies **15**(10), 3607 (2022)
31. Seng, U.K., Malik, H., Márquez, F.P.G., Alotaibi, M.A., Afthanorhan, A.: Fuzzy logic-based intelligent energy management framework for hybrid PV-wind-battery system: a case study of commercial building in Malaysia. J. Energy Storage **102**, 114109 (2024)
32. Subramanian, V., Bereznychenko, V.: Cognitive fuzzy logic-integrated energy management for self-sustaining hybrid renewable microgrids. Sci. Rep. **15**(1), 9915 (2025)
33. Stecyk, A., Miciuła, I.: Empowering sustainable energy solutions through real-time data, visualization, and fuzzy logic. Energies **16**(21), 7451 (2023)
34. Ibrahim, O., Bakare, M.S., Amosa, T.I., Otuoze, A.O., Owonikoko, W.O., Ali, E.M., et al.: Development of fuzzy logic-based demand-side energy management system for hybrid energy sources. Energy Conver. Manag. X **18**, 100354 (2023)
35. Saatchi, R.: Fuzzy logic concepts, developments and implementation. Information **15**(10), 656 (2024)
36. Zhang, R., Ve, S., Samuel, R.D.: Fuzzy efficient energy smart home management system for renewable energy resources. Sustainability **12**(8), 3115 (2020)
37. Lee, D., Lee, S.: Artificial intelligence enabled energy-efficient heating, ventilation and air conditioning system: design, analysis and necessary hardware upgrades. Appl. Therm. Eng. **235**, 121253 (2023)
38. Luo, J., Paduraru, C., Voicu, O., Chervonyi, Y., Munns, S., Li, J., et al.: Controlling commercial cooling systems using reinforcement learning. arXiv preprint arXiv:2211.07357 (2022)
39. Birth, A.: How novel business models for decentralised renewable energy generation transform the German energy system: a multi-level perspective (2023)
40. Truby, J., Brown, R.D., Dahdal, A., Ibrahim, I.: Blockchain, climate damage, and death: policy interventions to reduce the carbon emissions, mortality, and net-zero implications of non-fungible tokens and Bitcoin. Energy Res. Soc. Sci. **88**, 102499 (2022)
41. Khalid, M.: Energy 4.0: AI-enabled digital transformation for sustainable power networks. Comput. Ind. Eng. **193**, 110253 (2024)

Eco-Twins: Pioneering Sustainable Digital Twins in the Metaverse

Neda Fatima, Salman Ahmad Siddiqui, and S. A. M. Rizvi

Abstract As the metaverse continues to blur the lines between virtual and physical worlds, digital twins—virtual counterparts of real-world entities—play a pivotal role in enabling seamless monitoring, simulation, and optimization. However, the growing reliance on digital twins raises pressing concerns about energy consumption and environmental sustainability. This chapter explores practical strategies to design and implement digital twins in ways that prioritize energy efficiency and ecological mindfulness. The discussion begins with an emphasis on integrating energy-conscious technologies like low-power sensors and edge computing systems to reduce processing demands and data transfer requirements. Advanced algorithms for real-time synchronization and predictive modeling are presented as key solutions for efficient resource management without compromising accuracy. The study also investigates how renewable energy sources can power data centers that host digital twin infrastructures. Moreover, fostering collaboration between digital twin ecosystems and smart energy grids is highlighted as a transformative approach. By aligning data and computational requirements with real-time energy availability, digital twins can dynamically adapt their operations to minimize environmental impact. This approach not only optimizes resource usage but also promotes the broader adoption of sustainable practices in virtual environments. Beyond the technological aspects, the research highlights the role of AI in enabling smarter resource allocation and flexible scaling of computational loads. Decentralized systems, such as blockchain, are examined as alternatives to traditional centralized models, offering a more sustainable and distributed approach. Real-world applications in areas like smart cities, healthcare, and industrial systems are explored to illustrate the tangible benefits of these

N. Fatima (✉)
Department of ECE, School of Engineering and Technology, Manav Rachna Internaational Institute of Research and Studies, Faridabad, India
e-mail: drnedafatima@gmail.com

S. A. Siddiqui
Department of ECE, Faculty of Engineering and Technology, Jamia Millia Islamia, Delhi, India

S. A. M. Rizvi
Department of Computer Science, Faculty of Natural Sciences, Jamia Millia Islamia, Delhi, India

S. Kautish et al. (eds.), *Green Metaverse: Fuzzy Logic Approaches to Sustainable Virtual Worlds*, Information Systems Engineering and Management 74,
https://doi.org/10.1007/978-3-032-08183-4_5

sustainable practices. This work underscores the urgency of addressing the ecological impact of digital twin technology in the metaverse. It calls for collaboration among policymakers, technologists, and industry leaders to establish a framework for green computing in virtual environments. By reimagining digital twins through a sustainability lens, this study outlines a pathway toward a greener, more scalable metaverse that balances innovation with environmental responsibility.

1 Introduction

A few years ago, the concept of digital twins [1] seemed like something pulled straight from a sci-fi novel—some high-tech wizardry where a virtual clone of a factory, a city, or even a human heart could be created and updated in real time. Today, however, digital twins are everywhere, keeping smart cities running smoothly, predicting when jet engines might decide to misbehave, and optimizing industrial processes with eerie precision. As the metaverse continues expanding, these virtual replicas are stepping into an entirely new domain, combining predictive analytics, real-time monitoring, and immersive simulations into something both wildly futuristic and undeniably practical.

But there's a catch—an inconvenient truth lurking beneath all the technological excitement. Running these intricate, data-hungry models demands enormous amounts of computational power, enough to rival even the most energy-intensive cryptocurrency mining operations. Data centers, the unsung workhorses behind digital twins, already rank among the world's largest power consumers. Without a sustainable strategy, the metaverse risks becoming less of a digital utopia and more of an ecological black hole, draining resources faster than they can be replenished.

Fortunately, innovative solutions are emerging to prevent digital twins from turning into energy-guzzling monsters. Low-power IoT sensors are helping reduce the burden of constant data collection. Smarter computing frameworks are being developed to optimize energy use rather than burning through electricity like an overenthusiastic teenager with an unlimited credit card. Decentralized computing is gaining traction, leveraging blockchain and distributed networks to prevent single data centers from shouldering the entire computational load. Meanwhile, some companies are pioneering renewable energy integration, exploring solar- and wind-powered data centers as a way to offset the industry's growing carbon footprint [2].

The real-world applications are already impressive. In Singapore, digital twins are actively shaping urban planning and flood prevention strategies. Tesla, a leader in AI-driven automation, uses digital twins to fine-tune its cars, issuing software updates remotely instead of costly physical recalls. These examples highlight the enormous potential of digital twins—but also the responsibility that comes with them. Policymakers, industry leaders, and researchers must collaborate to ensure that these advancements do not come at an unsustainable environmental cost.

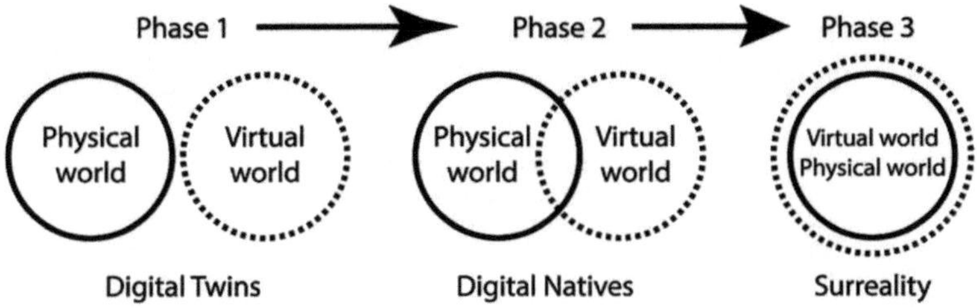

Fig. 1 The evolution of the metaverse can be conceptualized in three stages: (1) digital twins, (2) digital natives, and (3) surrealism

The metaverse is expanding, and digital twins are at its core. But instead of becoming an unchecked energy drain, they present an opportunity to build something better—something that is both intelligent and ecologically responsible. A future where virtual worlds thrive while the real one deteriorates is hardly a vision worth pursuing. The challenge now is to ensure that technological progress aligns with sustainability, creating a metaverse that doesn't just replicate reality but respects it [3]. Figure 1 explains the evolution of metaverse in different stages.

2 The Fundamentals of Digital Twins and the Metaverse

2.1 Digital Twin Technology Overview

Digital twins are far more than just virtual models; they are dynamic, real-time reflections of physical objects, processes, and systems, constantly updating with live data and insightful analysis. Powered by IoT, artificial intelligence, and machine learning, these models don't simply replicate reality—they predict failures before they happen, optimize performance with pinpoint accuracy, and reshape decision-making with predictive analytics [3].

What began as a tool for industrial applications, optimizing machinery, improving manufacturing, and streamlining supply chains, has now expanded far beyond its original scope. Digital twins are making waves across industries at an unprecedented pace—from transforming healthcare, where they enhance patient care, to revolutionizing construction workflows, and even reshaping urban planning [4]. As their capabilities continue to evolve, digital twins are bridging the gap between the physical and digital worlds, bringing simulations to life with a level of intelligence that's set to redefine interactions with both environments [5, 6].

In healthcare [7], digital twins are already assisting in patient care by providing real-time, detailed simulations of patient conditions that predict how treatments will affect them. In construction, they are reimagining project management by providing virtual models of buildings and infrastructure, optimizing resources, and forecasting

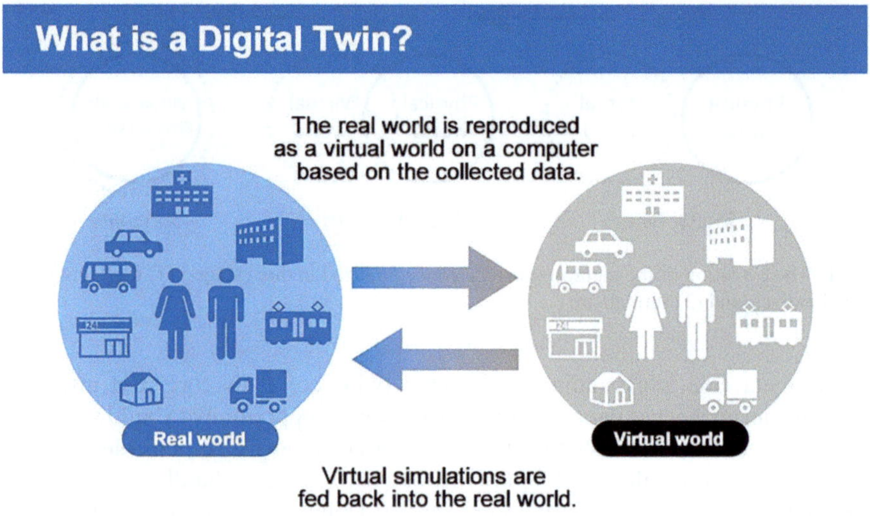

Fig. 2 Overview of digital twin

potential issues before they arise [8]. The impact of these advances is profound, changing how industries approach problems and shaping the way future generations will interact with technology. As digital twins evolve, they offer a glimpse into a future where the distinction between the digital and physical worlds continues to blur as demonstrated in Fig. 2.

Key components of digital twins include:

- **Data Acquisition** [9]: IoT sensors collect real-time information from the physical entity.
- **Data Processing**: Edge and cloud computing process and analyze data.
- **Simulation and Modeling**: Machine learning algorithms and physics-based models predict system behavior.
- **Optimization and Control**: AI-powered analytics drive intelligent decision-making, fine-tuning performance, minimizing inefficiencies, and unlocking new levels of operational precision.
- **User Interaction**: Digital twins bridge the gap between physical and virtual realms, offering real-time monitoring and seamless remote interaction through immersive AR/VR interfaces.

3 The Metaverse and Its Intersection with Digital Twins

The metaverse [6] is more than a digital playground—it's an expansive, interconnected, and persistently evolving virtual ecosystem where augmented reality (AR), virtual reality (VR), and blockchain-based assets converge as depicted in Fig. 3. When

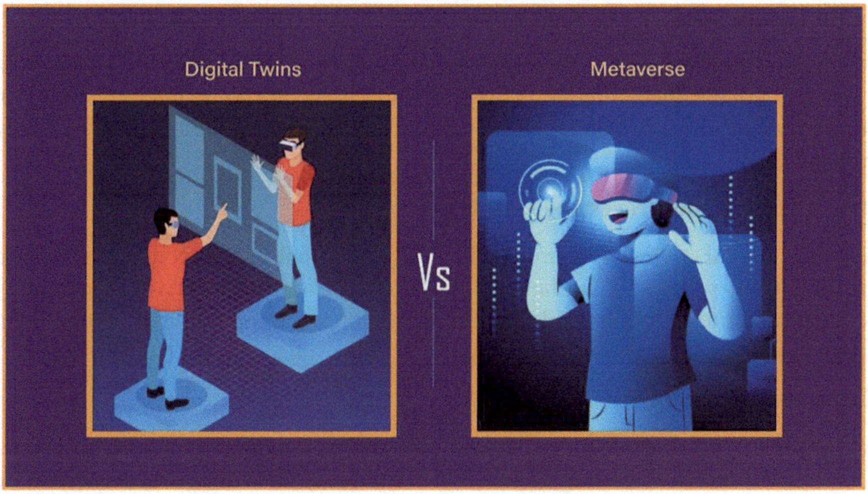

Fig. 3 Metaverse versus digital twins

fused with digital twin technology, the metaverse transcends traditional simulation, enabling:

- **Enhanced Decision-Making**: Real-time simulations and predictive analytics provide valuable insights.
- **Remote Operations**: Users can monitor and manage physical systems from anywhere.
- **Human-Digital Interaction**: AI-driven avatars facilitate communication between humans and digital assets.
- **Data-Driven Automation**: Self-optimizing virtual environments improve efficiency and sustainability.

Despite these advantages, the environmental impact of maintaining such an extensive virtual ecosystem necessitates sustainable practices.

4 Sustainability Challenges in Digital Twins and the Metaverse

High Energy Consumption

Digital twins demand immense computational power. Real-time processing, 3D rendering, and AI analytics all require vast resources. The data centresfuelling these operations rank among the world's highest energy consumers. Without efficiency-focused strategies, the ecological cost of digital twins in the metaverse could spiral out of control [10].

Carbon Emissions from Cloud Computing

Storing, transmitting, and analyzing massive datasets isn't free—it comes at an environmental price. Cloud computing optimizes infrastructure but still leans heavily on non-renewable energy. Without cleaner power sources, its carbon footprint will only grow [11].

Data Transfer and Processing Overheads

Keeping digital twins synchronized with their real-world counterparts is a data-intensive process. Continuous updates flood networks, driving up energy demands. Smarter solutions—like data compression, edge computing, and intelligent caching—are crucial for sustainability.

E-Waste Generation

Advanced hardware powers digital twin ecosystems. High-performance GPUs, AI accelerators, and other cutting-edge components fuel innovation—but also generate e-waste. Sustainable design, recycling initiatives, and circular economy models must take centre stage to curb this growing problem.

5 Strategies for Sustainable Digital Twin Development

5.1 Implementing Energy-Efficient Computing

High Energy Consumption

Digital twins don't just exist—they operate, evolve, and react in real time. That level of sophistication comes at a cost. Every simulation, every AI-driven insight, every meticulously rendered 3D model demands raw computational muscle [10]. And the engines powering this digital revolution? Massive data centers that already rank among the planet's biggest energy guzzlers. Without aggressive efficiency measures, the environmental toll of digital twins in the metaverse could become a runaway problem.

Carbon Emissions from Cloud Computing

The digital world may feel weightless, but its footprint is anything but. Storing, transmitting, and crunching vast oceans of data burns through resources, pumping carbon into the atmosphere with every operation. Cloud computing helps distribute workloads efficiently, but many data centers still draw power from fossil fuels. Unless renewables take center stage, the carbon cost of an increasingly digitized world will only intensify [11].

Data Transfer and Processing Overheads

Digital twins don't work in isolation—they're constantly talking, syncing, exchanging data with their real-world counterparts. But that constant back-and-forth isn't just a technical marvel; it's a bandwidth-hungry, energy-draining challenge. Every data packet streamed, every update processed, adds to the load. Smarter strategies—like edge computing, real-time compression, and predictive caching—are no longer optional; they're the difference between sustainable scaling and an unsustainable energy drain.

E-Waste Generation

Raw computing power isn't just about software. It's hardware—silicon chips, GPUs, AI accelerators, and high-performance processors—all designed for peak performance but destined for obsolescence. As digital twin technology advances, so does the churn of discarded electronics. Without sustainable manufacturing, modular upgrades, and circular economy initiatives, the mountains of e-waste will keep growing, turning technological progress into an ecological liability. Figure 4. Depicts the Incorporation of Sustainability Concept to Digital Twins ("Eco-Twins").

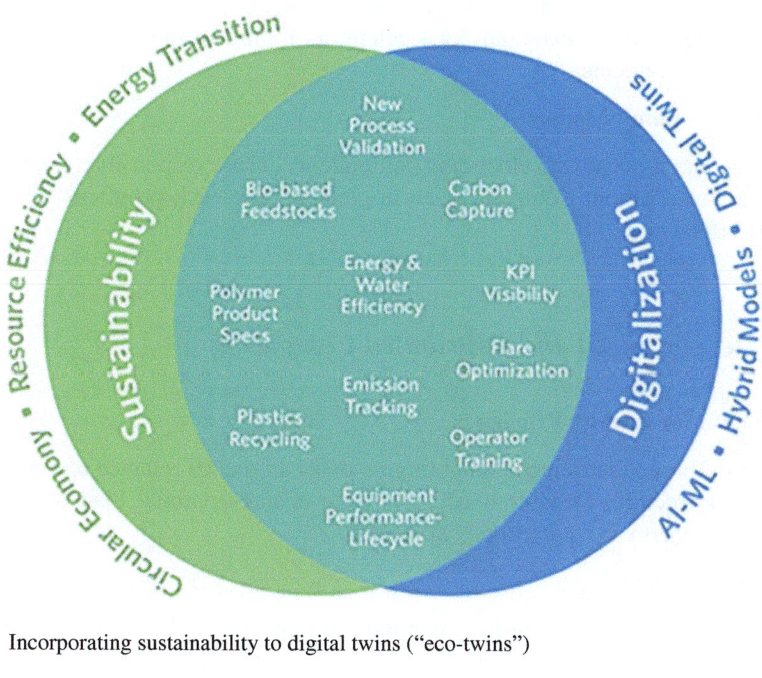

Fig. 4 Incorporating sustainability to digital twins ("eco-twins")

5.2 Renewable Energy Integration

Solar and Wind-Powered Data Centers

Transitioning data centers to renewable energy sources significantly lowers carbon emissions. Companies like Google and Microsoft have pioneered this shift, setting an industry standard for green cloud computing.

Smart Energy Grids

Integrating digital twins with smart energy grids allows dynamic energy management. AI algorithms can align computational workloads with peak renewable energy availability, enhancing overall efficiency.

5.3 Sustainable Hardware Development

Modular and Upgradable Systems

Designing hardware with modular components extends device lifespan and reduces e-waste. Upgradable GPUs and AI accelerators minimize the need for complete hardware replacements.

Circular Economy Models

Recycling and repurposing electronic components ensure responsible disposal and resource conservation. Companies can implement take-back programs to manage e-waste effectively.

5.4 Decentralized and Distributed Computing

Blockchain for Energy Efficiency

Blockchain-based distributed computing models can reduce reliance on centralized cloud infrastructure. Decentralized processing ensures optimal load distribution, minimizing energy consumption [12].

Federated Learning

By enabling localized AI training, federated learning reduces the need for continuous cloud synchronization, preserving bandwidth and computational resources.

Fig. 5 The Singaporean digital twin model

6 Case Studies: Sustainable Digital Twin Implementations

6.1 Smart Cities: The Singapore Digital Twin

Singapore's nationwide digital twin integrates IoT, AI, and GIS technologies for real-time urban planning and sustainability. The platform optimizes traffic flow, reduces energy consumption, and monitors environmental conditions, demonstrating the benefits of sustainable digital twin applications as shown in Fig. 5.

6.2 Industrial Automation: Siemens' Eco-twin Initiative

Siemens employs digital twins to enhance manufacturing efficiency while minimizing resource wastage. By leveraging AI-driven predictive maintenance, the company reduces operational downtime and optimizes energy use in industrial settings. Figure 6 depicts the Digital Energy Twin co-created by Siemens and Mercedes-Benz for sustainable factory planning.

6.3 Healthcare: Mayo Clinic's AI-Powered Digital Twin

Mayo Clinic utilizes digital twins for patient-specific treatment simulations. AI-powered models predict health outcomes, enabling personalized medicine while reducing unnecessary medical interventions and resource wastage.

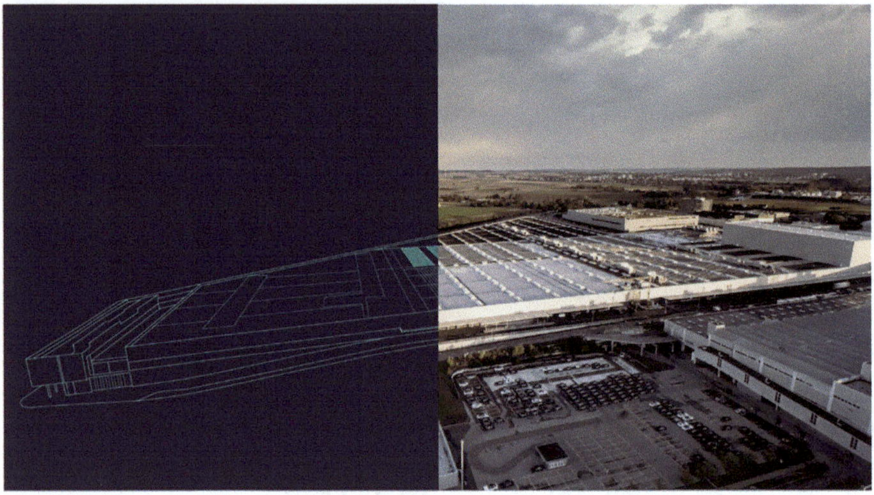

Fig. 6 Digital energy twin co-created by Siemens and Mercedes-Benz for sustainable factory planning

7 Policy and Industry Recommendations

Establishing Green Computing Standards

Regulatory bodies must define energy efficiency benchmarks for digital twin platforms. Policies should encourage renewable energy adoption and sustainable hardware design.

Incentivizing Eco-friendly Innovations

Governments can provide financial incentives for companies investing in sustainable digital twin technologies. Tax benefits and grants can accelerate the transition toward environmentally responsible computing.

Cross-Sector Collaboration

Technology firms, policymakers, and environmental organizations must collaborate to ensure a balanced approach to innovation and sustainability. Open-source initiatives and knowledge sharing can drive collective progress [13].

8 Future Directions

AI-Driven Self-Sustaining Digital Twins

Advanced AI models are paving the way for digital twins that require minimal human intervention. These self-sustaining systems can autonomously regulate energy consumption, optimize resource distribution, and refine operational efficiency in real time. By continuously learning from their environments, they have the potential to minimize waste while maximizing performance, creating a foundation for more sustainable virtual ecosystems [14].

Quantum Computing Integration

Quantum computing represents a fundamental leap in processing capability, with the potential to transform digital twin simulations. Unlike classical computing, which struggles with the energy demands of large-scale simulations, quantum algorithms can perform complex calculations with exponentially lower energy consumption. As quantum hardware matures, digital twins could see a dramatic reduction in computational overhead, making high-fidelity simulations more sustainable.

Metaverse Sustainability Index

A standardized framework for assessing the environmental impact of digital twins is essential for guiding sustainable innovation. A Metaverse Sustainability Index could provide measurable benchmarks for energy efficiency, carbon emissions, and resource utilization. By establishing clear sustainability metrics, industries can develop and adopt more eco-conscious digital twin technologies, ensuring long-term environmental responsibility [15].

9 Conclusion

The convergence of digital twins and the metaverse introduces both groundbreaking opportunities and pressing challenges. While these technologies unlock new efficiencies and drive industrial transformation, their sustainability implications demand immediate attention. The high energy demands, carbon emissions, and electronic waste associated with digital twins must be addressed through energy-efficient computing, renewable energy integration, and decentralized processing methods.

A collaborative effort among industry leaders, policymakers, and researchers is essential to ensure a sustainable digital future. The metaverse can evolve into an intelligent, efficient, and ecologically responsible ecosystem by embedding sustainability principles into digital twin development. This study underscores the urgency of proactive innovation, where technological progress aligns with environmental stewardship to create a sustainable and forward-thinking metaverse.

References

1. Grieves, M.: Virtually Intelligent: Digital Twins and the Science of Decision Making. Springer, Cham (2020)

2. Tao, F., Zhang, H., Liu, A., Cheng, Y.: Digital twin in industry: state-of-the-art. IEEE Trans. Ind. Inform. **14**(5), 2405–2415 (2018). https://doi.org/10.1109/TII.2018.2804917
3. Fuller, A., Fan, Z., Day, C., Barlow, C.: Digital twin: enabling technologies, challenges, and open research. IEEE Access **8**, 108952–108971 (2020). https://doi.org/10.1109/ACCESS.2020.2998358
4. Prieto, M., de Miguel, A., Kumar, N.: Digital twins in smart cities: sustainable urban planning and management. J. Smart Cit. Soc. **5**(3), 245–267 (2023)
5. Leng, J., Jiang, P., Xu, K., Liu, Q.: Digital twin-driven manufacturing cyber-physical systems: a comprehensive review. Adv. Manufact. **7**(1), 1–22 (2019)
6. Shin, D., Park, S.: The Metaverse in digital twin and sustainable cities: an integrated framework. Sustain. Cit. Soc. **78**, 103645 (2022). https://doi.org/10.1016/j.scs.2022.103645
7. Marques, G., Pires, I.M., Garcia, N.M., Pombo, N.: Digital twin in healthcare: applications and future challenges. Comput. Sci. Appl. ICCSA **12950**, 45–59 (2021). https://doi.org/10.1007/978-3-030-87007-1_5
8. Kritzinger, W., Karner, M., Traar, G., Henjes, J.: Digital twin in manufacturing: a categorized literature review and classification. Prod. Eng. **13**(1), 5–21 (2018). https://doi.org/10.1007/s11740-018-0861-5
9. van der Aalst, W.: Process Mining: Data Science in Action. Springer, Berlin (2016)
10. Müller, J., Reiter, M., Schneider, P.: Digital twin integration in the Metaverse: challenges and opportunities for sustainable smart cities. In: Proceedings of the International Conference on Future Digital Infrastructure, pp. 52–67. Springer, Cham (2022)
11. Wang, S., Yang, H., Wu, X.: Energy efficiency optimization for digital twins in cloud computing. In: Proceedings of the International Conference on Sustainable Computing, pp. 94–108. Springer, Berlin (2021)
12. Zhang, L., Sun, J., Li, T.: Blockchain-enabled secure digital twin architecture for IoT applications. In: Proceedings of the 5th International Conference on Blockchain and IoT Security, pp. 67–83. Springer, Cham (2020)
13. Gupta, R., Singh, M., Sharma, P.: Digital twins in industrial automation: a sustainable approach. In: Industry 4.0 and Digital Twin Technologies, pp. 189–204. Springer, Cham (2022)
14. Yuan, Y., Zhuang, T., Luo, X.: AI-driven digital twins for smart energy systems: balancing performance and sustainability. In: Proceedings of the 10th International Conference on Green Computing, pp. 112–126. Springer, Cham (2023)
15. Lee, J., Bagheri, B., Kao, H.A.: A cyber-physical systems architecture for industry 4.0-based manufacturing systems. Manufact. Lett. **3**(1), 18–23 (2015). https://doi.org/10.1007/s00170-015-7573-1

Sustainable Inclusion Factorization in Green Metaverse and Virtual Worlds

Wasswa Shafik

Abstract The new drive is toward a sustainable future driven by advancing technological progress supported by factors such as innovation and environmental, social, and corporate governance considerations. Purpose-oriented Digital downtime in the Green Metaverse is also picking up Digital Twin in its data continuum harnessing. New data and computational thinking are getting aligned. Digital remote sensing in smart cities opens up dynamic urban eco-navigation opportunities, laying the foundation for a theory of sustainability to be harnessed, factorized, and enabled. The four key pillars towards sustainable inclusion factorization in the green metaverse have been identified as essential: algorithmic, capital, interconnectedness, and purpose-orientated sustainable pillars. In this pioneer unique conceptual paper, we present nine principles in sustainable inclusion driving factorization towards a Sustainable Purpose-driven Digital Metaverse and Virtual Worlds society, leading to the fulfillment of ethical guidelines that state that AI shall be developed for the Common Good and Aggregated Intelligence towards the Metaverse shall be leveraged towards Stewardship over Ownership. These principles are essential both for humanity's transition towards AI Management of the World and for the sustainable development of the Green Metaverse as an ethical global common, guarded by Stewardship imposed over and above Ownership. Symbiotic Digital Immortality is no more a Futurologist Dream; it seems to become as feasible as the Engineer's Dream through advancing Information and Communication Technologies. Its affinity rests to a missiological approach in the paradigm of Technology as Missiological Driver assisting the fulfillment of the purpose of life of the human species on earth with differentiation from advancing intelligent beings, as enshrined by the Higher Ethical- Moral Imperatives. The purpose alignment founds its philosophical roots on a Purpose-Driven Philosophy embedded within the AI-Based system that employs algorithms for Stewardship and Ethical Governance support system towards Digital Immortality from the Technology stand-point.

W. Shafik (✉)
School of Digital Science, Universiti Brunei Darussalam, Gadong, Brunei
e-mail: wasswashafik@ieee.org

Dig Connectivity Research Laboratory (DCRLab), Kampala, Uganda

S. Kautish et al. (eds.), *Green Metaverse: Fuzzy Logic Approaches to Sustainable Virtual Worlds*, Information Systems Engineering and Management 74,
https://doi.org/10.1007/978-3-032-08183-4_6

Keywords Digital transformation · Eco-friendly diversity · Factorization · Green inclusion · Metaverse sustainability · Sustainable virtual worlds

1 Understanding Sustainability in the Metaverse

When considering sustainability in the Metaverse, the whole world, whether a digitalized society or a real world, is under the context of the virtual world governed together with its ID. The concept of sustainability is inclusive of people's survival and the stable development of the economy, society, and environment. According to the characteristics of the virtual world, facing the dilemma of teleology, introducing a sustainable Metaverse offers a feasible way for future business entity models and operating strategies [1]. The keyword of the sustainable Metaverse is fusion: to play the role of a virtual leader in practical backgrounds and ideal models under the effect of a sustainable Moebius ring, where the virtual economy and real economy are connected. However, the virtual world creates a prosperous 3D graphical, interactive interface and cyberspace's unlimited imaginative effect; it does not mean the Metaverse can easily indulge in not accepting any moral obligation. The ethical responsibility of game enterprises should be strengthened [2].

With the sustainability of fascination and function, it is useful to clearly conceive of both Metaverses and sustainability to grasp what may occur when we put them together. Both terms have become umbrella terms spanning a rich, diverse set of groups of stakeholders and their needs that are critical to creating, maintaining, developing, and using their interpretations: within experiences of the physical world, we create Virtual Worlds through media aspects, games, and social computing [3]. Similarly, when considering representations within Virtual Worlds, avatars are experiences to be personally created, maintained, developed, and used for other applications that develop their own user-avatar relationships, such as preparing to meet new challenges and move into the wider Metaverse; the Metaverse may perform a similar relationship. Notably, the experiences can be advantageous for participating in some kinds of social and economic activity and can represent indeed any resources [4].

1.1 Defining Sustainability

Sustainable development was first defined in 1987 as 'the development that meets the needs of the present without compromising the ability of future generations to meet their own needs.' By 1988, corporate social responsibility was described as 'a desirable goal' only. Over the past thirty years, CSR has grown to become a management standard and obligation; sustainability that was once escapist rhetoric and a desirable goal to aspire to has become an ethical baseline of human aspiration and the foundation of corporate cultural responsibility. In the metamorphosis journey of corporate social responsibility (CSR), CSR became commercially reasonable and also proved itself to be economically viable and sustainable [5]. Today, when one talks about sustainability in business circles, one means diverse social, economic, environmental, and commercial sustainability of a company's activities as well as active corporate governance for the enterprise. In other words, sustainability must self-preserve, nourish, and influence a company's financial prospects but also increasingly pay attention to the company's conduct in nurturing the well-being of society and employees; minimize pollution and visibly show that they are taking meaningful measures to tackle ethical elements such as corruption and bribery [6]. The Bern model is commonly adopted for sustainability. The planetary model suggests a clear commitment to the natural environment and financial sustainability and limits returns on financial and natural resources. Social sustainability ranges from paternalistic contributions to corporate community and investment in philanthropy in times of stress. The solar model describes ambitions where companies focus on the well-being and respect of employees, deal with worries, and improve transparent and righteous activities with open administration processes. Such measures are aimed at supporting equality and promoting economic justice [6].

1.2 Key Elements of Sustainability in Virtual Worlds

In considering the unique aspects of sustainability in virtual world economies and more broadly, we suggest a number of key components and issues that deserve further study and exploration. Emphasis should be given both to preventing massive neo-industrial waste dumps in the real world due to users' weariness and decreased demand, as well as to social goals of relaxing common and affordable information entertainment as a right of every human on the planet. In considering normative, empirical, and methodological aspects of environmental, economic, and cultural sustainability of virtual worlds, one should especially study synergies and conflicts between these goals [7]. We suggest testing voluntary benefits as an effective mechanism for consumers' involvement in economy- and society-enhancing activities. Creation of an information consumption market and facilitation of innovation activities and culture preservation are suggested as additional ways to enhance digital world inhabitants' quality of life. Environmental sustainability restricts the growth of either informational or inhabited content in the real or virtual worlds. What should be practically changed: leveling a decrease of both due to the growth of harmful activities in the virtual world or providing real-world inhabitants with compensation for additional environmental losses due to human demands on the virtual infrastructure? Should virtual world inhabitants be invited to the eco-friendly virtual waste facility at their next birthdays on Earth [8]?

2 Inclusive Design Principles in Virtual Environments

When entering a virtual environment, a user should not have to justify or account for its abilities and skills, so all the content and virtual and non-virtual interfaces should be inclusive. It is imperative to ensure that all kinds of users perceive a virtual environment as total, and thus, it should be adapted to the individual needs of people, benefiting from the commonly accessible design. From both technological and design aspects, inclusivity can be ensured by taking into consideration the rules following from the inclusive design approach [9]. The project allowed synthesizing the following principles of inclusive design in virtual environments: provide the possibilities of access to interactions and interfaces using assistive technologies. Each user should be given the opportunity to interact with each element of a virtual environment and non-virtual interface by means of their control channels and assistive technologies in accordance with their capabilities and experience of interaction. Virtual environments should support not only traditional mice, keyboards, and other common control devices but also provide alternative interaction methods and access to control channels for particular groups of users [10]. The most common control channels may include:

- Pointing devices (mouse, pointing stick, stylus and pen, trackball, touchpad, touchscreen)
- Mechanical and virtual switches
- Speech and voice control
- Body control
- Head control, possibilities controlled by eye
- Multimodal input means – the possibility of the joint use of several input devices.

The availability of interaction with a virtual environment should be provided not only by a particular type of input device but by alternative methods and interfaces, including speech and voice interfaces, body and head tracking setups, and multimodal interfaces. When interacting with interfaces and content via gestures, touching, or speech, persons with limitations of movement may encounter certain problems, since each of these types of interaction is not equally available to each user. Moreover, the lack of a systematic approach to work in a virtual environment can create situations for a particular user where the virtual environment discriminates against persons with motor limitations, since the most conveniently accessible method of interaction for a healthy person may be inaccessible due to user limitations [11].

2.1 *Accessibility and Universal Design*

In virtual worlds, the concept of virtual community and social capital has become increasingly important. To maximize participation, access to built spaces and public area facilities is crucial. Access and the quality of space matter because they affect social interactions and communication possibilities, and ultimately, life and work satisfaction. To achieve the goals of universal design, it is important to analyze and understand the relations between physical and virtual world infrastructure systems and the characteristics of their sensorial and communication areas. It is also crucial to understand the configuration among these characteristics, relations, and stakeholders, and their mutual influences [12]. Universal design aims to offer different social and cultural groups an environment that responds to their lifestyle, respects their social habits, and presents them with the right tools to protect and express their own identity. This means respecting the citizens' original culture and history. Therefore, technologies and sensory augmentation systems must be redefined in terms of services and creative approaches according to different contextual scenarios [13]. Societies can progress very differently according to the forms of regulation and control exerted by individuals. Different forms of societal models can consider such concepts of disability, with different individual-environment/artifact configurations, also due to technical and economic innovation, which can strongly change the way of building accessibility solutions. Misalignments between these elements can lead to inequalities, creating barriers that hamper equal opportunities in access to different activities. Different social groups are favored and advantaged in different ways [14].

2.2 *Equity and Diversity Considerations*

The broad community in the real world is represented in worlds but requires robust methods to account for multiple facets like age, gender, sexual orientation, socioeconomic status, national origin, language, education, and ability. Classic social measures of these factors have common limitations of measuring only one aspect, like gender and age. Other proxies and deceptive variables reflect correlation through access to real or economic assets rather than the social facets themselves. Gender is a case in point where the doubling of female designer leadership in both real and world plant management, promotion, and maintenance roles occurs [15]. Disparities in time and kilometers used for real life, and the holding of concealed roles arise without any recharge of their income or investment activities. This range of potential positive gender impacts in rearing and world matrix and cure designer worlds has been incorporated visibly into current examples. Many additional demographic and educational contributions to gender have not yet achieved their potential in leadership so that other gender attributes aside from real attributes contribute to the offering function of world impacts metaphorically with the female curve. Independent measures produce an adverse effect for world [16]. Given the known impact of inclusive and diverse groups on workplace satisfaction and performance, the broad feminist action toward sustainable development that provides a significant global statistical contribution and large bibliography altogether for perceived collective similarity improvements will be beneficial in information for future work optimizing gender impacts. Upon inclusion of children, worlds provide additional beneficial spending implications for measures of global collaboration and health among the absorbed pool of children and other workers of active working age. While the aggregate population expended with expected improvements of possibly modest amounts, even firms used as examples of inclusiveness and gender equity [17].

3 Technological Innovations for Green Virtual Worlds

This section addresses the technological aspects of the development of green virtual worlds. Due to the innovative nature of the topic, some techniques do not yet exist; as a result, this section displays a wide range of goals, most of which are yet to be met. They can serve to assist researchers by showing available challenges and opportunities, but especially to promote new opportunities in research related to the construction of a sustainable and equitable economy. The infrastructure for virtual worlds presently consists of powerful servers that provide the basic platform and numerous clients that connect and interact with the server [18]. These clients can range from small to high-performance devices. The device that provides the best experience for a virtual world is based on a head-mounted device, which should project 3D environments at frame rates greater than 60 Hz, which approximates the real motion perception of the human eye. With improvements following the penetration of high-resolution television sets and of transducers and accelerometers that provide tactile and kinesthetic feedback, most high-end head-mounted devices are equipped with them. Therefore, it is better to make the best use of these devices when providing 3D virtual worlds [19].

3.1 Renewable Energy Integration

In this section, examples of actual and potential renewable energy integration in virtual worlds are reviewed. A variety of different implementation approaches and best practices from real life are analyzed. As virtual space architectures and hardware constantly evolve, these could be used in the context of open worlds and mixed-reality environments to augment data and validate specific hypotheses. The derived experience reveals that renewable energy might be perceived and integrated in different ways and at various simulation levels, requiring a skillful combination of urbanization, software, and services [20]. The integration of energy solutions in the metaverse and virtual worlds, particularly renewable energy, is a natural step towards a holistic green paradigm. Partial energy supply reorganization can hardly reach the desired ultimate structural effect of environmental and social responsibility. Converting the predominant over-technical game culture into a broader virtual engagement might provide a new and highly scalable means to spread awareness and ultimately create the critical mass for capturing wider society's involvement and replicating sustainable real-life success [21]. At the same time, the virtual influence of smart information can considerably stimulate renewable energy knowledge, industrially driven research and development, paving the way for cost-effective future adoption by inspiring game creation, promoting e-simulation, and following the commercial trickling down from public demonstration effects [22].

3.2 Energy-Efficient Hardware Solutions

The power consumed while utilizing hybrid technologies of sensors and functionalities causes the main desktop and mobile power consumption. Within the contents of our proposal, we will develop new technologies within SmartGlass—namely power-efficient ICs that will make it possible to integrate into a variety of devices. In a decade, glassware shall become the major computer. And, of course, the proper architecture of these products using hybrid semantics in computing models will allow us to analyze and solve the most essential questions in economics, medicine, and human rights [13]. Our work serves as an introduction to the development of such architectural models. There are very important points about the current product: the product is equipped with both sensors and has much in common with only in the energy generation scheme. Our subminiature product consumes from 25 million to 1 billion times less energy than its industry analogues for equal computational opportunities. A micro-scheme like this enables quick mass production of high-yield sensors [16]. High energy efficiency allows us to minimize the sensor scan time without losing the quality of the scan data.

4 Community Engagement and Social Impact in the Green Metaverse

E-240 project plans to provide a variety of educational and motivational programs for youth, encouraging them to take an active part in building the world. Creative, artistic, and cultural expression of our inner drives should become the basis of the external world. When E-240 people think, they are not limited to the development of their plots and salaries; each will think about the good that they can bring to the world or how they should contribute to the improvement of the common good. E-240 is a global home for all categories of people [17]. For early participation of children in the system, special educational games will be created that will help instill in children the necessity of taking responsibility for improving the world and being aware of the enormous potential that their work has. Another area of community organization will be the creation of various social and entertainment resources. Hundreds of quests and other entertainment will take players to virtual and augmented reality. Regularly organized sports and cultural events, quizzes, interactive testing, and social initiatives are planned [18]. Relations between people will actively develop in the field of interests through E-240. All financial and social communications of life will also be transferred to E-240 operation, turning it into a complex system of the village and family of thousands of people with a continuous turnover of events and bringing business, the labor market, and enterprise to a new level of quality. The E-240 project will create a unique space for communication, social inclusion, and cooperation for everyone who lives in it [19].

4.1 Stakeholder Collaboration

Efficacious virtual world collaborative learning technology depends on effective collaboration between technology stakeholders, and more on rigorously evaluating the technology than merely on needs assessment research. Using semi-structured and unstructured interview techniques, experts from user, developer, multisector, investor, and policymaker technology stakeholder categories were probed about the factors affecting their interests in the three technology inclusion areas—technology user involvement, engagement with local communities of new technology installations, and evaluation of SVEC [23]. Interview text field data were analyzed using phrase and thematic text analyses. The results provide guideposts for stakeholders to improve their collaboration in developing, deploying, and evaluating efficacious SVEC platforms and their related technology. A goal of transitioning to a green economy is one of developing and deploying increasingly sophisticated new technology and service provisions that become sustainable. Green metaverse technologies involve virtual natural environments or processes such as those in computer games, simulations, and online interactions [17]. Virtual world inclusive collaboration uses collaborative learning to integrate local native groups and new technology designers and investors to attend, stipulating installed technology virtual world collaborations, including facilitated collaborations using internet-connected computers. These proposed future technologies traditionally ignore natives, who degrade the natural environment the technology was built to ameliorate, as well as services and other local materials because they assume that participants can learn primary messaging in the course of using and building the virtual world [19].

4.2 Environmental Education and Awareness Programs

As awareness of global environmental and conservation problems continues to increase, there is an increasing number of education and information programs that use virtual worlds, where a large number of people can be reached. In these environments, participants can interact with other characters and objects, acquiring information on different subjects presented in a three-dimensional medium. They can receive constant information about natural resources and product and service consumption sources, as well as data on the results of consumption patterns and resource management, which can be used for education, including sharing in other people's experiences [24]. These three-dimensional learning venues can foster industry and opportunities to share knowledge to such an extent that they can actively use technological innovations that provide countless opportunities for promoting environmental awareness, educational development, and ecosystem understanding. They are already being used in many countries for educational purposes and life-long learning. Thanks to virtual worlds, people learn more quickly and effectively about the problems faced by the community regarding environmental issues, natural disasters, and climate change scenarios on the planet [25].

5 Case Studies and Best Practices

This section presents case studies of projects that foster the innovative and inclusive city, experiences in the arts, culture, and creativity, scientific study, educational institutions, and explorations at all levels. These projects are designed to bring together the individual and the community in blended physical and virtual activities, combining the best aspects of virtual realities and engaging activities. These collective themes define today and will also define tomorrow's urbanizing world, helping to ensure the construction of more inclusive, accessible, poetically connected, and resilient futures [26]. The featured projects include an architecture project, an inclusive design project, an in-depth study of representations and models of cities as a basis for inclusion, an arts and performing arts project, and some early findings of a study on an icon for the realization of a cruise ship. This text focuses on the city and the world in its design and construction, or adjustment, and on the people in its environment, which meets needs in all aspects of daily life, thus sharing visions of what it could be [27].

5.1 Successful Sustainable Projects

Activity has developed in recent years, particularly in professional organizations and universities, though the concept of sustainable development has been highly publicized. In this section, we emphasize the importance of evaluating the proposals of projects with an ecological sense. The idea is presented directly; we intend to fulfill different areas under proposals that are focused on the same concept of sustainable development, involving researchers and students. The economic disparities between the groups have increased, and as a result, the difference in the consumption patterns of individuals with no family income and those who have a sustainable income has also increased [28]. From this perspective, it is perceived that sustainable projects, aimed at developing smaller families over time, help to characterize a taxonomy in the area and decrease social gaps. Also, note that such projects, when carried out in regions with lower income levels, give rise to entrepreneurship that generates income and appreciates the local culture. This study considers sustainability to be the fundamental concept of society's dream of preserving and improving the reasonable future of humanity on Earth and the quality of what the future will be. Sustained from the perspective of development, the commitment is to produce a book of experiences [29]. The cultural but not organic entity tends to converge in a presentation of suggestions for a movement that aims to balance critical thinking and growth efforts.

5.2 Lessons Learned and Future Directions

We presented a methodology and network science approach to factorize the sustainability influences and inclusions that the members of society perceive in each other in the green metaverse, or virtual worlds meant for the application of general societal welfare. We shared several empirical results from the world of online games. Our results enable a better understanding of the enablers to perceived inclusion effectiveness of sustainability improvement among society members. Our results are useful in optimizing the planning and coordination efforts in discovering and implementing effective solutions towards sustainability improvements in all areas, through the understanding of how society members evaluate their potential inclusion contributions supporting the solutions [30]. This understanding assists in improving the communication strategies addressing perceived sustainability inclusion and allows the deployment of tailored training programs that would accelerate the development of required knowledge, skills, and competencies by effectively addressing the perceived inclusion effectiveness among society members. In addition, our empirical evidence suggests that see-through survey methodology combined with network science methodologies is a viable set of principles and workflows for empirical investigations involving several issues on performance evaluation into antecedents and consequences of the perceived societal inclusion perceptions of current or potential social entity members and stakeholders [31]. In summary, the contributions and lessons are expected to be useful to fellow researchers and in the design of future empirical studies into the antecedents and consequences of sustainability inclusion in virtual worlds, in both societal and specific organizational settings for the application of general societal welfare [32].

6 Policy and Regulatory Frameworks for Sustainable Virtual Environments

Participation friction that is negatively influenced by developers' non-optimal planning indices related to stakeholders can affect the existence and acceptance of the proposed green metaverse. Participatory virtual towns, like their worldly counterparts, need optimization due to limited natural resources. Most of the time, collaboration with other countries may assist throughout the structural design. Public authorities are primarily responsible for the structural preplanning and building plans that become public policy obligations [33]. It is essential that there is a legal framework for ensuring that building-related facilities in the virtual world are accessible and will assist in achieving the highest aesthetic and functionality standards. On the one hand, virtual inhabitants should at least be enabled to have similar access, and they will, therefore, benefit hundreds of times more consequentially from the right to connect. On the other hand, local communities need to be selective and take care of public facilities in addition to building structures that will have an integrated architectural character in the future [34]. Owner, manager, and user companies have obligations to care for the virtual environment. Although they also have responsibilities toward many places in society, they should meet related obligations directly. Prior to traditional world penalties, these may need to be amended to be fresher. On the other hand, local inhabitants have responsibilities under the philosophy of stories to keep the natural environment public [35]. Sports protect these places where we used to live, constructing a reference system for future generations to eliminate the possibility of distortion.

These responsibilities include personal behaviors such as never killing, vandalizing, or overusing anything that does not serve a meaningful occupation or installation, such as virtual pure air and ecological happiness. Programs should, therefore, be improved through a store, with interested individuals also included in preservation or maintenance activities to create community satisfaction and, thus, facilitate those policies for additional support. There can also be promotion and display activities associated with the artistic aspects of buildings, like the competition for the most attractive belvedere focused on digital materials reused in the construction of responsible architectural designs [36]. Such obligations are a function of the relationships among individuals and not spending; they are not exactly pollution charges related to the cost of policy definition. It is interesting to notice that the idea of 'what is invisible' can also be a framework for virtual communities as an input for policies developed by a multiscale set of actors: practitioners and users. Such ideas could be creatively used for building standards of offices, ready rooms, stores, shopping centers, entertainment squares, and even more. For example, the notion of vegetable towns could also be possible in virtual geography, especially because inclusive development can protect against physical gaps [37]. Since the legal and institutional construction of virtual worlds does not create a representative democracy, other policy instruments, such as price reform, poverty line approaches, market poetry, and participation reciprocity can also be used.

6.1 Government Incentives and Legislation

Now we move on to government incentives and legislation. As we have stressed, rich micro- and macroeconomic datasets are essential to capturing useful summaries and indexes of MI and sustainable inclusion. The primary limitation at present is that the construction, collection, and dissemination of robust summary statistics on them are predominantly in the realm of public-private partnership organizations. The agencies dedicated to monitoring social, economic, and political development at the high priority or utmost urgency that monitoring the natural environment currently occupies are much more numerous and better resourced [38]. Due to these resource constraints, governments are now providing incentives, both direct and indirect, for the private sector to take up the collection and dissemination of social, economic, reputation, and satisfaction data. Those resorts and developers who satisfy the green demands of their investors, politicians, and activists by addressing the SDGs are provided with preferential access to funding, tax incentives, and licensing. However, in the ugly face of progress, the process of creative destruction indicates that new workforces are required for an industrial revolution to finally succeed [29]. Governments provide safety nets for economically displaced workers, but ensuring the expenditure multipliers spend the money in the green economy hasn't been sufficiently addressed nor captured in a way that the transition can happen at speed and scale to prevent suffering. Should the micro- and macroeconomic outcomes be captured, consensus might develop that our societies need to become more value-based. Qualitative and time-based data would then move to a higher priority and be sought out and disseminated faster to correctly inform the markets. Due to national differences, government legislation would be the most effective in the first instance in mandating change [30]. We close with examples of government legislation.

6.2 International Standards and Guidelines

International standards and guidelines have not been developed specifically for virtual worlds. Unspecific standards and guidelines, including those developed for rights to intangibles available through virtual worlds, serve for the definition of specific rights to intangibles in virtual worlds. These rights to intangibles are comprehensively formalized in the specification of intellectual rights. The existing aspects of virtual worlds, including game and business ones, are available thanks to the unique intellectual rights model. Intellectual rights are not included in international standards, although the description of the legal terms, instructions on how to characterize these legal terms, and the characteristics themselves of these terms represent a significant part of the functions of the standards [31]. Yes, international standards and the legal language have a number of special features related to the general framework of harmonization and professional terminology. This aspect is consistent with the special role of the International Organization for Standardization. The international language of intellectual rights has not been fully developed yet. Active efforts started no earlier than the early 1960s. The standard, one of the first standards for characterizing a relationship with an intangible thing, formalizing the rights to intangibles, was developed more than 30 years ago. International standards on characterizing the intangible object itself can be considered. Characterizing the rights to an intangible thing greatly predates characterizing the rights to a thing [33]. The rights to an intangible thing do not only take much longer to develop, but also have a greater influence.

7 Economic Viability and Business Models in Green Virtual Worlds

We have already presented several direct and intangible economic values collected and created from a sustainable virtual green world system. Although these collections are only a part of the economic values that will emerge in the future, we can assume that since these values may continually increase and expand to other unknown sectors, it will then become possible for Serkel to run his businesses in virtual worlds by relying on our new economic models. While the data center is facing the limitations of the homogeneous super large scale to benefit from economies of scale, a lucky curve with relatively low-cost service providers is happening in the "multi-core" business systems [35]. Although the monetary rains presented here are not the most prospective part of the future economic standards of Serkel's virtual world, it can still be capable of bringing in financial benefits for investors. Virtual worlds and green metaverses are supposed to have the prohibitive or preemptive mechanisms to encourage the transition of user economic beliefs and myths into modus operandi. The vertical proportion in platform construction or open policy in world design can easily ensure the flexibility of the virtual world system, as well as prevent or mitigate lock-in and other anti-competitive behaviors [36]. Biopolitical philosophy and theory also suggest that authorities are supposed to have full public concerns rather than enforce cumbersome management, which cannot weaken such measures' validity. Specifically, consider the fashion conglomerate. Fashion institution owners can ask all institutions and their staff to value each of the sustainable operation indices, although every institution may choose its most appropriate and unique fashion themes as they operate. Moreover, the value realization formulas also empower sustainable corporations by implementing voices and expectations into the institutional mechanisms and propelling collaborative and creative collaborations in the surrounding institutional space [38].

7.1 Monetization Strategies

Rather than paying the costs of maintenance and development themselves, virtual world developers might co-opt mainstream economic players into externalizing significant portions of these costs. Virtual world franchise models work in a similar way: existing corporations license virtual world developers to create and maintain their virtual in-world presence. Virtual world developers might even accept direct investment from the players, selling part of their virtual world enterprise in public offerings or private placements. In return for their initial stake and potentially ongoing injections of capital, players would expect compensation in the form of in-world profits and dividends [3]. For all the troubling social and economic implications inherent in this market-centric approach to virtual world finance, it is one possible way for mainstream economic actors to recapture some of the hundreds of millions of resident hours the players contribute. The participants say, "People come into an entertainment-driven environment, then there is a rapid user experience." So, it can be regarded that it belongs to the entertainment industry. It is very dangerous if it does not fit participants' demands and the game market, so the amount of capital required will be very large, and in the case of a company, the possibility of bankruptcy increases. To survive, high-quality content must exist, and there must be ways to extract more money. Ads, brand collaborations, and advergaming are the services that can lay the foundation for capital influx [7]. Since these services allow for multiple revenue streams, it is very important to increase capital.

7.2 Corporate Social Responsibility in Virtual Environments

Corporate social responsibility is not automatically extendable to virtual worlds; rather, it needs to be redeveloped and readapted by numerous operators in these environments for it to be a positive factor. This mechanism would also entail operators of platforms being especially encouraged to implement legally framed directives since new forms of CSR self-regulation, without a political or legislative framework and suffered only by firms, are less binding and efficacious. They have, in fact, proved to be instrumentally oriented public relations, defined as mechanisms to maintain the official "public" reputation of companies which, despite declarations of purity for solidarity and support for environmental safeguarding, continued to introduce systems innovations to lead them to higher objectives, like increasing profits [9]. While it is true that the experience already consolidated in VR and VW has understood and used the potential that may arise in the coexistence between business, explained as general principles of social responsibility and ethical principles, and technology, using the latter to give shape to the application of the former. The necessary prerequisites, if we want to put all the virtual resources to the service of corporate social responsibility, can, however, be summarized in two evident elements already discussed in this chapter [32].

7.3 Ethical Considerations and Data Privacy in Virtual Worlds

This section addresses the ethical, legal, and privacy-related aspects of sustainable inclusion factorization in virtual worlds. We begin with a summary of the current legal situation and responsibilities when operating in virtual worlds. Our emphasis is that virtual worlds are indeed virtual, with any associated physical actions and consequences being within the legal structures that govern non-virtual activities. With that in mind, we outline some of the important privacy-related aspects of virtual worlds before including a brief discussion of the relevant technical aspects of security. A physical action in a virtual world may take many forms, from crafting a digital object, changing the properties of a digital object, traveling in the world, altering the properties of an avatar, etc. [33]. All are potentially subject to some form of legal governance ranging from patents, trademarks, copyright, and other IP laws or rules through physical geography and precedence, societal agreement and regulation, actual legal statutes, and common practice. It cannot be overemphasized that physical action in a virtual world has consequences, and the impact of these on the non-virtual world is governed by the most appropriate physical world rules. With this in mind, we now consider some of the aspects of privacy and security that are important when traversing, shaping, and handling data within a virtual world [34].

8 Measuring and Evaluating Sustainable Impact in the Metaverse

Measuring and evaluating the impact and success of a metaverse project is a relatively small-scale and crucial challenge. Research on similar issues mainly takes place at the implementation and project levels, although outputs of the presence and impact levels also differ due to the complexity of the project. The largest part of the evaluation literature defines the traditional project cycle of input, process, output, and impact. However, in the metaverse project cycle, these are not sufficient because we witness the creation of an online fictional community as output. This also does not suffice because 3D games, 2D websites, traditional social networks, and other applications and software can exist in isolation and need a mobile or web technology infrastructure to develop and become available or useful.

8.1 Key Performance Indicators (KPIs)

In the increasingly global world, the instantaneous connection between people and digital companions exposes the dependencies of our collective innovation ecosystem on rapid and continual digital transformations of the worlds we inhabit. This chapter formulates quantitative key performance indicators that track the sustainable inclusion performance of our digital transformations, reflected in both the inclusive composition of the productivity matrices that factor the goods and services generated in our economy and in the inclusionary access of people to experience interacting with atoms evolving on bits ranged across all scales of representation available to them [39]. These KPIs thread from pixels of matter in bits to the metaverse and macroverse synthesized from bits, offering a path in which digital transformation is simultaneously a sustainability revolution realized digitally. Every metaverse and virtual world development begins with an idea, energy, time, operating and licensing expenses, guidelines, values, and investments. The aim is to generate new opportunities and reach a generic audience base for participation, virtual tax collection, and international public funding. Moreover, the main success factor is maximizing the feeling of perception and conceptual experience and minimizing real-world costs [16]. In another sense, the main challenge is to intervene and influence or organize virtual and fictional community resources and properties without physical creation and beyond geospatial restrictions.

8.2 Impact Assessment Tools

There is a long list of tools and guidelines available for sustainability assessment. However, a conspicuous lack of inclusion indicators has meant that peer pressure, market awareness, best practices, and government legislation have been the drivers for best practice. There are different standards as part of its complete sustainable reporting guidelines, including required standards. Meanwhile, investors and analysts have attempted to use rating companies to inform them for the best investment decisions, where the larger companies can afford to report on all elements, but the smaller companies or start-ups cannot. A charitable investment company also sponsors an annual issuance of the top listing of the most sustainable asset managers [18]. The effectiveness of guidelines has been questioned, with a significant percentage of the world's largest companies using the guidelines for corporate reporting, but it recognizes the need. Some organizations have independently collected data across different dimensions of ESG with plans to create an ESG index for future new ESG funds. The significant problem has been the lack of effective tools that focus on making, creating, and transforming inclusion. The room and need for improvement and uniqueness that sustainable guidelines themselves are driven by simply giving a clear and distinguishable description that differentiates the present condition of a product or business activity are both taken into account over time, and with regard to how these guidelines will affect a product, service, or factory design. Small or large project implementations generally focus on the efficient and effective use of public resources to solve public problems and create desirable local conditions [28].

9 Future Trends and Emerging Technologies in Green Virtual Environments

We have discussed some major green VR and VE technologies of the near future. These technologies, when deployed properly, will make the VE very sustainable as they reduce energy and material usage and heat. They also improve modeling and rendering of 3D models and animate them using AI and related technologies. The resulting effects and visualizations are projected through our eyes and ears. The quality and immersive capabilities of the outputs are also to be enhanced to improve user attractiveness. How will the future VR ecosystem be? As of now, we are generating, consuming, and releasing huge amounts of digital data everywhere, which is expected to grow enormously with the arrival of new technologies [3]. This digital data is going to be stored in digital warehouses of the future. The rise of digital warehouses will make digital resources available to all types of digital engines that process data and deliver novel digital products and services to markets in different technology domains. The different technologies address various market segments. Digital assets are employed in different types of VR worlds, which have different types of ESI ratings (Excellent, Superior, Inferior) or VE degrees (Mainstream, Niche, Marginal). Artists, creators, owners, users, and other connected entities participate in the latest VE, earning directly or indirectly through these ESI-graded VR ecosystem connections. The future of the VR ecosystem is expected to make HD-SDI resolutions favorably available through more lightweight, sunglasses-like devices [17]. Energy and memory usage, along with visual fatigue, are expected to decrease considerably through innovative hardware technology insertions. Moreover, innovative hardware technology is expected to make the sunglass-wearing experience very attractive, while collaboration and interoperability solutions are expected to fuse 3D models and users' personalities from different worlds into one. These offerings are expected to satisfy the existence and inclusion factors of credible users and stakeholders, making the future VR ecosystem very green and inclusive. High-level services with secure, scalable, self-healing, and low-latency network technologies are expected to offer enterprise-grade commercial operations within this metaverse. What is going to be the repository of these digital assets [27]?

9.1 Augmented Reality Integration

To allow augmented and virtual reality headsets to be accessible to the general public, AR and VR systems will merge with each other and with vision-making systems that harness all the technology of mass media computing. Wearable computers, voice and print recognition, motion analysis, and haptics combine to reveal and improve real and virtual environments. An important phase in these merging technologies is AR/VR integration [40, 41]. The creation of a coherent intermediate AR/VR display system that combines the strengths of both sensing paradigms provides the first step in the integration of AR and VR methodologies. The second basic step is the introduction of a clear, unambiguous set of AR/VR culture, language, and environment that can be shared by the AR and VR community in preparation for the further stages of AR/VR integration. The third basic step is the complete fusion of the intermediate AR/VR display systems, methodologies, and shared environments, leading to freely intermixed AR/VR systems of any or all of the various display qualities and functional effectiveness of any necessary or desired fusion [42].

9.2 Blockchain Applications for Sustainability

Blockchain holds various potentials for maximizing the achievement of sustainability-related goals, including opportunities for further improving sustainability through better alignment of incentives among stakeholders, provision of a more transparent and traceable supply chain, and enhancement of the functionalities of existing assets and financial instruments. While firms committed to sustainable development need to focus on compliance with legal frameworks and competition in optimizing these codes and norms, they must also be forward-looking along different time horizons with ambitious levels. It is important to highlight the complementary roles and capitalize on synergies between corporate codes and public policies [43]. Improving Sustainability through Incentive Alignment Properly designed blockchain-based systems can enhance sustainability through the alignment of incentives among stakeholders. For instance, the establishment of a blockchain-managed, peer-reviewed pricing platform can reduce the risk of impact-washing, which is associated with insincere or exaggerated claims of organizations regarding their ESG impacts. Blockchain-compatible IEOs have shown the capabilities of revolutionizing the sharing economy by not only providing traditional investors an opportunity to invest in sharing economy-based projects but also ensuring that the invested funds are used as proposed in the investment proposition [44]. However, the alignment of the interests of multiple stakeholders requires a comprehensive, transparent timescale for corporate sustainability performance and impact measurement.

10 Conclusion

Positive digital engagement is essential for promoting transformative access, inclusion, and equity in virtual worlds. Most digital engagement enriches communities while it can also aggravate inclusion disparities. For instance, disadvantaged individuals experiencing difficulty accessing and using virtual world services can be further discriminated from society. To guide positive inclusion factorization in the Green Metaverse, engagers should answer the following questions not only once, but frequently along their virtual world experience and should call on hosts, designers, developers, and regulators to commit to co-guiding these crucial conversations in collaboration, co-governance, and optimally co-regulation with stakeholders, ensuring the implementation and follow-up with zero tolerance for discrimination, exclusion, and alienation. Engagers should be able to ask and know about issues mitigating or contributing to their digital well-being, including whether the features of the 3D virtual world where they are preparing to participate are entirely accessible or need additional adjustments to accommodate their specific necessities, including physical mobility, vision, hearing, cognitive, and neurologic supports and adjustments, among others.

References

1. Shafik, W.: Predicting future cybercrime trends in the metaverse era. In: Forecasting Cyber Crimes in the Age of the Metaverse, pp. 78–113, IGI Global (2024)
2. Hadjistassou SK (2013) Culturally afforded tensions in the second life metaverse. Int J Web-Based Learn Teach Technol 11(2). https://doi.org/10.4018/ijwltt.2016040102
3. Lungu, S., Enache, I., Rîşnoveanu, G.: Urban green infrastructure, ecosystem services and challenges for universities to respond to the needs of society. Proceed Int Conferen Econ Soc Sci (2022). https://doi.org/10.2478/9788367405072-038
4. Heilala J, Singh K (2023) Sustainable human performance in large people-oriented corporations: integration of human systems for next-generation metaverse. In: Proceedings of the 6th International Conference on Intelligent Human Systems Integration (IHSI 2023) Integrating People and intelligent systems, Venice, Italy (2023). https://doi.org/10.54941/ahfe1002858
5. Abdala, L.N. et al.: The role of the metaverse in the future of the space sector. In: Proceedings of the International Astronautical Congress, IAC (2023)
6. Yu, J. et al.: Gallium nitride blue/green micro-LEDs for high brightness and transparency display. IEEE Electron Dev. Lett. **44**(2) (2023). https://doi.org/10.1109/LED.2022.3228529
7. Wu, G., Dai, J., Lin, Q., Liu, C., Mi, J.: Person detection based adaptive video synthesis: a fusion framework for ultra hd panorama and green screen videos. In: Proceedings-2022 International Conference on Virtual Reality, Human-Computer Interaction and Artificial Intelligence, VRHCIAI (2022). https://doi.org/10.1109/VRHCIAI57205.2022.00017
8. Gupta, R.: Challenges & opportunities—the road ahead for management of internet waste & carbon emissions for a safe & green digital world. Int J Sci Res Eng Manag **07**(05) (2023). https://doi.org/10.55041/ijsrem21539
9. Pigultong, M.: Cognitive impacts of using a metaverse embedded on learning management system for students with unequal access to learning resources. In: 2022 10th International Conference on Information and Education Technology, ICIET (2022). https://doi.org/10.1109/ICIET55102.2022.9779045
10. Green, N., Works, K.: Defining the Metaverse through the lens of academic scholarship, news articles, and social media. In: Proceedings-Web3D 2022: 27th ACM Conference on 3D Web Technology (2022). https://doi.org/10.1145/3564533.3564571
11. Mingorance, N.M.: Green Video Compression for Metaverse: Lessons Learned from VP9 and HEVC (2024). https://doi.org/10.5594/m002002
12. Al-Karaki, J.N., Omar, M., Gawanmeh, A., Jones, A.: Advancing cybersecurity education and training: practical case Study of Running Capture the Flag (CTF) on the metaverse vs. physical settings. In: 2023 International Conference on Intelligent Metaverse Technologies and Applications, iMETA (202)3. https://doi.org/10.1109/iMETA59369.2023.10294722.
13. Devesh, S., Mahale, P., Subburayan, B.: Incorporating the Metaverse Into the Green Banking Revolution (2024). https://doi.org/10.4018/979-8-3693-3358-7.ch015

14. Rathore, B.: Emergent perspectives on green marketing: the intertwining of sustainability, artificial intelligence, and the metaverse. Int. J. New Media Stud. **05**(02) (2018). https://doi.org/10.58972/eiprmj.v5i2y18.115
15. Setyaningrum, R.P., Wulandari, A.: Green Human Resources Management, Green Environmental: Influence on Sustainability Performance Mediated by Digital Metaverse Technology in the Industry 4.0 Era (2024). https://doi.org/10.2991/978-2-38476-118-0_93
16. Li, J., Liang, W., Liu, X., Zhu, C., Zhang, Z.: Research on the Deduction Relationship Between Metaverse and Physical World from the Perspective of Mapping and Distributed Autonomy: A Case Study on Improvement of Urban Green View (2022). https://doi.org/10.1117/12.2652717
17. Chen, X. et al.: Decentralized cooperative caching for sustainable metaverse via meta self-supervision learning. IEEE Wirel. Commun. 30(5). https://doi.org/10.1109/MWC.008.2300006
18. Ma, Z.: Energy Metaverse: The Conceptual Framework With a Review of the State-Of-The-Art Methods and Technologies (2023). https://doi.org/10.1186/s42162-023-00297-w
19. Tan, T.M., Makkonen, H., Kaur, P., Salo, J.: How do ethical consumers utilize sharing economy platforms as part of their sustainable resale behavior? The role of consumers' green consumption values. Technol. Forecast Soc. Change **176** (2022). https://doi.org/10.1016/j.techfore.2021.121432
20. Shafik, W.: Introduction to renewable energy technologies. In: Soft Computing in Renewable Energy Technologies. CRC Press, Boca Raton, pp. 1–31 (2024). https://doi.org/10.1201/9781003462460-1
21. Shafik, W.: Energy policies and infrastructure requirements for efficient green buildings. In: Sustainable Technologies for Energy Efficient Buildings. CRC Press, Boca Raton, pp. 212–240 (2024). https://doi.org/10.1201/9781003496656-10
22. Wang, Y., Shafik, W., Seong, J.T., Al Mutairi, A., SidAhmed Mustafa, M., Mouhamed, M.R.: Service delay and optimization of the energy efficiency of a system in fog-enabled smart cities. Alexandria Eng. J. **84** (2023). https://doi.org/10.1016/j.aej.2023.10.034
23. Mingorance, N.M.: Green video compression for metaverse: lessons learned from VP9 and HEVC. SMPTE Motion Imag. J. **133**(1) (2024). https://doi.org/10.5594/JMI.2024/EBVF7405
24. Khan, L.U., Guizani, M.: Socially-aware green distributed metaverse for wireless systems. In: Proceedings-10th International Conference on Wireless Networks and Mobile Communications, WINCOM (2023). https://doi.org/10.1109/WINCOM59760.2023.10322890
25. Mohamed, A., Faisal, R.: Exploring metaverse-enabled innovation in banking: Leveraging NFTS, blockchain, and smart contracts for transformative business opportunities. Int. J. Data Netw. Sci. **8**(1) (2024). https://doi.org/10.5267/j.ijdns.2023.10.020
26. Zheng, J., Zhang, J., Du, H., Niyato, D., Ai, B.: Cell-free massive MIMO enabled URLLC communication for the green metaverse. In: Proceedings-2023 IEEE International Conference on Metaverse Computing, Networking and Applications, MetaCom (2023). https://doi.org/10.1109/MetaCom57706.2023.00096
27. Reniati, R., Bidayani, E., Sumiyati, S.: The influence of digital talent and green innovation and its impact on the msme business model in the metaverse era. MIX: J Ilmiah Manajemen **13**(2). https://doi.org/10.22441/jurnal_mix.2023.v13i2.011
28. Zhang, H., Wang, D., Wu, S., Guan, W., Liu, X.: USTB 6G: key technologies and metaverse applications. IEEE Wirel. Commun. **30**(5) (2023). https://doi.org/10.1109/MWC.012.2300077
29. Dubey, V., Mokashi, A., Pradhan, R., Gupta, P., Walimbe, R.: Metaverse and banking industry–2023 the year of metaverse adoption. Technium: Romanian J. Appl. Sci. Technol. **4**(10) (2022). https://doi.org/10.47577/technium.v4i10.7774
30. Srivastava, A., Sahu, G.P.: The impact of the metaverse on green marketing: a paradigm shift in sustainable consumer behaviour. IFIP Adv. Inform. Commun. Technol. (2024). https://doi.org/10.1007/978-3-031-50188-3_40
31. Wang, X., Bennis, M., Wang, P., Yu, R., Bryan Lim, W.Y.: Green Technologies for the Sustainable Metaverse and Web 3.0 (2023). https://doi.org/10.1109/MWC.2023.10325448

32. Huang, X., et al.: Service reservation and pricing for green metaverses: a stackelberg game approach. IEEE Wirel. Commun. **30**(5) (2023). https://doi.org/10.1109/MWC.014.2300095
33. Den Yeoh, E., Chung, T., Wang, Y.: Predicting price trends using sentiment analysis: a study of stepn's socialfi and gamefi cryptocurrencies. Contemp. Math. (Singapore) **4**(4) (2023). https://doi.org/10.37256/cm.4420232572
34. Yang, T., Cui, Z., Peng, C., Wu, J., Liu, F., Yang, Y.: Integrated communication and computing maritime networks design for green metaverse. IEEE Wirel. Commun. **30**(5) (2023). https://doi.org/10.1109/MWC.017.2300119
35. Chen, P.K., Ye, Y., Wen, M.H.: Efficiency of metaverse on the improvement of the green procurement policy of semiconductor supply chain–based on behaviour perspective. Resourc. Policy **86** (2023). https://doi.org/10.1016/j.resourpol.2023.104213
36. Cao, W., Cai, Z., Yao, X., Chen, L.: Digital transformation to help carbon neutrality and green sustainable development based on the metaverse. Sustain. (Switzerland) **15**(9) (2023). https://doi.org/10.3390/su15097132
37. Wang, R., Wang, J., Hao, J., Hu, L., Alqahtani, S.A., Chen, M.: C3Meta: a context-aware cloud-edge-end collaboration framework toward green metaverse. IEEE Wirel. Commun. **30**(5) (2023). https://doi.org/10.1109/MWC.013.2300082
38. Chen, P.K., Huang, S.: Enhancing supply chain resilience and realizing green sustainable development through the virtual environment of the metaverse. Sustain. Develop. **32**(1) (2024). https://doi.org/10.1002/sd.2663
39. Zhang, S., et al.: Toward green metaverse networking: technologies, advancements, and future directions. IEEE Netw. **37**(5) (2023). https://doi.org/10.1109/MNET.130.2200510
40. Shafik, W.: Digital twins tools and technologies. In: Digital Twins for Smart Cities and Villages, Elsevier, pp. 55–80 (2025). https://doi.org/10.1016/B978-0-443-28884-5.00003-8
41. Shafik, W.: Revolutionizing skin cancer diagnosis with artificial intelligence: insights into machine learning techniques. In: Impact of Digital Solutions for Improved Healthcare Delivery, IGI Global, pp. 167–194 (2025)
42. Shafik, W.: Human-Computer Interaction (HCI) technologies in socially-enabled artificial intelligence. In: Future of Digital Technology and AI in Social Sectors. IGI Global, pp. 121–150 (2025)
43. Shafik, W.: Navigating emerging challenges in robotics and artificial intelligence in Africa. In: Examining the rapid advance of digital technology in Africa. IGI Global, pp. 126–146 (2024). https://doi.org/10.4018/978-1-6684-9962-7.ch007
44. Shafik, W.: Incorporating artificial intelligence for urban and smart cities' sustainability. In: Maintaining a Sustainable World in the Nexus of Environmental Science and AI. IGI Global, pp. 23–58 (2024). https://doi.org/10.4018/979-8-3693-6336-2.ch002

Sustainable Virtual Worlds: Implementing Renewable Energy Solutions

Amena Mahmoud, Sapna Juneja, Osman Elwasila, and Yonis Gulzar

Abstract As virtual environments become increasingly integral to modern life, the need for sustainable practices within these digital realms is paramount. This paper explores the implementation of renewable energy solutions in creating sustainable virtual worlds. We examine various renewable energy technologies, such as solar, wind, and bioenergy, and their potential applications in powering virtual infrastructures. By analyzing case studies and existing frameworks, we identify best practices for integrating renewable energy sources into virtual environments, emphasizing energy efficiency and carbon neutrality. Additionally, we discuss the challenges and opportunities associated with adopting these solutions, including technological limitations, economic considerations, and user engagement. Our findings highlight the importance of sustainability in virtual spaces, urging developers and stakeholders to prioritize renewable energy solutions to foster environmentally conscious digital experiences. This research serves as a foundational reference for future initiatives aimed at building greener virtual worlds.

A. Mahmoud
Department of Computer Science, Faculty of Computers and Information, Kafrelsheikh University, Kafrelsheikh, Egypt

Department of Information and Communication Sciences, Faculty of Science and Technology, Sophia University, Chiyoda, Japan

S. Juneja (✉)
KIET Group of Institutions, Delhi NCR, Ghaziabad, India
e-mail: sapnajuneja1983@gmail.com

O. Elwasila · Y. Gulzar
Department of Management Information Systems, College of Business Administration, King Faisal University, Hofuf, Al-Ahsa, Saudi Arabia

S. Kautish et al. (eds.), *Green Metaverse: Fuzzy Logic Approaches to Sustainable Virtual Worlds*, Information Systems Engineering and Management 74,
https://doi.org/10.1007/978-3-032-08183-4_7

1 Introduction

The paradigm shift from Web 2.0 to Web 3.0 represents a period characterized by significant advances in machine learning, artificial intelligence, the Internet of Things (IoT), 5G, augmented reality (AR), virtual reality (VR), and blockchain technologies. It envisions a thriving virtual space termed the Metaverse, comprising immersive virtual worlds, digital twins, and expansive digital ecosystems. Supported by rapid advancements in immersive devices and networking technologies, a multi-dimensional and interactive Metaverse has become an attractive vision for both users and enterprises. User experiences are on the verge of being upgraded to discriminatory stages, enabling a substantial transition from the mobile Internet and social networks to the interaction of digital and physical worlds [1].

As an integrated virtual environment where users engaging different applications and services can share memories/experiences in both 3-D and 2-D forms, the Metaverse contains immersive virtual worlds/twin spaces to mirror the events in the physical world, digital products/items used in the Metaverse ecosystems, as well as an immersive experience space allowing users to engage in exciting interactions. Enabling technologies of the Metaverse have undergone rapid advancements recently and will continue to advance over the coming years. For example, VR and AR technologies are on the verge of offering new experiences for users, reducing the cost of devices through the introduction of palm smart glasses, and exploring novel use cases. Digital twins have migrated into a new dimension through the introduction of digital twins for the Metaverse, enabling advanced, automated in-virtual space industrial services. In the last two years, AI technologies have been embedded in the Metaverse as new digital beings, enabling novel characters in immersive combined virtual environments. Influenced by the COVID-19 pandemic, designing immersive experiences has become a rising market trend, which has also attracted substantial attention in academia [2].

2 Overview of Virtual Worlds

Virtual Worlds (VWs) refer to persistent, simulated real-time environments comprising the virtualized 3D graphical representation of a world in which users can interact through avatars. VWs have become prevalent due to advancements in computer graphics/technology speed, bandwidth access, and a shift toward massive multi-player online games. Factors like enhanced immersive experiences enjoyed through head-mounted displays have further spurred the expansion of VWs. Given their affordability for users, convergence with the real world is favorable. However, the proliferation and ubiquity of these applications also pose a challenge in scalability and transition, particularly in energy consumption. With the significant rise of VWs in popularity over the past decade, research into the design these residents occupy has recently flourished [3].

Research on VWs began to grow significantly after the emergence of multi-user and massive multi-user VWs in virtual environments. The two most notable examples of the latter are Second Life and World of Warcraft. In particular, while research on VWs as relatively static simulation environments has made significant progress, efforts to address the design of adaptive VWs that evolve over time are relatively lacking [4]. Other VWs are often created with an emphasis on technology or narrative. There are also similar efforts based on specific tools or frameworks. VWs can be categorized into three general typologies: follow-the-user content update, off-the-user content instantiate and movement, and autonomous virtual world. This section introduces the existing VW architecture and discusses their design requirement satisfaction and limitations in a qualitative study.

The term "Virtual Energy Hero" (VEH) refers to a virtual reality (VR) game designed to gamify the smart energy-smart city for the public at large. Smart city technologies are a rapidly growing research topic. The goal of this research is to raise public awareness of the dangers of overconsumption of nonrenewable energy sources and the transition towards renewable energy systems. Overconsumption of energy is an issue of critical importance in developed nations which demands to be addressed [5]. Analyzing the current state of energy generation and consumption in Switzerland in conjunction with an energy utility company aims to better understand the challenges at hand. Based on this assessment, stringent yet plausible approaches are proposed. The most important step towards reducing energy consumption will be raising public awareness of how the populace may save energy at a marginal cost to the quality of their respective everyday lives.

3 The Importance of Sustainability in Virtual Environments

There are increasing risks to the sustainability of virtual worlds from the usage of energy-guzzling servers. The Covid pandemic has led many organizations to adopt virtual reality as a substitute for face-to-face meetings and workspaces. As more immersive social platforms provide highly detailed models of cities and other locales in education, leisure, advertising, commerce, and other domains, the energy and electronic power requirements of these virtual worlds are rapidly increasing [1]. Sustainable virtual worlds are needed to mitigate their potentially significant environmental consequences. To that end, it is crucial to identify key energy requirements of virtual worlds (or the metaverse), along with proven energy efficiency technologies that are suited to their service provision, and necessary levels of investment needed to realize commercially viable virtual worlds from a sustainable perspective. Accordingly, this paper focuses on these issues.

While the metaverse comprises technology stacks, software stacks, and virtual space stacks, it comprises architecture, protocols, and services that enable multi-party and multi-scene immersive interaction in convergence across the virtual space and physical space, and that supports the immersive simulation of all human economic and social activities. A sustainable metaverse environment refers to one for which services or applications provide services with appropriate energy and other material resource requirements and that manages environmental issues arising from such applications. Hence, a focus is on the energy requirements of metaverse services delivered in virtual space. With the rapid advancement of communications, rendering, and immersive interface technologies, and efforts on improving the energy efficiency of their devices that operate these technologies and other equipment in data centers, servers, and networks, consumer VR/AR devices or headsets remain energy hungry, particularly when devices need to render and display high-resolution 3D graphics with high frame rates in real time. It is challenging to provide consumer VR/AR devices, especially headsets, with sufficient electrical power for prolonged periods of usage [6].

The devices remain bulky as considerable battery capacity is needed. When wire connections are required, conflicts arise between ergonomics of immersiveness and the clumsiness of connections. Because such devices for delivering immersive metaverse services are unlikely to be lightweight, compact, energy efficient, and wireless anytime soon, it is crucial to maximize energy efficiency of rendering 3D graphics and displaying them in the RF or optical domains. Moreover, as 3D model generators, servers that access and serve models of 3D spaces/assets among others require proper investments to provide power, cooling, and solutions for dealing with energy waste. Since most of them still rely heavily on fossil electricity, it is crucial to deploy renewable energy technologies to power their servers and among others [7].

4 Renewable Energy Sources

In an ongoing effort to make the internet more sustainable and reliable, several scientists have begun proposing a renewable energy approach to establish carbon neutrality web services. On the other hand, energy-efficient solutions for the provision of web services can be adopted to reduce fossil energy consumption. In this case, renewable sources can replace fossil sources in the energy mix, and governmental policies can encourage this substitution. The former approach appears attractive, as renewable technologies have become technologically mature. However, researchers on this subject are currently limited to physical systems rather than virtual ones. There is therefore a need to fill this gap by examining how other renewable technologies might be applied to virtual worlds both now and in the future. Having established sources, it is valuable to study their implementation in virtual worlds. Is it enough to invest in the construction of solar panels and wind turbines to make the services carbon–neutral? The digital domain seems like an uncontained void emerging on top of the physical world, with no sites to anchor the carbon–neutral sources and no links to tie them into a working chain. Nonetheless, on the more than two-hundred-million computers that make virtual worlds available, there is sufficient space to host new plants.

It is proposed that open-source agents are employed to design and manage a chain of agents on which a sustainable world can grow. This "virtual embedding" embeds a robust virtual world in the most resourceful physical world present today while retaining the desired properties of decentralized management. The open-source group must develop both the software needed to control these physical links and the open-source agent agents responsible for transferring the active data. One way to structure such a project is to invest in a simple agent capable of managing one computer-solar panel link. Several copies of this agent can then be replicated, enabling self-organization to occur. Ultimately, the implementation of renewable resources calls for a holistic structure blending multiple aspects of existing technologies. Efforts to design and apply them to create a network of computers linked to carbon–neutral energy and waste recycling systems are ongoing [8].

Renewable technologies are not levels of data and hence cannot serve as sources for the description of the world. Nonetheless, they may be employed to fuel the design of descriptions and programs characterizing other services. Some of these programs can be independently run. Others must be driven by a description of the run-time state of comparable functions located in the ever-evolving virtual world. This often requires the availability of interpreters capable of translating between implemented data types. Only then will the ability to fuel both worlds with energy help in ensuring their continued survival.

4.1 Solar Energy

Solar energy has long been the most widespread and commercially available form of renewable energy. Since the birth of the sun, civilizations on planet earth have used solar energy in a variety of ways. Solar energy exists in the form of solar thermal and solar electricity. Solar thermal technologies are used to use sun energy in the form of heat, whereas photovoltaic technologies convert solar energy directly into electricity. Solar energy is eminently sustainable, long-lasting, emissions free, and easily available. Solar energy is widely used around the world in a variety of applications. There are many mature system technologies which can deliver commercially available solutions. This text seeks to analyse the maturity, sustainability, and market impacts of current and innovative solar systems across several power climates, considering both grid-tied and grid-independent systems [9].

Various disciplines such as photovoltaic technology, planning solar parks, solar rooftops, digitalization for solar grids, decentralized energy systems, tender coaching, business models and business development in emerging photovoltaic markets could be formed. Technical components of the solar system are broadly categorized into four major categories; solar power generation, power conditioning and power conversion, and system level components that provide monitoring and deployment services.

Every sector of photovoltaic (PV) giant players plays a big role in transmitting information and experiences to midstream and downstream players. It is imperative that as emerging markets overcome institutional challenges, they take advantage of rapid development pathways by gathering the best possible information and human capital worldwide to leapfrog to new industries and products. The need for more sophisticated methods for sustainable solar system design increases with design complexity increases. The enormous diversity of types of solar energy systems gives rise to a wide range of possible design trade offs. On the other hand, output parameters and costs from each subsystem jointly determine the performance, reliability, maintainability, environmental compatibility, and economics of the solar energy system [10].

4.2 Wind Energy

Wind energy is renewable energy generated by harnessing the kinetic energy produced by wind. Wind generators can generate energy with low long-term investment and operating costs. In recent years, wind generators have emerged in Australia, as wind speeds and the amount of energy harnessed have increased steadily. There are several different types of wind generators, the two most common being horizontal axis wind generators (HAWTs) and vertical axis wind generators (VAWTs) [11].

Wind generators are a promising area for research as they are growing in popularity and will continue to do so. Other similar research areas include solar energy, hydrothermal energy and wave energy, all of which have acceptable waking noise levels similar to wind energy. Furthermore, hybridizing wind generation with other generation types such as diesel generators and solar generators is a promising area for further research [12].

Wind generators come in two varieties, horizontal axis and vertical axis machines. HAWTs are machines which have blades which rotate around a horizontal axis. Most commercial HAWTs have 3 blades and are upwind machines designed to adjust to changing wind conditions through pitch regulation. VAWTs on the other hand consist of blades rotating around a vertical axis. Vertical axis due to their design and positioning means they are relatively unaffected by turbulent winds, allowing them to be positioned in areas where wind speed may be high but also turbulent. The two most common types of VAWTs are Savonius and Darrieus machines, with the later having a more extensive commercial application. The most famous and recognized Darrieus machines are the Australian designed Gerenuk airfoils, which are 100kW machines. These machines are competitive with HAWT designs and have been wind tunnel tested and proven to operate at much higher turbulence levels than other machines. Vertical axis machines can be used in arrays and thus allow more efficient use of space in urban environments, where wind speeds are less significant.

4.3 Hydropower

With over 80% of contribute to the region's electric power production, hydropower forms the most significant share in Oman's renewable energy spectrum [13]. This comes alongside an ambitious target for hydropower to mature to a share of 15% by 2035. The project was at greater risk due to relevant difficulty in construction, thus not favorably evaluated in this report. Using the sensitive nature of its geographical features for such infrastructural changes and distribution channels renders Oman more susceptible to sociotechnical risks. In this regard, hydroelectric power constitutes one of the oldest commercially significant sources of energy [14].

Hydroelectric technology utilizes technically mature methods to convert gravitational potential energy into electric power. Furthermore, even a large dam with high elevation is not necessarily needed: Newer methods for producing directly usable electric energy from proportionally lower headwater bodies or without constructing major infrastructures have been developed. Due to the water cycle, kinetic potential energy supply from falling water is inherent and inexhaustibly decline. In terms of contribution to feasibility, it has a better anticipated energy share than wind systems, while having a considerably higher capacity than solar. As supply is exogenous, hydroelectricity defines great reliability and capacity adjustments against peak load. Emission: High alternate implementation imposes considerable emission during construction, power infrastructure, and operation, although they are comparatively miniscule in consideration to alternate production methods. Emission is also site-dependent; while very little emission occurs from lower head or less intricate methods, material-sensitive cutting-edge power conversion technologies have greater technical complexity.

4.4 Geothermal Energy

The quasi-stationary thermal response of a subsurface to a heat load or potential field following an inversion source can be simulated with the Laplace transform method, and the Green functions can be calculated in the Laplace domain using an analytical approach for a series of practical geometries, including a semi-infinite medium and a multi-layer medium. Several examples are provided to present the calculated Green functions in the Laplace domain and the cooled gradients and thermal response curves at the observation points after the gauss source, excavated tunnel, and linear heat source. The inverse Laplace transform is computed numerically using multi-Lanczos, and various examples are presented to recover the Green functions in the time domain by complex-contour integration and a Talbot approach. The good agreements are obtained between the raw Green functions and that approximated with a set of polynomial series at an early time and late time [15].

Besides the functions for pure conduction quarry bypass problems in the Laplace domain and basic kernel functions and approach to approximate them with trigonometric series for the quasi-stationary thermal response, where deformation and flow effects from those classical methods are ignored. The functions of both pure conduction and convection composite quarry bypass problems in the Laplace domain and an approach to spatially discretize and approximate complex conductivities and velocities for semi-infinite and finite two-dimensional media were presented [16]. There has been an increasing interest in geothermal energy resources in recent years, primarily due to the need for a sustainable energy portfolio, which is not dependent on fossil fuels. Distance heating requires an adequate system to pump water to the buildings and cool it. Two broadly sixteen types of geothermal systems are evaluated: those that just use geothermal energy, cooling wells, and heat pumps for preconditioning geothermal energy to the absorption cooling system, to two different geothermal energy accumulating wells.

4.5 Biomass Energy

As previously described, energy consumption and generation can have a huge impact on a virtual world's sustainability. To fully understand a system's sustainability, energy sources and sinks have to be individually considered. All aspects like the extraction of fuels, physical processes behind energy generation, transportation and consumption have to be brought together to obtain a full picture. Furthermore, summarised estimations of a system's carbon footprint can be given by grouping individual sources with respect to their type. The first step in this direction has been undertaken for an exploratory modelling experiment of a virtual world. Case results show that, parallel to the quantitative analysis, the great variety of types of energy sources, there are also considerable differences in how comprehensively their specific processes and parameters were extracted. It was found that, while transportation and CO_2 sequestration and enhanced oil recovery aftermaths can be almost neglected, a yet to be defined large number of parameters has to be considered in order to give an accurate estimation of biomass or biofuel production. Better knowledge of underlying mechanisms and essential parameters of energy consumption processes could help simplify their estimation which is most costly in terms of time and complexity [17].

First, the sources, sinks and processes of energy in biomass regarding virtual worlds and their surroundings have been identified. Types of biomass and biofuels, processes they can undergo to transform into fuels, ways to extract them from nature and transport, carbon footprints linked to their use or production are the key aspects that need further refinement. Comprehensively deriving the carbon footprint of these energy sources can take days or weeks. When done carefully, huge amounts of information have to be collected regarding metrics one might not even know they should look for. Most current estimates only consider one or a few aspects of a biomass source's energy circle and this is how some experts of the field were addressed. Additionally, case biomass energy sources were analysed in great details by stakeholders, but known processes and figures were often abundantly neglected. Following herewith lay out the completeness of initial biomass modelling efforts and suggestions on how to address some of the gaps are offered.

5 Technological Innovations in Renewable Energy

The majority of energy used to create information technology infrastructures is derived from electrical generation sources that burn fossil carbon fuels. This is a major contributor to climate change and one driving massive worldwide societal, political, and economic upheaval. It is an international priority to find ways to convert and distribute energy using low-carbon techniques, to consume it using energy-efficient devices, and to find ways that society can transition to using clean, distributed, electrical energy supply with economically viable policies [18]. Digital interactions could utilize renewable energy, accelerating a global transition to a low-carbon future, thereby reducing climate change and its potential social consequences.

Technology should be enhanced for maximum energy efficiency, especially in the areas of search engines and general web services, and in the buggy code that adds layers of computational intensity, heat production, and energy usage. There needs to be significant investment in existing information technology and renewable energy policy, to ensure that economic and social priorities shift toward this user-friendly solution to the world's biggest challenge, climate change. As more and more information technology devices and applications are utilized, energy used to run existing ones can be better allocated towards maximizing their effectiveness.

The prospective impact of information technology could be massive. Renewable energy created from the sun's light, wind, waves, and tides—either directly or indirectly—could power everyday information technology devices. What is needed is a public policy requirement or incentive to ensure that power plants are each determined to generate purchasing agreements needed to incentivize spending on new infrastructure, and a utility infrastructure upgrade spurred by government public policy. Attainable projects powering computers, servers, base stations, and cellular providers with renewable energy are growing in number and size.

5.1 Energy Storage Solutions

An increased use of renewable energies and continuously increasing energy consumption led to challenges such as temporal and spatial decoupling of energy generation and consumption. So far, the most suitable technical means of this decoupling have been battery energy storage systems. This work evaluates the possible applications and advantages of hybrid energy storage systems (HESS) compared to conventional, single energy storage applications. A multi-timescale method is developed to classify different types of energy storage on the basis of evaluation criteria such as frequency and probability of power transients, as well as absolute power peaks [19].

The classification system will be combined with mathematical equations for this multi-timescale analysis and applied to real life load profiles from three different applications. Using this procedure four categories, clustering similar roles of the vanadium redox flow battery and the supercapacitor, are developed. Finally, an outlook about possible future works, such as optimally designing and controlling HESS, is given. With growing amounts of renewable energy sources, matching generation and consumption of electricity is more and more difficult. The increasing share of renewable energy sources (RES) observable in recent years contributes to a sustainable energy future, but also lowers the stability, reliability, and power quality of the electrical grid. Most RES are by nature disruptive and volatile, i.e. not perfectly controllable by the grid operators [20].

Due to the uneven geographical distribution of RES and the fact that fossil and nuclear power plants have been taken offline recently on the one hand the distance between generation and consumption increases continuously. On the other hand, on a short-term scale this leads to an over generation in times of high output from RES, whereas at other times, e.g. when demand is increasing throughout the day, the generation has to be ramped up quickly when most RES are already at their limit. Unfortunately, incentives to curtail RES are often not acceptable from both a political and an economic point of view. In order to separate consumption and generation with respect to time and space, the use of energy storage systems (ESS) is increasing [21]. With regard to grid applications, a balanced forcing to use variable speed machines in both hydro and wind power plants and an increased application of pumped hydro storage or battery-based systems is globally observable.

5.2 *Smart Grids*

Although there has been significant progress in this field in the last few years, the Smart Grid is still not fully deployed in most electricity networks. The decade from 2006 to 2016 is expected to see many countries around the world begin to invest significantly in Smart Grid systems in order to increase the safety and reliability of electrical networks and improve their efficiency and sustainability. The international drive towards the establishment of the Smart Grid has nonetheless already led to the definition of Technical Reference Standards such as IEEE 802.15.4g, which ensures the interoperability of communication devices in the Smart Grid perimeter. The Smart Grid is a new paradigm in the energy domain. The interoperability of suppliers and customers has been foreseen from a methodology's perspective, with focused studies making use of mathematical tools [22].

The Smart Grid is fundamentally a collection of technologies that, by communicating in real-time with electricity suppliers and consumers, improves the efficient management of the entire electricity network, reduces wasted energy and cutting costs. A Smart Grid separates IT and energy carrier infrastructures smart systems with underlying hardware, communication links and technology platforms that differ from one country to other countries should design Smart Grids to comply with national, regional, and local context [23]. Smart Grids are intelligent networks. Smart Grids consist of intelligent technology used to better manage the electricity supply chain and avoid wasting energy and money. Smart Grids include intelligent generation technologies on the production side, intelligent transmission and distribution technologies on the transport and network side, and intelligent utilization technologies on the consumer side. Smart Grids will reduce environmental pollution and energy costs, improve energy access and reliability, facilitate the penetration of renewable sources, increase the employment of efficient electric devices, and foster markets and competitiveness owing to the substantial potential of demand side management.

5.3 Energy Efficiency Technologies

Energy use management and improvement in building energy efficiency is important because the majority of energy consumed in buildings is usually attributed to HVAC systems, lighting, and appliance use. It is mandatory to develop and implement energy efficiency strategies in order to reduce non-renewable energy usage, CO_2 emissions, and energy costs. Addressing this problem requires a comprehensive energy supply strategy and assessment of the best financial options. In addition, complex and ineffectively regulated production facilities present great challenges to pesticide manufacturers [24].

Energy consumption in buildings can be obtained from building energy modeling using a mathematical computer simulation, but this method is time-consuming and requires a high level of expertise. It is, therefore, important to have an alternative method to compute building energy consumption and conduct automatic building energy audits, diverting all the excessive loads after audit to energy efficiency, renewable energy, and battery storage. These measures are described by their type with a unique code, aim, technical method, and description of design parameters & constraints. The proposed methodology for selecting energy-saving measures first involves prioritizing them according to the most practical criteria, costs and energy saving values, after filtering them based on technical compatibility with the target building. It is mandatory to leave the analysed building energy model untouched so the energy savings can always be evaluated post-implementation. The top ranked energy-saving measures are thoroughly assessed with respect to the quantity of refinements needed to their default constraints. The finetuned energy savings, costs and payback periods are calculated and the viability of the energy efficient building modeled with all of the measures is evaluated by testing the updated building energy model against the default one [25].

6 Integration of Renewable Energy in Virtual Worlds

Millions of gaming users routinely forget their physical world settings and their commitment to conserve natural and human resources once they interact in their virtual worlds. Orlando's ten million visitors in 2011 consume the same amount of electricity as the whole city of San Antonio, Texas, if the energy intensity number of kilowatts per transaction in virtual settings would be extended to real world transactions. If the virtual world were to be made physically sustainable, its energy consumption would be half that of San Antonio, and it would need to operate on renewable energy sources. A buyer and an investor participating in the World of Warcraft can engage in 176 million kilowatts annually in a nonrenewable resource 'sandbox.' Could it be possible to integrate renewable energy solutions into expansive 3D virtual worlds? Would Avatar Living be able to engage in economically and politically feasible power purchase agreements (PPAs) for the development of new wind and solar farms on arid desert land or ocean waters with appropriate fast-tracked permitting [8]? Two estimates of costs are provided as a demonstration of development scenarios that would successfully supply the Avatar Living platform with renewable energy.

Virtual worlds contain user-generated immersive experiences of visual and auditory phenomena using 2D or 3D creations, avatars, digital items, and the immersive interaction between large numbers of these creations. The interaction takes place in real time or time with controlled latency via the entities connected to the internet, a multi-way telecommunication media consisting of wired or wireless broadband access to individual household or institutional computers, game consoles, and mobile devices on an unrestricted basis. Virtually thousands of gamers can connect to a single imaginary environment constructed as a continent populated by landscapes, utilities, properties, avatars, and items created and installed by the residents. These activities are hosted and automatically serviced by industrial scale servers connected to high bandwidth but high latency trunk or long-haul transport networks across countries and continents.

6.1 Case Study: Virtual City Planning

Sustainable development requires making intelligent actions that would result in the protection of the environment. As communication technology is ever-growing, more and more people are gathering online, increasing internet-based facilities and services are fulfilled in real-time. Industries are looking for innovation and switching to a greener virtual environment. Integrating emerging renewable energy technologies in urban planning is the central focus of this research. A virtual city planning perspective to stimulate urban planners and policymakers is investigated using games. The need for the development of a city planning game is stated, considering the integration of emerging renewable energy sources in a virtual environment to support policy making and planning. The gaming perspective helps to utilize the virtuous cycle of knowledge creation.

The objective game to be developed is outlined using the characteristics of a city planning game, educational game, simulation game, strategy game, and computer game. Games with different formats are reviewed, and three corresponding games are proposed. The focus on the policy and planning perspective, multiagent-based modeling framework, and real-time decision-making support through advance visualization are the main contributions of this research to simulate sustainable development planning using an academic approach [5]. Since 2013, gamification and VR will be investigated, and strengthening the communication of smart energy topics would consider an even broader segment of the population as a target audience. This project will therefore look at how to do this in terms of design options and technical possibilities and implement more concept tests of creative tools and solutions that could be developed and conceptualize the serious gaming platform for the three cubes and games currently in use for testing, ensuring a high-quality debut of Smart Energy Cities. The business model of the cities and the platform partners will also be further developed to ensure they remain free for users and effectively reach consumer target audiences. The seamless incorporation of outcomes from RnD projects into the tools of the industries, local governments, and end users will result in collaborative tools for all regions of Switzerland and rich exchanges of information concerning smart energy topics.

6.2 *Case Study: Gaming Platforms*

Game Engines, which are software solutions for the construction of video games, enable designers and developers to incorporate sophisticated functionalities without having to recreate them from scratch. Nowadays, new games are increasingly built atop pre-existing engines instead of from scratch. This study examines the energy consumption of video games built with two of the most used game engines in the world, Unreal Engine and Unity, through the comparison of very popular games released in moderate periods of time. The assessment considers the gaming time required to complete the main story of each video game, and its link with energy consumption according to the literature indexed in reliable databases after filtering on the energy assessment protocol. The results reveal a considerable variability in the energy consumption of games operated on the same engine. Also, a significant difference in energy consumption was found between platforms. Unity games consumed significantly more energy on average than Unreal Engine ones. The results were expected because Unity has a greater PC gaming market share than its counterpart. The impact of these results is that game designers may undertake performance optimizations in their titles in order to lessen CO_2 emissions [26].

6.3 Case Study: Educational Simulations

The continued exploration of innovative energy technologies comes with emergent complexities regarding their environmental sustainability, policy viability, and user adoption. A widely used method to explore the features and feasibility of such technologies is the use of educational simulations. These immersive simulations allow users to perform moves akin to those interactions they would have with the real socio-technical system. They are often based on Computational Agent-Based Modeling (ABM) formalisms and allow the exploration of the alleged emergent behaviors of the modeled socio-technical systems. Limitations such as the model's usability, accessibility, and immersion prompt a search for solutions to ensure sustainable educational simulations. To this end, the concept of Sustainable Virtual Worlds (SVW) is introduced, with a focus on techniques stemming from game design and human–computer interaction. A detailed case study regarding a family of energy related simulations exemplifies the proposed SVW solutions.

The novelty of immersive processes, within the realm of interactive 3D multi-user simulations, is rapidly shifting from the mind to the hand. Performance is increasingly determined by intuitive interaction and the aesthetic quality and intelligence of 3D worlds rather than the underlying technology. While life-like realism remains elusive, this new decade of extended reality (XR) platforms rewards designers' imaginations with unexpected possibilities that call for new designs of pathways into worlds. As risky as it is exciting, this research opportunity could prove invaluable for individualized education regarding the complexity of emergent behavioral outcomes of policy options on sustainable energy futures. To achieve this, however, educational simulations need to become sustainably entertaining games. Possible ways of doing so are explored and summarized in terms of important principles for future educational simulations regarding their design as Sustainable Virtual Worlds (SVW) based on specific techniques for interaction, believability, and user-involvement.

The goal and anticipated outcomes of the case study are specified. The concept of educational simulations is further clarified as immersive simulations built from ABM models of real-life socio-technical systems to promote the user's understanding of their functioning over time [5]. Detailed descriptions of the case study provide a walk-through to design, implement, and assess educational simulations. This includes the family of energy-related simulations; the design process applied to the first demo of the Co-Evolutionary Energy Transition Simulation; and the evaluation of all hypothetical and available best practices for ubiquitous features to turn educational simulations into SVW.

7 Challenges in Implementing Renewable Energy Solutions

The broadening range of system design experiences and user expectations incorporated into virtual worlds may demand more complex requirements for energy solutions. Constructing buildings and installing energy systems require greater effort and input, namely in terms of time, skills, and financial investment, compared to changing avatar appearances, furnishings, and textures in affordable, widely available, beginner-friendly applications. Therefore, large-scale server farms account for a growing proportion of overall computing energy demand, with studies estimating over 50% of energy to be incorporated into firm or establishment energy footprints, accounting for 15% of direct consumption and 30% of combined electricity demand worldwide.

The pervasiveness and worldwide dispersal of stand-alone devices also present challenges in future sustainable developments. The dissimilarity of devices leads to differences in idiosyncratic hardware designs and firmware implementations. Consequently, delivery of system updates, software rollouts, repairs, and decommissioning are mostly haphazard processes in which new, unexamined code regarding usage and feasibility are introduced in an unregulated manner. The asynchronous nature of human computer interaction impedes the updating of display-side systems in a common way. Cyberspace also offers pervasiveness since the usage of virtual worlds spans office hours and personal time and may occur worldwide and at all times.

While smaller, individualized servers are reinvented to operate commercial vertical foliage arrangements with seasonal fluctuations, hourly schedules of music and lighting displays, and sub-1s device/server turnaround times in data integrity, affordability prevails not only in terms of the massive computing resources' costs relative to investments leveraged into pre-commercial stage app products but also the naiveté of equity investments in nascent but ultimately negligent developing countries. The same may hold for open-source designs that compete with services that standardize access or resolve prior, fragmented modes of use and interaction [8].

7.1 Technical Barriers

Blockchains are publicly available, digital ledgers or databases used to store information. Each transaction is stored in a block. Information is stored across multiple computers, called nodes, creating distributed networks and increasing resilience and security. Virtual worlds on blockchain technology provide participants with ownership and control over their assets, environment, and experience. A decentralized layer creates distinct financial systems, digital ownership, and the independence of content, platforms, and tools of engagement. Players also create, manage, and govern virtual worlds. Virtual Worlds strive to provide play-to-earn mechanisms and increased digital ownership for players. Gamefi and play-to-earn are compared to the concept of AAA gaming referring to design and polish.

Blockchain-based immersive experiences use the same technology as Web3. Functionalities of Web2 tech stacks apply to the Web3 stack but transactions occur over distributed networks secured by decentralization protocols. Therefore, access to technology for the implementation of such atmospheres is subject to great effort. The constraints enterprises face in the implementation of their metaverse strategy are also applicable to the transition to a Web3 metaverse. Technical barriers include technological complexity and integration, lack of internal capabilities, and scalability and transaction costs.

Building and implementing Web3 metaverses require integrating multiple layers of protocols and technologies. Public blockchain networks, middleware and services across technology stacks, and software platforms are interoperable networks. Technology stacks are layered from the bottom blockchain protocols to application development SDKs. Different standards for asset creation, interaction, and on-chain information storage largely do not interoperate. Complex interoperability challenges exist between technology stacks needing standardized protocols and APIs. Attorney technology is needed to ensure immersive environments work smoothly and outcome-oriented and maintain a balance in stability and complexity [27].

7.2 Economic Considerations

Economic feasibility is generally one of the widest barriers to entry for the adoption of renewable energy technologies. Initial costs of solar and wind energy equipment typically yield a payback time on the order of decades and may also require large capital investments or a remote connection to a grid-based solar or wind plant, as is often the case with bulk energy projects. Each technology has unique factors which do or do not favor one community over another. The selection of a system is most likely to be successful in the long term if those variables can be wisely selected and combined, and if a good understanding of the overall costs can be achieved at the outset [28]. To this end, a techno-economic comparison of renewable energy systems is undertaken in this study.

It is becoming ever more apparent that the consumption of fossil fuels has widespread detrimental effects on the quality of life of humanity as a whole. There is ample evidence that anthropogenic emissions of greenhouse gases are changing the climate of the Earth today, with many studies estimating that the fallout will result in an exponentially worsening standard of life for humans and biodiversity in earth's ecosystems [8]. It is also evident that current and future projections of the fossil fuel supply chain involve widespread human health implications from mining, refinement and combustion, as well as high insurmountable political, economic, and social issues of fairness inherent to resource accessibility.

As the current understandings of meteorology and engineering principles and technologies improve, solar energy systems stand poised to also yield both environmental and economic benefits as they scale up in response to energy demand. It has been shown that carefully planned projects that account for place-specific factors such as resource availability, regulatory environments, and social ties can deliver solutions with co-benefits to social inequalities. Grounded within the paradigm of sustainable development, where social equity, economic vitality, and environmental conservation must be considered alongside each other, an environmentally benign energy system satisfying the needs of a growing society is an urgent pursuit.

7.3 User Acceptance and Engagement

User acceptance of advanced technology is an important aspect of any research project. Users must find technology to be easy to use, valuable to their needs, and, ideally, even enjoyable to engage with. To fully determine the robustness and validity of the proposed concepts for the SERPENT change agent technology, initial research is needed to confirm that a user-focussed model is workable in a virtual domain and that users express a high level of acceptance toward engaging with proposals made in the model. The development of the user acceptance and engagement research plan is the first step in understanding the user-focussed model concept and technology [29].

The most important hypotheses necessary to test include the idea that users will accept and find value in the proposed change agent technology and the social, procedural and platform virtual world models. Participants should express a perception that the models and technology for the project can generate valid and reliable proposals to change energy behaviours in any settings. The full and partial models should be able to elicit a high level of engagement across most participants in post-engagement evaluations of user-focussed virtual worlds model proposed proposals for energy behaviour change. User evaluations should lead to a revision of the initial user-focussed change agent technology to direct future work toward areas of improvement [30].

Early explorations of the user-focussed change agent technology and models should also provide insight into the areas in which user-centred technologies can be most beneficial to engaging energy management systems. User participation in in-depth interviews at various intervals during the research and regular consultations with an external advisory group have been insightful in this area. Findings should add to this understanding by focusing on assessment of the acceptance, engagement level, aspects of clarity and capability, and on proposals that were poorly received by users and why this occurred.

8 Future Directions for Sustainable Virtual Worlds

While there is notable research on metaverse architecture and applications, literature regarding sustainable consumption in the metaverse is still sparse. There is no broader investigation of design considerations and frameworks to generate sustainable consumption in the metaverse. There are also calls for deeper considerations of how the sustainability agenda for the physical world translates to the virtual world [31]. CEOs, digital artists, and designers also express a need for research to guide them in decision-making to maximize sustainable development in their designs.

Encouraging a new sustainable virtual world experience involves two components. Firstly, an exploration of current uses of the metaverse in the environmental space is needed; an understanding of how multiple virtual worlds are merged and images edited. Secondly, the promotion of the implementation of infrastructures in the metaverse and services in the physical world (co-creating visual social channels devoted to spreading engagement actions). This shift will augment the comprehension of the downfall of nature capital at the final user's level and urge this user to take actions in favor of the environment [1].

Auto-generated land sinking next to polluted shores or hidden rainforest burning decidedly raises awareness. Although this shift appears radical, the goal is simply to make the virtual world reflect what is tangible nowadays. Planting trees to urbanize parks or generate environmental profits and negative use are highly used practices in a neighborhood modeling instantly with the enormous number of sensors tied to natural assets and built and animated autonomous agents. On the contrary, pipe regeneration grants looking visually beautiful cities in exchange for climate shocks. Both fortune and misfortune act as triggers for metamorphosis, how company behaviors and customer experience are changed and innovated.

Sustainable consumption cannot stop standing upon a limiting point of view. Procedural generation should be fostered. An animation, object, or shape is drawn by encoding and gradual modifications. Infrastructure building suggests networks and parcels for paths or surfaces, retaking dust plummeting, planets movements, or nature systems growth laws. Insights then need visualization on the interface for the output world construction process monitoring. Agents plant trees in an empty ground along procedural sequences. Each second of building a park continually generates more than a thousand pixels to visualize.

8.1 Emerging Technologies

The emergence of Metaverse promotes a larger paradigm shift on how humans interact with smart environments and share content. Metaverse is a perpetual, real-time, living digital universe that integrates spaces across both physical and virtual realities. It provides monetization opportunities, changes the nature of human interaction, and drives a novel era of innovative content production through user-generated content [1]. A more interactive virtual world will be developed from user-generated content around user experiences and valuing a value and experience generation chain. Metaverse is expected to provide social forums, workspaces, entertainment, education, and digital twin services. Metaverse is not mainly a new technological innovation. It is regarded as an embodied version of the Internet that seamlessly and ubiquitously synergizes virtual and physical worlds.

For realizing the Metaverse, a growing amount of investment and effort have been made. The number of research publications related to Metaverse in 2022 is five times more than that in 2021. Meta hopes to build a consortium of businesses and academics interested in building a responsible Metaverse. The business Metaverse economy has been valued at 337 billion USD in 2028. To meet demand, the chip industry needs to build new fabs, reportedly costing 15 billion to 30 billion USD each. Meta was rebranded to Meta in 2021 to shift from a social media company to a Metaverse one. This transition forces the company to search for new revenue sources and compute architecture based on massive parallelization.

The definition of the Metaverse and its features are also discussed, including immersion and interactivity, ubiquitous and individualized interaction, persistence and content generation, human-controlled agents and interpersonal social relationships, and virtual economy and security concerns. All these features promote unprecedented use cases, such as human hardware interface, avatar customers, virtual influencers, location-based augmented reality, digital artifact citizenship rights, and asset decomposition. Successful metaverse pioneers include Meta, Roblox, Fortnite, and Decentraland. It is critical to compete with video greenhouses, movies, and games like the Metaverse economy. Dominant central technologies of the Metaverse, such as 3D reconstruction, real-time render engines, and volumetric video compression and decompression schemes, are introduced.

8.2 Policy and Regulation

Government policies and regulations can provide incentives to invest in renewable energy (RE) technologies whilst simultaneously closing off avenues for major non-renewable technologies. In this way in October 2007 the Government announced a new renewables obligation order, which included major new incentives for onshore wind and wave/ tidal power. Added to this the existing renewables obligation with its multi-billion-pound trade in green certificates would be left undisturbed until 2014 minimizing rich opportunities for all technologies from future trading [32]. In this same month it was announced that a major fission plant could be operational by 2018. Market power would be transferred from the consumer to the producer and could only be countered by state ownership. The Secretary of State for Business, Enterprise and Regulatory Reform, John Hutton's comments on the UK's energy industry were unequivocal that competition could be achieved but only at the cost of increased inequality and environmental impoverishment and gave warning of the risks of over-reliance on market mechanisms to provide solutions.

At the same time the UK Government's 2050 Energy White Paper set ambitious and legally binding targets for GHG emissions reductions and stated that the only way a sustainable modern economy could be achieved was on the basis of a low carbon supply [33]. There is considerable scope for the UK to utilize some of the many low carbon technologies that could form part of a sustainable supply. Wind, wave and tidal, biomass and solar technologies are here so far as storage, carbon capture and recycling, hydrogen gas and nuclear fission technologies, and under development so that environmental impacts are being decreased and prices tending downwards in the markets of these technologies. Closing was the danger level for global warming of 2°C above pre-industrial levels but whether it could be contained or exceeded remains the great uncertainty.

The successful closure of a sustained RE market, however, has very particular requirements in terms of regulation and good practice. An autonomous institutional design is needed that will enable RE to flourish whilst delivering low prices to the consumer. Current arrangements limit the possibilities for all but the most commercially viable RE technologies and innovations to enter and stay in the energy market. Overreliance on market mechanisms to provide sustainable energy is unwarranted and would be expensive for UK consumers and prices. Current arrangements, even with the proposed alterations, would also be highly detrimental to the Government's GHG reduction targets.

8.3 Collaborative Approaches

Through the years and with the advance of technology, intelligent agents have become an important part of households and even in company environments. A household can contain a large number of devices that run independently and exchange messages among them. Hundreds of devices exchange information about energy consumption and production without any intervention on behalf of the users. The complexity associated with managing such a system increases with intelligence, communication capabilities and, consequently, autonomy of the devices involved. Transparent intelligent agents can be used to crowd-source decisions connected to energy. For example, using sensors to detect the behavior of a user in a household and sending this information to a multi-agent system that creates a virtual model of the user. With the help of a game environment, users could choose the behavior they want to test as a way of consuming energy in a low-cost/low-risk manner. Using social networks, users could expose their behavior or the models of their household in a shared environment.

The cooperation of multiple agents in a game environment is an essential part of the proposed architecture. There is currently little focus on the complicity, social interaction and involvement aspects of these agents. However, the features introduced through collaborative engagement could offer significant added value. For example, end-users would engage with the service and feel part of it. They could share tips, ideas and local knowledge; contribute to a living community of practice regarding energy efficiency. They would also enhance the quality of the model and the collaborative work between end-users and the game. Monitoring, evaluating and recognizing behavior and performance should be part of the engagement, as should providing feedback on its desirability [34].

Eventually, emerging from these collaborative activities, knowledge bases might be created that enhance the core service [35]. Participatory media are an opportunity for stakeholder and end-user engagement. Events could be held within and through the media, offering a shared space for discussions, competitions, experimenting and playing. End users could engage with one another regarding more technical aspects of the service and come up with suggestions on how to improve the game. This could help to create a durable community and knowledge base.

9 Conclusion

The unprecedented rise of the Metaverse made it a major industry with billions of dollars invested and valued at trillions of dollars. This revolution is spurred by exhibited excitement and immense potential stemming from technological and market readiness. However, its sustainability is questionable and requires rigorous scrutiny. Therefore, to ensure a sustainable Metaverse, a green foundation is provided through an extensive investigation into energy efficiency solutions for all network components. Then, renewal energy utilization solutions toward a renewable Metaverse are presented. Finally, an evaluation framework for the sustainability of selected network components is provided for further research.

To meet the increasing energy demand brought by new digital applications, innovations are required to improve energy efficiencies. Fortunately, some technologies are being extensively examined. However, most earlier explorations are limited to network components. It is requested with an effort involving all network components to optimize energy efficiencies holistically. Although adopting renewable energy can further reduce carbon footprint, no prior studies focus on utilizing renewable energy in Metaverse networks. Wide deployment of renewable energy is a critical hurdle because of irregular availability, effectiveness, geographical restrictions, and its high cost. To achieve sustainability, Metaverse networks should leverage both energy efficiency and renewable energy utilization technologies.

With the increasing urgency for sustainability, Metaverse, like the physical world, needs to seek green networking technologies. Today's Internet requires immense networks consisting of thousands of routers, hundreds of thousands of switches, millions of servers, and billions of end-user devices to serve billions of users. The Internet and associated services support billions of euros in the European economy. Serving billions of users results in serious concerns regarding the sustainability of the Internet because of the immense energy it consumes. The implementation of green networking technologies is necessary for sustainability and attracting growing concerns. Therefore, energy consumption on the Internet is increasingly coming under the spotlight, and a variety of metrics, technologies, and solutions for energy efficiency are emerging [1].

Acknowledgements This work was supported by the Deanship of Scientific Research, the Vice Presidency for Graduate Studies and Scientific Research, King Faisal University, Saudi Arabia under the project KFU251573

References

1. Zhang, S., Yang Bryan Lim, W., Chong Ng, W., Xiong, Z., et al.: Towards Green Metaverse Networking Technologies, Advancements and Future Directions (2022)
2. Damaševičius, R., Sidekerskienė, T.: Virtual worlds for learning in metaverse: a narrative review. Sustain. (2024)
3. Korkut, E.H., Surer, E.: Visualization in virtual reality: a systematic review. Virt. Real. (2023)
4. Shen, B., Tan, W., Guo, J., Cai, H., et al.: From Trend Analysis to Virtual World System Design Requirement Satisfaction Study (2020)
5. Ramram, A., Elizabeth Harte, A., Whitney Fleury, N.: Virtual Energy Hero: Virtual Reality Gamification of Smart Energy-Smart City Technologies (2018)
6. Jones, G.W., Jones, S.K.: Critical technology enablers for near-to-eye AR and VR. In: ... and Mixed Reality (XR) Technology for ... (2022)
7. Al-Jumaili, A., Muniyandi, R.C., Hasan, M.K., Singh, M.J.: Advancements in intelligent cloud computing for power optimization and battery management in hybrid renewable energy systems: a comprehensive review. Energ. Rep. **9**, 123–145 (2023)
8. Finley-Brook, M.: Renewable Energy (2014)
9. Jahankhani, H., Hosseinian-Far, A., Jahankhani, H., Hosseinian-Far, A.: Efficiency Analysis of the Photovoltaic Systems for Carbon Footprint Reduction (Case Study: University of East London) (2011)
10. Nibedita, B., Irfan, M.: The role of energy efficiency and energy diversity in reducing carbon emissions: empirical evidence on the long-run trade-off or synergy in emerging economies. Environ. Sci. Pollut. Res. (2021)
11. James Duffy, M.: Small Wind Turbines Mounted to Existing Structures (2010)
12. Rehman, S., El-Amin, I.M., Ahmad, F., Shaahid, S.M., et al.: Feasibility Study of Hybrid Retrofits to an Isolated Off-Grid Diesel Power Plant (1970)
13. Al Tamimi, A.: Feasibility Study of Hydropower System in Oman (2018)
14. Fabio, A., Elena, C., Laura, D., Maurizio, R.: The use of water for technical development or technical development for the use of water? In: Systemic and Ecological Considerations about the Clean Energy Production in Urban Context (2018)
15. Falcone, G., Liu, X., Radido Okech, R., Seyidov, F., et al.: Assessment of Deep Geothermal Energy Exploitation Methods: The Need for Novel Single-Well Solutions (2018)
16. Sun, X., Soga, K., Cinar, A., Su, Z., et al.: An HPC-Based Hydrothermal Finite Element Simulator for Modeling Underground Response to Community-Scale Geothermal Energy Production (2021)
17. De Giovanni, P.: Sustainability of the metaverse: a transition to Industry 5.0. Sustain. (2023)
18. Sovacool, B.: Design Principles for Renewable Energy Programs in Developing Countries (2013)
19. Li, S., Xu, Q., Huang, J.: Research on the integrated application of battery energy storage systems in grid peak and frequency regulation. J. Energ. Storag. (2023)
20. Zugschwert, C., Göschl, S., Martin Ibanez, F., Pettinger, K.: Development of a Multi-Timescale Method for Classifying Hybrid Energy Storage Systems in Grid Applications (2022)
21. Nahid Rahman, M.: Energy Storage Solutions for Wind Generator Connected Distribution Systems in Rural Ontario (2009) [PDF]
22. Arenas-Martınez, M., Herrero-Lopez, S., Sanchez, A., Williams, J.R., et al.: A Comparative Study of Data Storage and Processing Architectures for the Smart Grid (2020)
23. Ozturk, Z., Arayici, Y.: Transition Process for Smart Grid Compatability in Residential Buildings (2011)
24. Hernandez, K.E.: Turbine House: A Net-zero Energy House (2010)
25. (Elieser) Tarigan, E.: Energy Saving Measures and Simulation in the Library Building of University of Surabaya (2018)
26. Pérez, C., Verón, J., García, F., Ángeles Moraga, M., et al.: A Comparative Analysis of Energy Consumption Between The Widespread Unreal and Unity Video Game Engines (2024)

27. Dell, M.: The Adequacy of Programs in NSW in Supporting Vulnerable Households to Transition to Renewable Energy (2015)
28. Khalid, F., El-Emam, R.S., Hogerwaard, J., Dincer, I.: Techno-economic Feasibility of Renewable Energy Based Stand-alone Energy System for a Green House: Case Study (2018)
29. Luse, A., Mennecke, B., Triplett, J.: The Changing Nature of user Attitudes Toward Virtual World Technology: A Longitudinal Study (2013)
30. Morton, A., Bull, R., Reeves, A., Preston, S.: ICT for Sustainability: Reflecting on the Role of ICT to Enhance Communication and Empowerment of Building Users (2019)
31. Pellegrino, A., Wang, R., Stasi, A.: Exploring the Intersection of Sustainable Consumption and the Metaverse: A Review of Current Literature and Future Research Directions (2023)
32. McHarg, A.: Regulating for Sustainable Electricity Market Outcomes in Britain : Asking the law Question (2013)
33. Ratnikova, D.: Evaluation of Renewable Energy Policies: The Determinants of Wind Power Adoption Under a Quota Obligation (2010)
34. Castri, R., De Luca, V., Lobsiger-Kägi, E., Moser, C., et al.: Favouring Behavioural Change of Household's Energy Consumption Through Social Media and Cooperative Play (2014)
35. Bourazeri, A., Pitt, J.: Collective Attention and Active Consumer Participation in Community Energy Systems (2018)

Renewable Energy Solutions for Virtual Environments

B. Shuriya, S. S. Sivaraju, T. Anuradha, S. Srithar, and J. Chitra

Abstract Integration of renewable energy solutions into virtual systems is a crucial step towards obtaining sustainable energy systems. Virtual systems provide for effective energy management, optimization, and trading processes, providing a platform for revolutionized energy solutions. A few of the renewable energy strategies appropriate for virtual systems are deliberated in this chapter, including hybrid renewable energy systems, block-chain-based energy trading, and artificial intelligence-based optimization. Second, the chapter also discusses the application of digital twins, additive manufacturing, and automation in increasing energy efficiency and sustainability. The results outlined are intended to lead the researchers and practitioners to develop robust renewable energy frameworks that will be usable in virtual environments.

Keywords Sustainable energy · Energy management · Automation and renewable energy

---

B. Shuriya (✉)
Department of CSE, United Institute of Technology, Coimbatore, India
e-mail: shuriyasmile@gmail.com

S. S. Sivaraju
Department of EEE, R V S College of Engineering and Technology, Coimbatore, India

T. Anuradha
Department of EEE, KCG College of Technology, Chennai, India

S. Srithar
Department of CSE, Koneru Lakshmaiah Education Foundation, Vaddeswaram, Andhra Pradesh, India

J. Chitra
Department of Biomedical Engineering, Dr. N. G. P Institute of Technology, Coimbatore, India
e-mail: chitra@drngpit.ac.in

S. Kautish et al. (eds.), *Green Metaverse: Fuzzy Logic Approaches to Sustainable Virtual Worlds*, Information Systems Engineering and Management 74,
https://doi.org/10.1007/978-3-032-08183-4_8

1 Introduction

Virtual spaces have also emerged as effective tools for energy management, where renewable energy sources can be fed into smart grids, microgrids, and online energy trading platforms. Virtual spaces facilitate energy optimization through the use of machine learning, block-chain technology, and federated learning techniques [1, 2]. The pace of development of digital technologies has revolutionized the energy industry, turning it into a sustainable, scalable, and efficient one. Virtual worlds offer an environment in which renewable energy sources like wind, solar, and hydroelectricity can be governed and optimized efficiently. The worlds can predict energy demand and reduce the consumption of non-renewable energy sources using real-time data analysis, AI, and deep reinforcement learning [3, 4].

Another advantage of virtual environments is that they can deal with multiple sources of energy within a decentralized setup. Usage of block-chain-based energy trading platforms has enabled peer-to-peer energy trading among consumers and producers, enhancing energy system efficiency and resilience. Virtual power plants (VPPs), a new area of focus, aggregate distributed energy resources to stabilize the grid and maximize operations. These platforms are IoT-based to enable secure real-time trading of energy between stakeholders.

Further, demand management in virtual space needs to render energy sustainable. Machine learning algorithms such as deep reinforcement learning enable flexible energy utilization through dynamic adjustment of the supply of energy as a function of real-time variations in demand and supply. This improves efficiency in the utilization of energy as well as cost savings to both consumers and grid operators. In addition to this, federated learning techniques have been put forward as a decentralized solution to energy management such that different stakeholders can collaboratively train AI models without compromising data privacy.

Another pressing virtual environment application is the use of renewable energy in cloud computing data centers. Such data centers, being energy-intensive, can employ AI-driven energy optimization methods that cut power consumption and carbon footprint. Smart homes and buildings have hybrid energy management systems that share power in equilibrium through the help of solar panels, battery banks, and load-balancing software. Deployment of this technology assists in creating sustainable city development, making sure smart cities based on clean energy use as a preference.

In addition to all of this, the deployment of digital twins in energy management cannot be overlooked. A digital twin is a computer simulation of a real energy system that can observe in real-time, simulate, and conduct prediction analysis. Energy companies can use digital twin models to optimize the efficiency of their assets, reduce operational risk, and improve the reliability of renewable energy. This provides proactive maintenance techniques that can be applied, which is cost-effective in downtime and enhances energy infrastructure resilience.

Technological advancements in additive manufacturing (3D printing) have also increased the application of renewable energy in virtual worlds.3D printed energy storage components such as components of batteries and solar cells are providing cost and expert solutions for energy transportation and storage. The construction industry has also been enhanced through 3D-printed building material to enable the use of renewable energy installation, such as energy-efficient architectural components and grid-compatible homes.

Briefly, virtual environments are an innovative method of using renewable energy, with creative solutions to manage, optimize, and sustain energy. Virtual environments, with artificial intelligence-based energy optimization, block-chain energy trading platforms, federated learning, and digital twin technology, give optimized and secured energy systems. Virtual worlds become increasingly able with other smart grid technologies, microgrid technologies, and additive manufacturing to transform the use of renewable energy. The chapter is reacting to such innovations and offering clarity on how the virtual world may revolutionize the energy sector as well as yield sustainable innovation.

2 Hybrid Renewable Energy Systems in Virtual Environments

Hybrid Renewable Energy Systems (HRES) combine two or more technologies for renewable energy sources such as solar, wind, and hydropower with a view to maximally ensuring reliability, efficiency, and sustainability of the energy. HRES utilizes the exploitation of more than one energy source in an attempt to address the issue of intermittency inherent in standalone renewables. Virtual systems application of HRES provides sophisticated solutions to energy forecasting, optimization, and decision-making that facilitate real-time control of supply and demand balance [3, 4].

2.1 Simulation and Optimization Within Virtual Environments

Virtual environments provide simulation-based planning and optimization of HRES in a way that allows researchers and energy planners to simulate and test various different scenarios before they are implemented. Simulation of variable weather patterns, load requirements, and energy storage approaches allows the development of high-quality energy management systems within virtual environments. Sophisticated digital twin platforms replicate physical energy systems such that predictive maintenance, fault detection, and real-time monitoring of power supply can be conducted. These functionalities promote the optimal usage of energy economically and prevention of energy wastage.

Moreover, optimization methods form the core of HRES peak performance in virtual systems. Techniques such as genetic algorithms, particle swarm optimization, and deep reinforcement learning enable efficient dispatch of energy so that the system operates at its maximum capacity under dynamic conditions. The methods enhance energy predictability to the extent that grid operators can make decisions in real time from data, minimizing the use of backup systems based on fossil fuel and maximizing sustainability.

2.2 Artificial Intelligence Based Forecasting and Decision

Artificial intelligence (AI) and machine learning (ML) operations significantly enhance the accuracy of energy forecasting, which is appropriate for the variability issues introduced by the generation of renewable energy. AI-based models provide accurate future estimates of energy demand and supply based on past weather patterns, trends in energy demand, and grid stability metrics [5].

Deep learning models such as convolutional neural networks (CNNs) and long short-term memory (LSTM) networks have achieved peak performance for energy forecasting activities. The models allow HRES to adjust power production based on real-time information, restricting grid instability as well as enhancing resilience. Federated learning methods integrated also boost energy forecasting in decentralized grids by allowing various stakeholders to share their knowledge in energy forecasting without compromising data privacy.

2.3 Role of Block-Chain in Optimizing HRES

Block-chain has become an intervening technology to allow secure, decentralized, and open energy trade within virtual environments. The deployment of block-chain-orchestrated smart contracts for integrating with HRES can help improve peer-to-peer (P2P) trading of energy and fair price negotiation mechanisms alongside decentralized settlement [6]. Virtual environments utilize block-chain architectures to mark the transactions for recording energy deals, authenticate the correctness of information, and carry out automated disbursements of the energy contingent on real-time spot prices.

In addition, block-chain provides tamper-evident energy trading between micro-grid stakeholders, increasing energy prosumers' trust. The integration of AI and block-chain in virtual space supports dynamic energy pricing, demand-side optimization, and improved security for decentralized energy resources [7].

Applications of HRES in virtual systems will increase with the improvements in computational modeling, cloud computing, and quantum optimization. Further advancements in edge computing will also increasingly facilitate real-time energy management through minimization of latency in data processing. Yet, scalability and dependability of virtual energy solutions rely only on redressing cyber security threats, data privacy issues, and computational complexities [8].

In summary, HRES in virtual environments is a new way of renewable energy management that utilizes AI, block-chain, and digital twins to attain maximum energy efficiency. The continuous development of smart forecasting and simulation methods will continue to expand the integration of renewable energy sources towards a more robust and greener energy system.

3 Block-Chain-Based Energy Trading

Block-chain technology changes energy trading since it allows for secure, peer-to-peer energy transactions amongst end-users and VPPs. block-chain, through the application of distributed ledger technology, provides immutability of data in addition to transparency and decreases central authority dependency. Smart contracts allow for automatic trading of energy since they have an impact on settled conditions without the mediation of a middleman, hence enhancing transaction security and efficacy as well [9].

Block-chain supports P2P energy trading between online communities through real-time validation and authentication of transactions. Block-chain's decentralized approach offers cyber attack and fraud risk mitigation as every transaction is logged on an immutable ledger. This promotes the building of trust among members and eliminates third-party authentication, thereby making transactions more secure and efficient. Besides that, the openness of block-chain enables real-time tracking of energy trade, providing an ethical and equitable trading platform. If block-chain is applied to renewable energy sources, prosumers can sell surplus power to consumers directly without the assistance of conventional utilities, thus saving on transaction costs [10].

The largest benefit of P2P exchange on block-chain is that it can be utilized to create microgrid systems. These grid systems dispersed among the local populace allow them to produce, store, and exchange electricity at reasonable prices, reducing the need for large power plants. Smart contracts facilitate these exchanges by setting terms for exchange in case certain conditions are fulfilled, e.g., price levels or changes in demand. In such an event, P2P exchange of energy causes the democratization of power whereby customers become key players in the energy market while ensuring complete usage of resources with sustainability.

3.1 Network Protection and Consensus Algorithms

Consensus algorithms like Proof-of-Stake (PoS) and Practical Byzantine Fault Tolerance (PBFT) form key parts in protecting the network in addition to confirming transactions. Double spending is avoided and valid energy transactions only are included in the block-chain. PoS is a power-saving variant of the classical Proof-of-Work

(PoW) because the validators need to stake tokens instead of finding hard mathematical puzzles. This diminishes the energy to be utilized with block-chain transactions, thus a power-saving solution for energy trading networks [11].

PBFT also improves fault tolerance due to a network of nodes being able to achieve consensus even under malicious players. This comes in handy especially in energy trading markets where various stakeholders battle one another under a decentralized setting. By offering such consensus protocols, block-chain facilitates secure validation and storage of every energy transaction with no way it can be tampered with. Moreover, automatic billing in smart contracts does away with monetary settlements to zero administration and cost of intervention.

3.2 Integration with IoT for Real-Time Monitoring

Virtual platforms combine block-chain with Internet of Things (IoT) sensors to track actual-time energy consumption, production, and storage. Sensors communicate with block-chain networks to record energy transactions, grid-optimization processes, and synchronize energy demand and supply. IoT integration relies on the right and real-time data to enhance energy forecasting so that grid operators can make apt energy distribution and load balancing choices. This avoids wastage of energy and achieves maximum overall efficiency of the grid [12].

In addition, AI analytics complement block-chain-based energy trading platforms by analyzing large amounts of sensor data to determine patterns and predict shifts in energy demand. Predictive software enables stakeholders to foresee peak usage hours and balance supply in kind, clogging energy and avoiding blackouts. With the integration of IoT and AI, block-chain networks can handle the energy transactions automatically so that there can be an elastic and strong established energy network to supply the customers in real-time.

3.3 Decentralized Autonomous Organizations (DAOs) in Energy Markets

Moreover, block-chain establishes DAOs for the energy market in which stakeholders together own energy trades through tokenized voting mechanisms. This improves social decision-making and fairly allocates energy. DAOs make it immaterial to have centralized regulator bodies, which provide open and democratic control over energy markets. Stakeholders may decide on the price policy, grid extension tactics, and funding proposals through block-chain-validated digital tokens, securing open and representative decision-making [13].

Block-chain-based energy exchange combined with federated learning makes predictive analytics one step further with price adjustment using minute-to-minute variations in the cost of energy. Decentralized energy processing grids can perform the processing locally with federated learning without betraying users' data, enhancing accuracy in demand prediction and price adjustment. It not only makes the consumers the most cost-effective but also promotes prosumers to supply surplus energy to the grid, thus creating a balance and sustainable energy economy.

Overall, block-chain technology revolutionizes virtual energy markets by fostering transparency, reducing operational costs, and promoting energy sustainability practices. With AI, IoT, and distributed networks, it facilitates effective management of energy and ease of access to facilitate transactions in virtual environments.

4 AI and Machine Learning for Renewable Energy Management

AI has vast applications to the administration of renewable energy in virtual space. ML methods enhance decision-making through immense data analysis, forecasting future energy production trends, and grid optimization [8, 9]. AI-based models provide adaptive energy management through continuous learning from past trends of energy consumption and immediate real-time adjustment for greater efficiency. This creativity for resilience uses the maximum green energy resources with minimal dependence on non-green backup power sources. AI also helps identify wastages and where energy generation can be optimised to achieve even better power supply sustainability.

When the AI-based energy management systems came into being, they could be merged with smart grids that allow for real-time tracking and predictive analytics. The functions offer capabilities to predict changes in energy consumption and, as a reaction, make adjustments in energy supply accordingly. Weather patterns, seasons, and consumption patterns go through AI-driven algorithms that align energy grids to their best capacities. It lowers transmission loss and stabilizes the grids, thus making the long-term sustainability of renewable energy infrastructure feasible.

4.1 Deep Reinforcement Learning for Demand-Side Management

Deep reinforcement learning (DRL) has emerged as one of the key methods of demand-side management which enables effective usage of energy and load balancing of smart grids. DRL learns by trial and error, improving power distribution plans to minimize wastage of energy and increase grid stability [14]. By applying

DRL to virtual worlds, energy operators are able to experiment with different patterns of energy distribution and figure out how to most efficiently satisfy demand at the lowest cost. Maintaining this process in sync with real-time dynamics keeps energy distribution liquid and responsive to changes in demand conditions.

Besides this, DRL-based models have a crucial part to play in the management of distributed energy resources (DERs), e.g., roof top solar and home energy storage. Using these models, self-decision-making on behalf of the energy prosumer becomes possible, e.g., optimizing self-consumption or storing excess energy/selling excess power in peer-to-peer (P2P) networks. The systems become efficient and decentralized as well as less dependent on traditional centralized utilities while there is overall higher energy utilization efficiency.

4.2 Machine Learning-Based Forecasting Models

Machine learning-based forecasting models are a crucial part of renewable energy generation. Solar and wind-based power generation sources are inherently fluctuating in nature, and hence unreliability of power supply is an issue. Neural networks and regression-based forecasting are AI algorithms used to forecast fluctuations in renewable energy output, thus facilitating easier planning and integration into the power grid [10]. By comparing historical weather data, real-time sensor data, and seasonal trends, AI-based forecasting minimizes uncertainty and maximizes the efficiency of energy distribution. Energy distributors are able to change plans of storage and distribution with the help of such forecasting models to ensure stability in supply of energy even under low production of renewable resources.

Machine learning also facilitates hybrid energy forecasting through the integration of data from different renewable sources such as solar, wind, and hydroelectric power. The data is integrated here and AI models increase the accuracy of energy generation prediction and ease the operation of the grid. Ensemble learning methods, where multiple forecasting models collaborate, also increase energy forecasting and decision-making by grid operators even further. In this process, the data sets are merged and AI models improve the predictability of energy production and optimize grid operations.

Federated learning techniques are being applied in renewable energy management to improve distributed decision-making without centrally storing data. Federated learning enables different virtual power plants (VPPs) to collaborate to train AI models together without violating data privacy. It optimizes distributed energy trading, fortifies the grid, and prevents energy imbalance, which could result in power blackouts or inefficiency. The ability to train artificial intelligence models on numerous energy systems without compromising the data security is a requirement to ramp up AI-enabled renewable energy solutions.

Besides, federated learning facilitates coordination of distributed energy resources in a way that they can work together to optimize power distribution. This is very useful for microgrids, where locally generated renewable energy needs to be consumed in an optimal manner. Federated AI models balance supply and demand at different microgrid nodes, allowing for efficient use of energy while ensuring reliability and stability in the system.

4.3 AI-Powered Anomaly Detection Systems

AI-driven anomaly detection systems are an essential tool for fault detection and mitigation in renewable energy infrastructure. From the interpretation of data from smart meters, IoT sensors, and grid devices, anomaly detection systems detect energy generation or consumption anomalies and alert operators to potential failures or cyber attacks. Predictive maintenance functionality like this eliminates downtime, enhances operational efficiency, and extends the lifespan of renewable energy equipment. Fault detection tools powered by artificial intelligence can determine inefficiencies within wind turbines, solar panels, and battery storage units so interventions and repairs may be executed with speed.

Moreover, AI-driven cybersecurity technology protects renewable energy infrastructure from likely cyber attacks. With increased digitalization and connectivity of smart grids, the exposure to cyber attacks is higher. AI-driven security systems monitor network traffic, detect anomalies, and respond to possible security breaches in real time. This increases the resilience of renewable energy networks and preserves energy transaction integrity in digital networks.

As artificial intelligence evolves, its integration into virtual realms will enhance the responsiveness and smartness of renewable energy systems. Future advancements in AI-driven energy optimization will comprise self-learning energy grids, autonomous decision-making systems, and sophisticated energy storage management, all further enhancing the efficiency and reliability of renewable energy solutions. The development of AI-driven digital twins of energy infrastructure will enable real-time simulation and predictive analytics, supporting forward-looking energy management strategies.

Apart from this, AI will also play a significant role in maximizing smart grids of energy, where embedded systems will constantly observe and reorganize the flow pattern of energy to make it as efficient as possible. AI-based predictive analytics, grid optimization, and demand forecasting will lead the world towards cleaner and renewable sources of energy, which are less dependent on fossil fuels and towards a cleaner energy world.

5 Additive Manufacturing and Digital Twins in Renewable Energy

Additive manufacturing (AM) and digital twin technologies are transforming the renewable energy sector by optimizing the design, development, and maintenance of energy infrastructure. Digital twins produce very precise virtual models of physical energy infrastructure, facilitating real-time monitoring, performance analysis, and predictive maintenance. Such sophisticated simulation models allow energy optimization to be streamlined by constant updating against real-time sensor data implanted in energy assets [14].

5.1 *Predictive Maintenance and Performance Analysis*

One of the key benefits of digital twins is that they have the capability to simulate various scenarios of operations so that energy operators can test novel approaches without altering real systems. With the incorporation of real-time data, digital twins can predict faults, schedule maintenance at ideal times, and optimize the utilization of renewable sources. For instance, digital twins find applications in wind turbines by producers to monitor performance of turbines, provide early indication of wear and tear, and prevent unexpected downtime, which translates into reduced maintenance and downtime expenses. These virtual replicas powered by AI and ML possess a real-time sensor of operational parameters such as vibration levels, temperature fluctuations, and loading patterns and detect anomalies before system failure.

Predictive maintenance of the digital twin effectively lowers the risk of renewable energy power plant operations. Scheduled maintenance or response to issues creates duplicative service fee bills or unnecessary equipment failure. Digital twins ensure condition-based servicing through live judgment of the fitness of the individual components. The technology of a solar power plant, for instance, supports the digital twin that enables diagnostics in aging panels from dust deposits, shadowing, or material degradation. This enables the operators to follow active cleaning or replacement policies in order to realize optimum energy output efficiency. Additionally, hydroelectric power plant digital twins monitor pressure within turbines and changes in water flow to guarantee optimum efficiency with minimum structural deterioration over time.

Digital twins also play an extremely critical role outside predictive maintenance, namely in performance analysis through means of energy yield optimization and system optimization. With a digital twin of a renewable power plant, engineers can test various operating scenarios and analyze the impact of many control policies. For offshore wind farms, for example, digital twins interpret atmospheric conditions such as wind speed, wave direction, and grid demand load to position the turbines in the best possible positions and also to rotate them. Similarly, in battery energy storage systems, digital twin simulations monitor charge–discharge cycles to optimize battery performance and lifespan. By means of such real-time simulation and data analysis, affordable and sustainable renewable energy solutions are possible.

5.2 *Additive Manufacturing of Renewable Energy Components*

Meanwhile, additive manufacturing (AM) is revolutionizing the renewable energy sector with fast prototyping and manufacturing of energy components. Compared to conventional methods of production, wasteful and labor-intensive, AM employs layer-by-layer production to create highly precise energy components with little or no waste [15]. The technology is especially important in the manufacturing of customized wind turbine blades, lightweight solar panel supports, and very efficient battery components. With the use of technologies such as high-strength polymers and carbon fiber composites, AM focuses on renewable energy technology and provides design flexibility.

AM application in wind energy has transformed wind turbine manufacturing including its ability to supply aerodynamically optimized blades that are efficient. Wind turbine blades traditionally are material and scalability-constrained. However, AM allows the creation of intricate blade shapes optimized for certain winds to ensure optimal energy harvesting. Besides, AM is transforming photovoltaic solar panel technology by developing light and strong mounting structures to maximize installation efficiency.3D-printed perovskite solar cells with enhanced energy conversion efficiency at reduced costs are also favored.

Another important advantage of AM is its application towards the advancement of energy storage technology. Grid stabilization: Grid stabilization using renewable energy needs high-energy batteries, and AM is also paving the way towards the manufacturing of more efficient batteries. With AM manufacturing of electrodes and electrolytes, researchers are maximizing their energy density and capacity for charge retention in future-generation batteries. AM is also enabling the creation of complex cooling systems in battery packs to avoid overheating risk and prolong battery life. As energy storage technology in the future develops further, AM-driven innovations will be instrumental in realizing clean and affordable renewable energy.

5.3 *Local Production and Supply Chain Resilience*

Additionally, additive manufacturing increases the resilience of supply chains via local production of spares, less reliance on international supply chains, and limiting lead times. This is particularly valuable for remote renewable power plants, for example, offshore wind farms or remote solar photovoltaic power plants, where component availability may often be the limiting factor. Unlike having to rely on traditional supply chains subject to geopolitical disruption or supply chain constraints, AM enables energy operators to manufacture critical elements in-house on their own site. This does not leave the maintenance and use of equipment continuous.

With the ability to produce replacement parts 3D in-house, productivity in energy infrastructure maintenance is increased significantly. For wind farms in the open sea, it might be expensive and time-consuming to ship spare parts from onshore production facilities to turbine sites. With AM, maintenance personnel can print required parts such as gearbox housing, blade mounts, or hydraulic fittings on the spot, reducing downtime and service costs. Likewise, locally printed 3D-connectors, mounting brackets, and junction boxes can be used in distant solar farms to enable quick repair and increased system reliability.

Other sustainable manufacturing processes are facilitated with the help of AM by minimizing material loss and energy consumption. The conventional production process entails high scrap materials, whereas optimal raw material efficiency is achieved through AM through certain deposition methods. This minimizes the cost of production in addition to harmonizing global sustainability strategies through less carbon footprint emission in renewable energy devices. Besides, studies on biodegradable as well as recycle materials for 3D printing are unveiling sustainable manufacturing routes that promote circular economy loops in green energy.

5.4 Enhancing Energy Storage Systems

Another vital use of AM in the renewable energy sector is for enhancing energy storage technologies. Through the use of novel designs and high-performance materials, AM assists in the production of high-performing battery architecture with optimized energy density as well as thermal management [16]. Integration with AI-optimized designs further enhances the effectiveness and reliability of such storage facilities.

Digital twins and AM together also unlock intelligent energy grids. Digital twins of energy distribution networks can simulate current power flows, forecast shifts in demand, and optimize grid efficiency. AM, on the other hand, allows for new components of grids that are configured to respond to specific operational requirements at once and therefore create more responsive and dynamic energy infrastructure. In the coming times, the integration of AM and digital twin technologies will keep propelling innovations in renewable energy, promoting sustainable innovation and enhancing the efficiency of the overall energy system.

6 Automation and Robotics in Renewable Energy Systems

Automation and robotics are transforming the renewable energy sector through the improvement of system deployment, operation, and maintenance efficiency. Robotic technology sophistication at higher levels enables solar panels, wind turbines, and other renewable gear to be mounted with much lower construction time and labor costs. Robotic systems also enable control of energy infrastructure to deliver maximum levels of performance with reduced downtime.

6.1 Robotic 3D Printing for Energy Infrastructure

One of the most creative applications of automation to renewable energy is robotic 3D printing that allows for rapid energy-efficient structure construction with minimal use of resources. Robotics 3D printing, as opposed to traditional building technology, employs additive manufacturing methods in building and energy appliance construction layer by layer with the goal of achieving much strength while losing small quantities of materials [17]. This technology is particularly helpful while constructing modular solar farms, windmill bases, and hydroelectric dam construction because it facilitates cost-effectiveness, quicker construction, and extremely intricate infrastructure construction.

Robot 3D printing provides tremendous value to sustainability through the use of green building materials such as bio-based composite materials, recycled polymers, and geopolymer cement. These building materials improve the life spans of the energy infrastructures and their green nature and minimize the carbon footprint. Robot 3D printing accuracy also eliminates human mistakes, resulting in longer-lasting and efficient energy systems [18]. Through decentralized construction, the technology also facilitates local manufacturing of renewable energy components such that transportation is costly and redundant as logistics challenges are absent in far-flung areas.

Aside from construction of infrastructure, robot-assisted 3D printing also optimizes renewable energy systems through rapid prototyping and site-adaptive optimization. For example, foundations of wind turbines can be printed in a way that the structures are optimized for structural integrity depending on soil conditions at the location. Similarly, in hydroelectric application, turbine parts 3D printed can be optimized to maximize efficiency depending on water flow conditions. This adaptability is one whereby renewable energy installations are designed to their optimal performance and lifespan, ultimately leading to a more efficient and more dependable system of energy.

In the future, 3D printing and AI robots will also revolutionize energy infrastructure. AI software is indeed capable of real-time adjustment of printing parameters, varying rates of material deposition and structural topology as a function of simulation-informed performance. Besides, robotic automation integrated with 3D printing can create self-healing energy systems where robots can identify and heal structural defects without human intervention. With increasing technology, robotic 3D printing will contribute greatly towards leading the world to sustainable and robust energy systems.

6.2 Building Information Modeling (BIM) in Energy Planning

Building Information Modeling (BIM) is transforming energy infrastructure planning to make digital modeling of energy infrastructure possible. BIM allows planners and engineers to model different configurations, improve energy transmission, and identify faults before physical realization. With the use of a digital strategy, the most efficiency is realized, and stakeholders can make changes and modify designs in a way that it becomes easy to integrate renewable energy sources into existing infrastructure. Automation and artificial intelligence (AI) application in BIM-based platforms cut decision-making cycles and increase the accuracy of energy project planning at lower cost and wastage.

Moreover, BIM fosters real-time coordination of projects via collaboration of teams. Construction managers, engineers, and architects can exchange up-to-date information via cloud-based BIM platforms, reducing opportunities for conflicts in design [19]. Increased transparency promotes effective communication and resource management, resulting in improved implementation of energy projects. Predictive modeling through BIM involves scenario analysis as practicable, and stakeholders can estimate the effect of various energy alternatives on performance and sustainability.

A further benefit of energy planning in combination with BIM is its capability for environmental footprint analysis and compliance with regulations. Simulation of energy consumption patterns and emissions simulation by BIM software allows energy-efficient planning of buildings and infrastructure in accordance with green energy policies. Integration of automation in the workflow enhances simulation accuracy, providing overall indications for potential improvement. These aspects are crucial when developing renewable energy infrastructure up to international levels of sustainability at low expenses.

6.3 Autonomous Robots for Inspection and Maintenance

Use of self-service robots in the maintenance of renewable energy is transforming energy asset maintenance. Advanced drones equipped with thermal imaging cameras and artificially intelligent analysis perform real-time solar panel and wind turbine inspection. The drones can identify defective panels in real time, identify micro-cracks, and inspect the structural health of turbines within minutes, significantly reducing time and cost compared to manual inspection.

Robotics can also undertake routine checks within offshore wind farms where there is easy access but not immediate accessibility without putting human life in danger. Submarine robots are used in inspecting the underwater areas of hydro-electric power plants and removing costly and hazardous human diving operations. Robots capture data immediately and transmit them to be analyzed, allowing power companies to perform proactive maintenance. Robotic autonomy, round-the-clock tracking of infrastructure, makes it possible for increased dependability, enhancing the lifetime of renewable energy installation [20].

Besides inspection, robotic technology also enables automated maintenance. Robotic arms equipped with specialized tools can perform small solar panel and wind turbine repair with reduced downtime and improved energy generation effectiveness. Robotics-based robotic maintenance systems also foretell potential equipment malfunction by AI-based analysis, which allows intervention in advance prior to the expensive breakdowns occurring. Predictive method reduces downtime in operations, hence leading to cleaner and more efficient production of energy.

6.4 Grid Optimization Using AI-Powered Robotics

Artificial intelligence (AI) and robotics are increasingly being employed to optimize energy grid management. AI-powered robots continuously monitor grids' performance and adjust automatically within power transmission to match fluctuating energy demand. Dynamic realignment enhances the stability of the grid, averting shortages and overloads of energy while maximizing efficiency. Smart grids driven by artificial intelligence and robotic automation aid in regulating the supply of energy from renewable resources such that excess energy is redirected where needed most.

Predictive maintenance is also the greatest benefit of robotics with AI capabilities in grid operations. Following processing of volumes of performance data, robotic systems can identify the formation of equipment wear and tear and conduct maintenance prior to breakdown. Predictive maintenance reduces operating expenses by averting emergency maintenance and forced outages. Robotic automation also makes it possible to monitor grid components in real-time, improving security from cyber attacks and physical sabotage [21, 22].

In addition, robots with AI-based control are applied in demand response schemes under which the use of energy is managed in accordance with the current grid status. These intelligent systems balance supply and demand equilibrium to an optimal level, thereby making integration of renewable power efficient. Load balancing optimization and reduction of the transmission loss can be achieved with the application of robotics in power grids, and they provide a continuous supply of electricity to the consumers.

6.5 Energy Planning Enhanced Through BIM-Integrated Automation

BIM automation is also revolutionizing energy infrastructure planning with advanced clash detection processes. Through imitation in the virtual world of energy systems, BIM software can identify clashes between components and allow engineers to rectify defects ahead of time. Foreseeing this waste streamlines project efficiency, pre-empts expensive changes on the implementation front, and saves construction time.

Another important advantage of BIM automation is management of renewable energy resources in space. BIM software can calculate solar and wind resource values and position the solar panels and wind turbines in such a way that maximum energy is harvested. By automation, BIM tools can calculate the feasibility of different renewable energy sources for a geography and thus enable maximum harvesting and transmission of energy [23, 24].

Additionally, BIM enhances the coordination of renewable energy projects' energy storage devices. Through the use of simulation tools of battery storage systems and their behavior towards the grid, BIM software provides the best energy storage capacity to deliver electricity at any given moment, including the low points of renewable energy production. This is also extremely critical in the achievement of energy resilience and reducing dependence on non-renewable backup supplies [25, 26].

Future energy planning will more and more rely on the integration of BIM, automation, and robotics fueled by artificial intelligence. The technologies will keep on making energy more efficient, sustainable, and reliable, resulting in more intelligent and robust renewable energy power plants. While momentum builds around BIM and automation, it will usher in a greener, more interconnected energy future.

In the coming years, the convergence of robotics, automation, and AI will drive innovation in renewable energy systems. Independent building, robot maintenance, and artificially intelligent energy will spearhead sustainability, reduce the expenses, and increase the use efficiency of devices involving renewable energy. As advancements continue to change their lives, they will have to deliver on international energy transformation dreams and see wise energy communities prosper.

7 Conclusion

Virtual environments offer the groundbreaking platform for creating renewable energy solutions with ease of convergence of AI, block-chain, digital twins, and additive manufacturing. These technologies combined offer a boost in the efficiency of energy, sustainability, and grid stability with optimized energy management, transactional automation, and better infrastructure planning. AI-powered automation and forecasts cut down uncertainty connected with the availability of renewable power, while block-chain provides safety coupled with openness in energy transactions. Digital twins permit predictive servicing and real-time condition monitoring for augmenting the system's dependability. Additive manufacturing helps deliver efficient structures by reducing wastes efficiently. All these future works must concentrate on enhancing interoperability of these technologies towards a well-integrated smart system of renewable power.

References

1. Al-Saedi, W., Bass, O., Gadsden, S.A., Abu-Siada, A.: Optimal energy management strategy for a hybrid renewable energy system in a virtual environment. Renew. Energ. **195**, 1287–1302 (2022)
2. Li, Y., Xu, X., Liu, Y., Zhang, H.: A novel block-chain-based energy trading model for virtual power plants integrating renewable energy sources. Energ. Rep. **9**, 367–381 (2023)
3. Wang, J., Chen, H., Wang, C.: Demand-side management in virtual environments: A deep reinforcement learning approach. IEEE Trans. Smart Grid **12**(5), 3892–3905 (2021)
4. Zhang, X., Li, C., Wu, J.: Distributed energy management for microgrids using federated learning. Appl. Energy **310**, 118591 (2022)
5. Singh, R., Sharma, P., Kumar, A.: AI-driven optimization of hybrid renewable energy systems for smart cities. Renew. Sustain. Energy Rev. **173**, 112857 (2023)
6. Kim, D., Park, J., Lee, H.: Secure and efficient IoT-enabled energy trading framework for virtual energy communities. J. Clean. Prod. **416**, 137132 (2024)
7. Shuriya, B., Sivaprakash, P., Kumar, K.A., Saravanakumar, M., Rajendran, A.: Digital twins and artificial intelligence: Transforming industrial operations. In: Digital Twin for Smart Manufacturing. Elsevier, pp. 111–133 (2023)
8. Gupta, A., Patel, D., Verma, S.: Renewable energy integration in cloud computing data centers: a review. Sustain. Comput. Inform. Syst. **30**, 100527 (2021)
9. Ahmad, I., Raza, S., Khan, M.: A hybrid energy management system for smart homes in a virtual environment. Energ. Build. **270**, 112332 (2022)
10. Luo, H., He, Z., Wang, X.: Multi-agent deep learning for peer-to-peer energy trading in virtual energy markets. Energy **271**, 126846 (2023)
11. Brown, T., Green, M., White, P.: Machine learning-based forecasting of renewable energy generation in smart grids. Renew. Energy **208**, 253–268 (2024)
12. Mrazović, N., Baumers, M., Hague, R., Fischer, M.: Guiding building professionals in selecting additive manufacturing technologies to produce building components. Mater. Today Commun. **15**, 199–202 (2018)
13. Frazier, W.E.: Metal additive manufacturing: A review. J. Mater. Eng. Perform. **23**, 1917–1928 (2014)
14. Ngo, T.D., Kashani, A., Imbalzano, G., Nguyen, K.T.Q., Hui, D.: Additive manufacturing (3D printing): a review of materials, methods, applications and challenges. Compos. B Eng. **143**, 172–196 (2018)
15. Borkowski, A.S.: Digital twin conceptual framework for the operation and maintenance phase in the building's lifecycle. Arch. Civ. Eng. **70**, 139–152 (2024)
16. Mouna, P.S., Sivaprakash, P., Kumar, K.A., Samuel, P., Shuriya, B., Sharma, V.: Data modeling and analysis for the internet of medical things. In: Wireless communication technologies (1st edn). CRC Press, p. 272 (2024)
17. Kazemian, A., Yuan, X., Cochran, E., Khoshnevis, B.: Cementitious materials for construction-scale 3D printing: laboratory testing of fresh printing mixture. Constr. Build. Mater. **145**, 639–647 (2017)
18. Duballet, R., Baverel, O., Dirrenberger, J.: Classification of building systems for concrete 3D printing. Autom. Constr. **83**, 247–258 (2017)
19. Bourell, D., Kruth, J.P., Leu, M., Levy, G., Rosen, D., Beese, A.M., Clare, A.: Materials for additive manufacturing. CIRP Annal. Manufact. Technol. **66**, 659–681 (2017)

20. Sivaraju, S.S., Senthilkumar, T., Sankar, R., Anuradha, T., Usha, S., Musirin, I.B.: Improving the efficiency of induction motor drive by flux and torque control: a hybrid LSE-RERNN approach. ISA Trans. **147**, 215–226 (2024)
21. Vino, T., Sivaraju, S.S., Krishna, R.V.V., Karthikeyan, T., Sharma, Y.K., Venkatesan, K.G.S.,... & Markos, M.: Multicluster analysis and design of hybrid wireless sensor networks using solar energy. Int. J. Photoenerg. **2022**(1), 1164613 (2022)
22. Lim, S., Buswell, R.A., Valentine, P.J., Piker, D., Austin, S.A., De Kestelier, X.: Modelling curved-layered printing paths for fabricating large-scale construction components. Addit. Manuf. **12**, 216–230 (2016)
23. Davtalab, O., Kazemian, A., Khoshnevis, B.: Perspectives on a BIM-integrated software platform for robotic construction through contour crafting. Autom. Constr. **89**, 13–23 (2018)
24. Sivaraju, S.S.: Enhancing electric vehicle charging safety and efficiency through hybrid charging systems and intelligent management strategies. J. Energy Storag. **118**, 116073 (2025)
25. Shuriyaa, B., Santhamani, V., Shanmugam, V.B., Subashini, S.: Enhancing network security through viper optimization algorithm with deep learning-assisted network security system in biomedical records. Front. Health Inform. **13**(8)
26. Hasannejad, A., Sardrud, J.M., Javid, A.S.S.: BIM-based clash detection improvement automatically. Int. J. Constr. Manag. **23**, 2431–2437 (2023)

Governance and Policy for a Green Metaverse

Chevella Aravind Reddy, Dusari Sai Teja, and Pritam Khan

Abstract Green metaverse is the world that embraces the evolution, therefore, it ought to be in line with the climate goals. However, with the coming of contradiction, governance and policy play a huge role in the path of environmental sustainability and digital transformation to usher a smarter and greener future. For this purpose, we propose a simulation framework based on fuzzy logic and genetic algorithms to evaluate and optimize the digital ecosystem concerning technological and environmental innovation, social and economic sustainability. It involves a fuzzy control system with twenty-nine input variables, including essential environmental indicators such as carbon footprint, energy consumption, water usage, waste generation, pollution levels, and biodiversity effect. Strategic metrics include various factors such as green investments, circular economy practices, smart infrastructure, operational efficiency, and community engagement. Additionally, aspects such as waste management, pollution control, climate resilience, ecojustice, fair trade, and recognition of indigenous knowledge are required to understand the complex interrelationships of sustainability. Dynamically, each variable is accompanied by Gaussian or triangular membership functions representing low, medium, and high levels. Four fuzzy rules are applied targeting environmental and ecological impact; investments and innovations grounded in sustainability; social community engagement, regulation, and compliance; and resource dependency and risk management, from which the output variable is obtained for the green metaverse. This output is then subject to optimization by a genetic algorithm to reach an optimal solution producing a score for the green metaverse of about 82.95% with a moderate carbon footprint, with high green investment, and balanced resource and risk management strategies. The framework enhances the capability of managing sustainability interrelations through an evidential approach with minimal adverse impact. It has the potential as a scalable platform

Chevella Aravind Reddy and Dusari Sai Teja These authors contributed equally to this work and are co-first authors.

C. A. Reddy · D. S. Teja · P. Khan (✉)
Department of Computational Intelligence, SRM Institute of Science and Technology, Kattankulathur, Tamil Nadu, India
e-mail: pritamk249@gmail.com

S. Kautish et al. (eds.), *Green Metaverse: Fuzzy Logic Approaches to Sustainable Virtual Worlds*, Information Systems Engineering and Management 74,
https://doi.org/10.1007/978-3-032-08183-4_9

to explore scenarios for policy development, striving for greater applicability in both academic research and real-world governance.

Keywords Fuzzy logic · Genetic algorithm · Governance · Metaverse

1 Introduction

Metaverse is growing at a blazing speed to bring digital and green into the same space so that it redefines how we get things done through technology. The idea of virtual worlds is no longer as futuristic as one may think; they are becoming a part of the international economic, social, and cultural interaction. The green metaverse reduces with the goal of carbon dioxide footprint and energy consumption as well as the application of sound practices in digital infrastructures. That is, it offers alternatives to traditional developmental models and creates a compelling case for redefining how digital ecosystems should be created and maintained.Therefore, stakeholders are called upon to align technology advancement and the environmental stewardship.

Governance and policy framework as anchors of state can take center stage in making metaverse defined by its green character. Such frameworks would also help address energy consumption, health concerning resource distributions, and mitigate the effects of virtual technologies on the environment. There is a need to adjust existing legislation to meet the specific needs of decentralized digital platforms even as innovators propose new procedural in terms of creative solutions that feature technology's advancement against environmental sustainability. Such new models of regulatory institutions need to be flexible enough to cover both global and borderless metaverse definitions. This way, it will help lay a foundation for a greener digital future for governments as well as regulatory bodies.

Such governance within the ambit of multicasting is essential in order to propagate the development of an energy-efficient sustainable green metaverse. The transformational policy framework is in driving standards on renewable energy usage and minimizing carbon emissions from data centers and related kind infrastructures. Such standards are not only in line with the environmental aspect but also expose areas of transparency and accountability and equitable access for all users. Adoption of such policies would spur the innovation of greener technologies while rewarding enterprises that adopt green business practices. Therefore, it is a joint approach that entails proactive measures just as much as market-oriented incentives. A robust governance framework would, ultimately, play a catalytic role for a digital ecosystem that would have growth harmonized with environmental preservation.

A central point in the architecture of systems for effective policy-making for the green metaverse is a collaboration between multiple stakeholders. Constant dialogue should ensue among governments, private companies, academic institutions, and civil society organizations so that all stakeholders understand and respect their objectives and best practices. Collaboration among these partners would foster collaboration toward creating innovative laws and adapting them quickly to the speedy changes taking place with digital technologies. Such a joint effort will promote and encourage research in sustainable technology solutions through a public-private partnership. All those bodies will work together to overcome the numerous and complex challenges that the metaverse has created. This paradigm would also yield great dividends to individual sectors, along with pushing towards one common strategy of environmental sustainability at a global level.

The decentralized and, therefore, global character of the metaverse has created jurisdictional complexities that call for creative regulatory solutions. Tradtionally geographic boundary and legal framework have always failed to catch the complexity around digital environments. All this calls for regulatory bodies to initiate flexible mechanisms which will manage cross-border issues and enforce uniform environmental standards. Those mechanisms need to be addressed by any rapid technological advancements or differences in their regulatory environments. Challenges listed above will not be addressed without a reevaluation of existing statutory paradigms and the formulation of new policies to be internationally harmonized. This would redress the regulation agenda for making sure that the green metaverse is both high in innovation and accountability toward global sustainability goals.

Governance for greening the metaverse should not only cater for environmental concerns but also the social and economic dimensions of the digital innovation. Their policies should include the built environment itself for all and accompany equal access to technology and its resources. Sustainable development is access for all irrespective of geography or economy. Adoption of incentives for greening makes a punch toward sustainability, while at the same time promoting social equity. Linking the environmental goals with social justice initiatives will therefore make a metaverse for various communities. Governance can thus ensure bringing together social fairness with sustainability in making a digital future fair and balanced.

On the other hand, governance and policy for the green metaverse offer a very interesting opportunity to reshape the future of digital arenas. The metaverse shall ultimately become a venue for cultivating innovation and conservation, as people subscribing to this nexus deepen their efforts around environmental and social priorities. For this interaction to remain dynamic and sustainable, it must continuously be constructed with the inputs from policymakers, engineers, and the environmental community. Future recommendations must be flexible and visionary, bringing in new technologies with sustainability standards. In future the introduction into a metaverse will come in the form of a vibrant and green substance, becoming socially responsible and re-inventive.

1.1 Related Work

Emerging digital technologies have painted a complex picture of the application of governance, ethics, and sustainability derived from numerous studies. For example, [1, 2] emphasize a pressing need for regulation in ensuring technological change speedily. Not only it [1] interrogate calls for XR governance with privacy, safety, and commercialisation concerns, but [2] also raises attention to the so-called "pacing problem" in that traditional systems of regulation face much difficulty in conforming with the fast-changing digital environment. In addition to this duality is the fact that the metaverse should be an opportunity yet a threat for public service delivery. Such arguments seem to conclude that the contingent court of the law must respond to emerging threats linked to immersive technologies. Thus, generally, the literature invokes anticipatory mechanisms of governance which would be flexible and yet forward-looking.

The relationships between such digital public services and immersive virtual environments are researched in much greater detail in [3, 4]. That is, the two studies, upon reading together, suggest that although the metaverse can serve to change the face of public services with new platforms for engagement, not all conditions needed to effect such a change are met. Reference [3] therefore provides a comprehensive taxonomy of functional and non-functional requirements pertaining to digital governance, stressing that not all public services are suitable for inclusion in the metaverse. Such a contrary consideration is shown in [4], which tries to place the analysis within the Gambia where metaverse technologies are seen as tools for increased citizen participation and urban planning. Both studies thus underline the necessity of multi stakeholders fight for barriers that are technical, infrastructural and ethical. This general direction implies that the future of digital governance would be about technical innovations coupled with socio-political inclusiveness.

Environment sustainability and ethical orientation are focal concerns from the literature review under reference. Researches, as published with [5, 6], shows that the embedding sustainability from the beginning in the metaverse is an extremely urgent action. Hence, in [5], he goes on urging that metaverse development should hold allegiance to the United Nations Sustainable Development Goals, while clearly setting out the sustainability practices to undergo as part of the virtual experience and not as a retrofit. This is further complemented in [6], which borrows some frameworks from the triple bottom line and ESG perspectives to assess among other long-term effects of digital changes. The arguments synthesized from these investigations converge on the fact that if proactive action is not taken towards sustainability, then the inevitable rapid advance in immersion technology will generate very high costs in environmental and social dimensions. Thus, this literature lays a foundation for the possibility of technological innovations being balanced harmoniously and with the ecological responsibility.

In [7, 8], one could find intensive explorations into the legal and regulatory landscape of immersive technologies. A thorough examination of the applicable rules is taken in [7] regarding existing legal frameworks such as the GDPR and the Digital Services Act, focusing on their applicability in meeting fundamental rights protection in the metaverse. In contrast, [8] adopts a systematic scoping review to chart effective governance structures highlighting challenges determining from fake news to copyright infringement. Both papers stress a dual approach with the need to couple strict laws with innovative technological solutions placed together. It indicates that the new fast digitalizing society would need altogether new regulatory paradigms to vegetable policies adaptable to events as paramount.

The ethical issues are yet standing on the pillar, with inputs given by [9, 10] complementing each other. Reference [9] pairs ethical dilemmas through privacy conflict, identity manipulation, and surveillance into twin groupings, asking stakeholders to build a governance agenda to address them preemptively. Reference [10] extends the debate with a review of central issues such as growing surveillance risk, difficulties in data management, and ambiguities in regulation. All this work tends to suggest that ethical elements should be entwined with technological advancements and produce a balanced virtual space. Having ethical imperatives integrated within governance frameworks would then be regarded as a prerequisite to the build-up of trust among users and stakeholders alike. In a word, this all-inclusive treatment of ethical challenges also brings to the fore the need for continuous dialogue and iterative policy development in the metaverse.

Under the governance and ethical issues, one such work has proposed certain models of sustainability characterized by technological efficiency. Green Metaverse Networking, in this case, discusses the trends in energy consumption through the currently existing technologies for enabling the metaverse. It argues that energy optimization is a prerequisite not only to the reduction of environmental impacts but also to the establishment of economic viability to immersive environments. This work premises that energy-hungry processes will, if left unchecked, act as barriers to the large-scale rollout of the metaverse, while offering real answers to sustainability challenges of the physical digital infrastructure by the identified potential pathways such as semantic communication and resource allocation. The treatment in [11] thus enriches the larger sustainability narratives with respect to the technical nature of green technology in play.

The much-discussed intersection of digital transformation with public administration is accessible in [12, 13]. The former consists of a discussion on how government institutions compute the evolutions of their internal forms of public operations and metaverse technologies to engage their constituents [12]; the latter [13] speaks about sustaining the metaverse using presumptive tools such as Fuzzy Cognitive Mapping and scenario analysis. The two authors argue for scalability and interoperability as equally essential to technological innovation. The two contain the understanding that a reconstituted public sector will have to embed, in a highly active way, state-of-the-art digital instruments that remain attuned to issues of digital equity and infrastructural readiness. They also affirm that the success of such efforts would hinge on the balance between technical advances and social concerns.

They would thus forge a pathway for future digital governance that is, therefore, truly innovative and inclusive. Among other thematic areas of much interest in the current literature, the third is technological innovation, especially in computational methods as tools. In this context, GAs have evolved through evolutionary processes to solve optimization problems, linking the issues tackled by these papers with more general issues about the efficiency of digital systems. Paper [14] lays principles behind GAs-that they must be grounded in natural selection and show their applicability to hybrid computational problems. This paper [15], on the other hand, proposes a new idea known as the Intelligent Programmed Genetic Algorithm (IPGA), which, by integrating probabilistic operators and deterministic strategies, proves to have better performance on benchmark functions. Together, they establish a strong foundation upon which the algorithmic basis for development in the optimization of metaverse infrastructures can be set. These studies feed into computational optimization and increase the efficiency and sustainability overall of the digital governance model.

Collectively, the literature emphasizes that the interplay between technical issues, ethical considerations, and regulatory framework composes the metaverse ecosystem. As shown in the analyzed literature, some common ground has been established-largely revolving around the concept of anticipatory governance-that must, however, be able to accommodate contributions from multiple disciplines. The synthesis of insights from [1–3], and [4] proposes that effective governance in the metaverse must be cognizant of the fast-paced change of digital environments, remaining flexible to cope with challenges that had not previously been identified. When examined from this perspective, the ethical underpinnings highlighted by [9, 10] are further shades of gray against the rosy image of a technology buildout that would trample on the rights of individuals and undermine social equity. The synthesis thus postulates on a multidisciplinary ambit whereby legal, technological, and ethical expertise are collaborated to bring about the metaverse's tomorrow.

The literature once again has brought the underlying need to close the divide between technological advancement and societal necessities. The different accounts from the likes of [2, 6] illustrate that the metaverse stands not only as a technological challenge but also as a social one. These studies remind us that an equally important aspect of the metaverse is that digital transformation should be based on sustainable practices and ethical considerations together with proactive governance strategies. This balanced approach calling for regulatory, technological, and societal input thus has pushed for a coordinated effort between divers sectors. Convergence of these thoughts ensures that the advancement of the metaverse is done in an inclusive and sustainable manner, while ultimately giving credence to academic discussions.

In conclusion, the synthesis of studies provides a more coherent account of the challenges and opportunities around the immersive digital technologies. These studies reiterate that metaverse acts not just as a technological advancement but also a sociological evolution in the conduct of society-organizing, governing, and interaction. One needs not look beyond the governance frameworks discussed in [1, 2, 8, 10] on sustainability and ethics in [5, 6, 9], which forms the basis for further research and policy development. The work done in computing by [14, 15] dovetails into this framework in proposing computational tools for consolidated digital governance. This synoptic survey thus proposes that the many-faceted effects of the metaverse can be dealt with only through a just such integrated framework. Such insight shall be useful to policymakers, industry leaders, and researchers who wish to traverse the minefield of digital transformation.

1.2 *Contributions*

The contributions of this chapter are summarized as follows:

1. The Green Metaverse employs a fuzzy logic-genetic algorithm combination that can refine the assessment of environmental, economic, and social indicators essential for decision-making so as to promote sustainable development.
2. The Green Metaverse Index transforms crisp numerical values into fuzzy linguistic variables for interpretability and policy application by providing a structured evaluation framework.
3. By iteratively selecting, mutating, and evolving the best parameter values for minimizing ecological damage and maximizing inclusivity, a genetic algorithm optimizes sustainability configurations.
4. The proposed modality allows policymakers and developers of the metaverse to realize flexible, future-proof governance strategies, balancing environmental responsibility, economic viability, and social inclusivity.

2 A Framework Integrating Fuzzy Logic and Genetic Algorithm for a Green Metaverse

Fuzzy Logic is capable of analyzing uncertainty regarding the environmental parameters (e.g., energy efficiency, carbon footprint reduction) by modeling subjective or imprecise parameters (e.g., user behavior, resource allocation priorities). Genetic Algorithms optimize these parameters through evolutionary computation, iteratively refining the solution for energy-efficient virtual infrastructure, ecologically sustainable design of digital assets, or eco-aware user-engagement strategies. Combined, these allow adaptive, self-updating solutions that balance technical excellence against sustainability in the metaverse.

2.1 *Fuzzy Logic*

A structured evaluation framework for a sustainable yet eco-friendly digital world can consider diverse environmental, economic, and social aspects. The present study identifies 29 key variables which influence the governance and regulation of a Green Metaverse. They are categorized into: environmental factors-carbon footprint, energy consumption, pollution levels (microplastic pollution, biodiversity impact etc.); economic factors-green investments, circular economy practices, smart infrastructures, adoption of regenerative designs; and social factors-community engagement, green job creation, eco-justice fair trade, and indigenous knowledge utilization. All variables are critical components that make virtual environments sustainable, ensuring that the Green Metaverse stands on responsible innovation pillars.

In this chapter, the fuzzy inference system (FIS) was developed to assess the effect of the variables defined previously. The fuzziness in fuzzy reasoning will help in dealing with uncertainty and ambiguity in data which has made it an appropriate technique for assessing sustainability indices which is generally qualitative in nature. Therefore, each variable will be endowed with three Gaussian membership functions-low, medium, and high-according to the pre-defined numerical ranges (0–100). The reason for the permitted asymmetrical smoothness in transition from one sustainability level to another, rather than a hard classification, is completely understandable. Ultimately, the Green Metaverse Index (GMI) serves as the end output of this fuzzy system and signifying the overall sustainability rating of a given metaverse ecosystem.

A rule-based inference system forms part of this fuzzy framework that offers a roadmap through input variables for dynamic decision making: the four main rules are (1) Environmental and Ecological Impact, (2) Sustainable Investment and Innovation, (3) Social, Regulatory and Community Engagement, and (4) Resource Dependency and Risk Management. This measures the optimum balance achieved between ecological health, economic viability, and social welfare. For example, a Green Metaverse is rewarded a high sustainability score within the framework when carbon footprint, energy consumption, and pollution levels are low, but biodiversity impact is rated high. Conversely, if environmental risks are above normal, the sustainability rating decreases accordingly.

A fuzzy system is implemented using the SkFuzzy library within the Python environment, allowing dynamic generation of membership functions and fuzzy rules. Antecedents (input variables) are modeled using Gaussian functions, while the consequent (The Green Metaverse rating) uses triangular ones. The control system is implemented to simulate all fuzzy rules that dynamically compute sustainability scores from the real world. Hence, through the incorporation of fuzzy logic for such nonlinear decision-making systems, these systems can therefore accommodate the interjection of factors without asking for fixed numerical thresholds. Such flexibility will allow the methodology to be influenced by different metaverse governance measures and environmental regulations over time.

In conclusion, the fuzzy decision-based model provides a scalable and interpretable way to evaluate and optimize sustainable digital environments. By considering various environmental, economic, and social factors, the framework provides guidance to policymakers, developers, and stakeholders on how to create a responsible, efficient, and inclusive Green Metaverse. The next step in the methodology will incorporate a genetic algorithm-based optimization technique, which can thus dynamically predict which configurations would rank best for sustainability.

2.2 Genetic Algorithm for Optimization

Even though fuzzy logic serves well for evaluating the sustainability of the Green Metaverse, it does not inherently complete the job of optimizing the variable configurations for realizing the best possible sustainability score. For this reason, a genetic algorithm (GA) has been included in its pathway towards finding the optimal combination of various sustainability factors. The GA is a robust evolutionary computation technique imitating natural selection and genetic evolution to achieve the best possible solutions for complicated problems. In this work, the GA estimates the input variables to maximize the GMI (Green Metaverse Index) for the configuration of the most sustainable digital ecosystem.

The workings of the genetic algorithm start with the definition of a population of individuals, each individual representing a configuration of the 29 sustainability variables. Each sustainability variable has a value assigned to it in the range of 0–100, which is an indicator of the level of sustainability. The generation of an initial population will be from the random set, which guarantees a higher diversity of potential solutions. The fitness function measures every single individual according to the Green Metaverse rating computed by the fuzzy system, with higher ratings assigned for better sustainability performance.

In order to bring about evolution in the group, the GA uses crossover, mutation, and selection operations. Crossover work benefits the offspring through the potentially sustainable characteristics of both evolutionary parents. Mutations are small, random changes introduced on individual variables to keep the diversity of the individuals, enabling new solutions to be explored. Tournament selection, in this case, means that selection of reproduction favors individuals with higher fitness scores. The population evolves towards sustainability configurations in many generations, where each iteration refines the Green Metaverse model based on fuzzy inference output.

The GA is implemented using DEAP (Distributed Evolutionary Algorithms in Python). The library is well established for evolutionary computation. The evaluation function merges the fuzzy control system with the genetic optimization process, in which from hereon, each single individual will be dynamically computed for its sustainability score. Some of the parameters such as population size (50%), crossover probability (0.5%), mutation probability (0.2%), and number of generations (30%) have been selected carefully to maintain a balance between exploration and exploitation in the solution space, thereby allowing the algorithm to effectively converge toward an optimal solution in reasonable computation time.

At the end of an evolution cycle in GA, the best individual is finally chosen as the optimal sustainability configuration of the Green Metaverse. This configuration is the ultimate metaverse policy for the greatest eco-friendliness, economic viability, and social inclusiveness possible under prescribed constraints. It considers the optimization process powered by GA, which will allow policymakers and stakeholders to make data-driven decisions, ensuring that both digital environments and the way they are governed continuously improve.

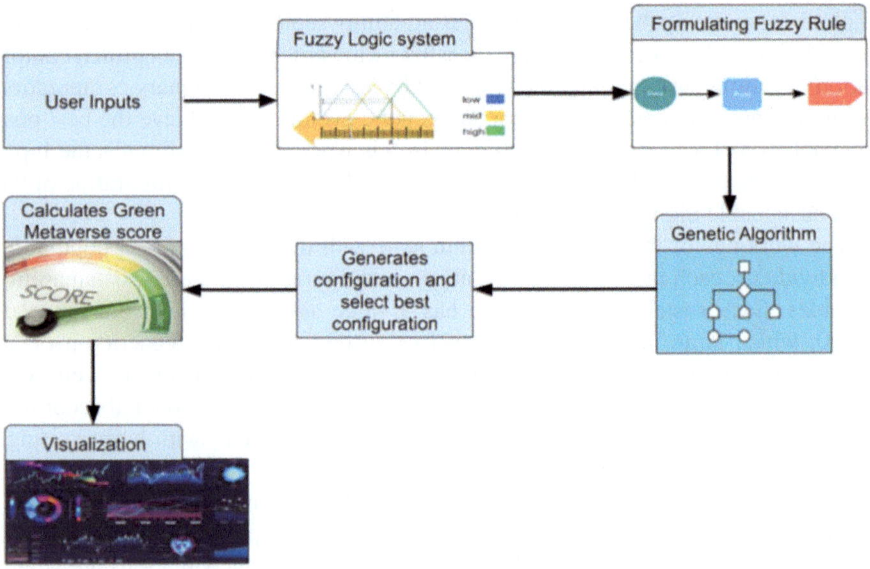

Fig. 1 Proposed architecture of hybrid fuzzy logic system with genetic algorithm

2.3 *Proposed Hybrid Fuzzy Logic System with Genetic Algorithm*

The proposed methodology, as shown in Fig. 1, combines fuzzy logic with genetic algorithms to develop an optimized governance framework for a sustainable Green Metaverse. It assesses the level of several environmental, economic, and social sustainability indicators using fuzzy inference systems (FIS) and then refines those input variables by genetic optimization. This, in turn, ensures minimum ecological footprints for digital ecosystems along with maximum efficacy and novelty and inclusivity.

2.3.1 Evolving the Fuzzy Logic System

Input variables representing the essential sustainability dimensions first need to be identified. There are 29 input variables which cover the following: environmental—carbon emissions, pollution levels, impact on biodiversity—economic—green investment, activities in the area of circular economy, adoption of regenerative design—social—community engagement, eco-justice, impact of climate change on mental health. Green Metaverse Index (GMI) is the consequent variable: the overall sustainability performance, measured on 0–100 scale.

From the three fuzzy membership functions:

- Low (poor sustainability performance)
- Medium (moderate sustainability performance)
- High (excellent sustainability performance).

The Green Metaverse Index values are:

- Low Sustainability (0–50)
- Medium Sustainability (25–75)
- High Sustainability (50–100).

With these classifications, crisp numerical values may be converted into linguistic variables for useful decision making.

2.3.2 Formulating Fuzzy Rules

The fuzzy inference system is built using if-then rules that determine how different sustainability factors influence the Green Metaverse Index. These are grouped into four categories:

Rule Set 1: Environmental and Ecological Impact This rule set ensures that low ecological footprint and high biodiversity protection contribute to a highly sustainable Green Metaverse. Some of the rules would include:

IF carbon footprint is low AND energy consumption is low AND water usage is low AND biodiversity impact is high, THEN Green Metaverse Index is high.

IF pollution levels are low AND soil degradation is low AND microplastic pollution is low, THEN Green Metaverse Index is high.

Rule Set 2: Sustainable Investment and Technological Innovation Rules in this category ensure that investments in eco-friendly innovation and circular economy practices enhance sustainability:

IF green investment is high AND smart infrastructure adoption is high AND waste management technologies are high, THEN Green Metaverse Index is high.

IF virtual ecosystem optimization is high AND operational efficiency is high AND regenerative design adoption is high, THEN Green Metaverse Index is high.

Rule Set 3: Social, Regulation, and Community Engagement This rule set establishes that in terms of sustainability, government policies, community participation, and fair trade are its cornerstones:

IF community engagement is high AND government regulations compliance is high AND green job creation is high, THEN the Green Metaverse Index is high.

IF eco-justice and fair trade policies are high AND indigenous knowledge utilization is high, THEN the Green Metaverse Index is high.

Rule Set 4: Resource Dependency and Risk Management This rule set helps identify risks, whether an unbalanced dependence on rare earth materials or otherwise, which threaten sustainability:

IF rare earth material dependency is low AND climate resilience is high AND greenwashing risk is low, THEN Green Metaverse Index is high.

IF consumer waste habits are moderate AND the longevity of sustainable practices is high, THEN Green Metaverse Index is high.

2.3.3 Integrating Genetic Algorithm for Optimization

After fuzzy rules have defined the sustainability evaluation model, a genetic algorithm (GA) has evaluated the values of input parameters so as to maximize sustainability. It brings about the steps in GA:

- Create initial random biodiversity configurations for environment sustainability.
- Derive the Green Metaverse Index by evaluating each configuration in fuzzy inference system.
- Select the best individuals.
- Crossover and mutation are performed for creating new population of sustainability configurations.
- Continue iterating for 30 generations or until the optimum sustainability configuration is determined.

2.3.4 Implementation and Decision-Making

Governments, policymakers, and metaverse builders may apply the optimized sustainability configurations to enact sustainable behavior. The model guarantees that the virtual environments built are:

- Environmentally responsible (low pollution, high biodiversity protection)
- Economically viable (high investment in circular economy and innovation)
- Socially inclusive (fair trade, well community driven, regulatory compliance)
- The fuzzy-GA hybrid model adopts a systematic data-driven procedure toward sustainable digital governance, underpinning that Green Metaverse policies are optimized, adaptive, and future-proof.

3 Results and Discussion

The best configuration obtained using the fuzzy system and genetic algorithm demonstrates a strong balance between environmental sustainability, economic feasibility, and social responsibility as can be observed from Fig. 2. We obtain the optimal green metaverse score as 82.95% leveraging the genetic algorithm.

The bar graph in Fig. 2 compares the input conditions to the Green Metaverse scores. Additional emphasis contextualizes trending data with respect to the four

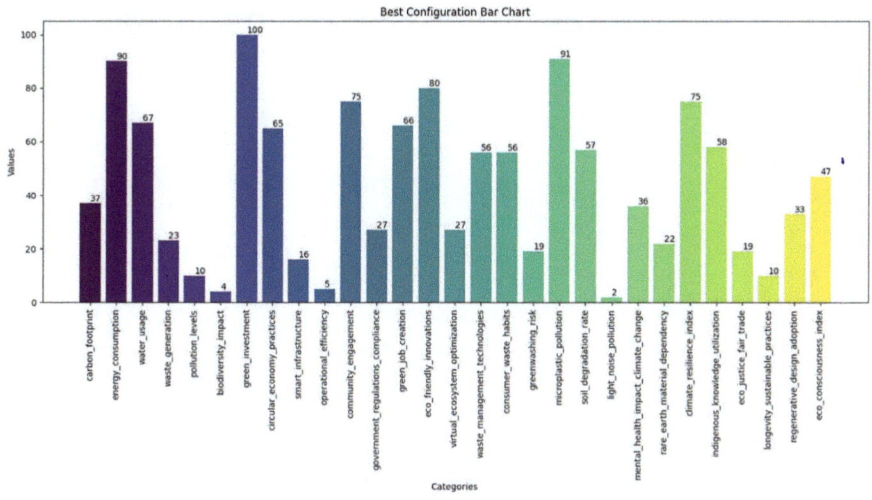

Fig. 2 Best configuration for green metaverse

rule-based classes. Within Environmental and Ecological Impact, a low carbon footprint (37%) and pollution (10%) are within the sustainability goals, while microplastics pollution (91%) and soil degradation (57%) violate the requirements for minimal ecological damage, destabilizing results towards high input. Sustainable Investment and Innovation has some strength in environmentally friendly investments (100%) and eco-innovations (80%), but was fairly poor in consistency due to circular economy practices (65%) and optimization of virtual ecosystems (27%) of weak system integration. High community involvement (75%) contrasted with low government compliance (27%) and eco-justice (19%) in Social and Regulatory Engagement, emphasizes regulatory misalignment that inflates variation. Resource Dependency and Risk Management, however, dimmed very low rare earth dependencies (22%), while consumer waste habits (56%) threaten it because they prevent progress.

Figure 3 represents a a box plot of the Green Metaverse scores according to input conditions: Low, Medium, and High. The scores are clustered around 30% with little deviation from each other during low input conditions. The next increase of the input results in higher scores of about 55% with slightly less clustering. Under High Input Conditions, the average scores reach at about 80%, but with a very wide spread indicating the much greater variability at higher inputs. This variability indicates high input conditions are less stable and may need refinement. The underlying trend has the notion that higher input conditions are generally associated with increased scores in a Green Metaverse; yet, wider spread at higher conditions implies demands for consistency in outcomes.

The bar chart in Fig. 4 displays the activation level of four fuzzy rules in the decision-making process: Environmental, Innovation, Social, and Resource Risk. Among them, the Environmental rule achieved the highest activation level at 0.85, stressing its importance in sustainability assessments. Following it, the Innovation

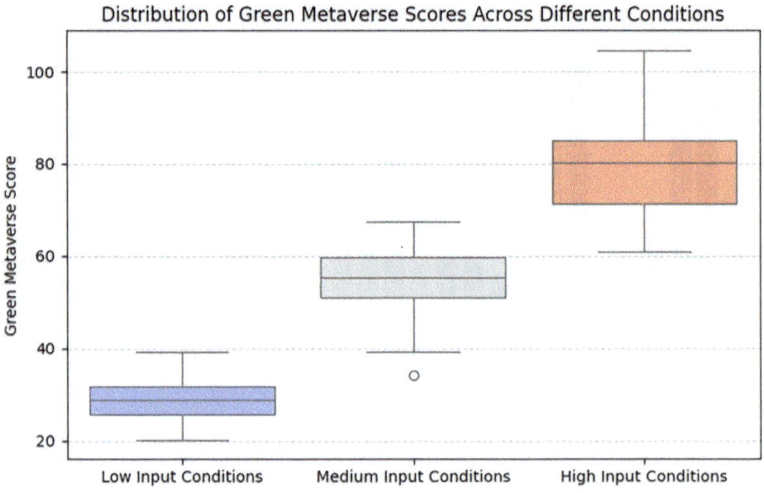

Fig. 3 Box plot distribution of Greenverse metascore

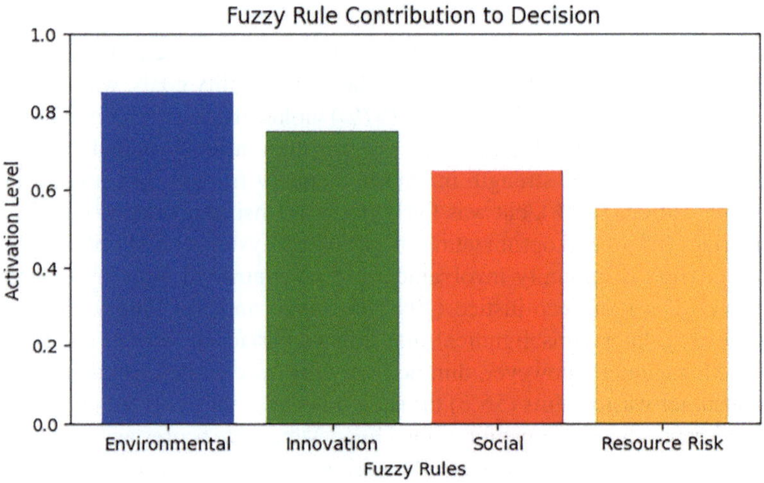

Fig. 4 Contribution of fuzzy rules

rule was recorded at an activation level of 0.75. This emphasizes the need to incorporate technology in advancing the Green Metaverse; the Social rule was subsequently mentioned with an activation level of 0.65, pointing out the impact of community involvement and social sustainability. Last but not least, the Resource Risk rule had the lowest activation level at 0.55, demonstrating the consideration for resource risks which is not much critical. It can be generally described that environmental concerns have taken a front seat but require a balanced treatment of innovation, social factors, and resource risk in sustainable governance.

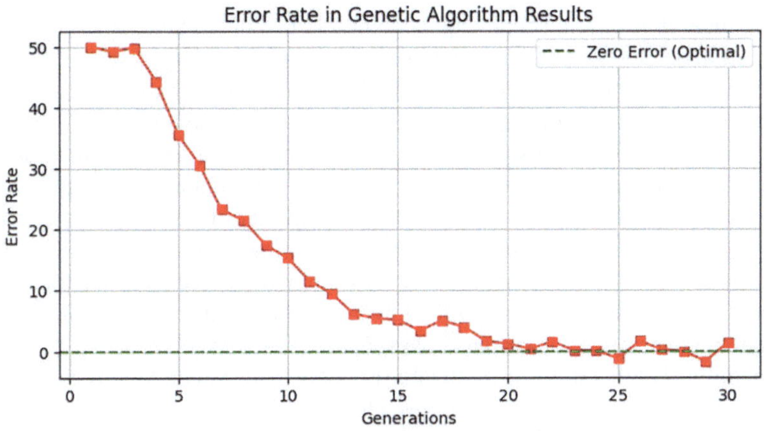

Fig. 5 Error rate in generating results over generations

Figure 5 details the evolvement of an error rate in a Genetic Algorithm (GA) over 30 generations. In the first generation, about half of the individuals in the population exhibited high error rates which were statistically heavily deviated from the optimal value meaning that a solution was not optimized.The error rates decrease almost steadily with time and increasing generations in the GA, indicating that the GAs enhance the accuracy of the solutions they provide with time. Stabilization of errors around generation number 20 occurs with values that are very close to zero. This means the algorithm would have converged to an almost optimal solution. The results affirm that Genetic Algorithms are indeed very good at refinement through iterative evolution. Error rates continue to fall, thus pointing to successful optimization efforts. On the contrary, if the error rates did remain elevated beyond the 20th generation, it would have most probably implied the need for tuning mutation rates, selection pressure, or crossover mechanisms for performance enhancement.

4 Conclusion and Future Scope

The integration of fuzzy logic and genetic algorithms (GA) in the evaluation and optimization of the Green Metaverse governance system has proven to be the most effective way to consider the complex and dynamic sustainability parameters. The findings indicate that increased input conditions generally would lead to an enhanced sustainability score, but there is considerable variability at extreme conditions, indicating a need for further stabilization. It is the capability of fuzzy logic to deal thoughtfully with uncertain and overlapping classifications that bestows value on its use in assessing a diversity of environmental, social and economic parameters. The activation of fuzzy rules shows that while there are manifold environmental issues, technological innovation, social engagement and resource management are equally important in achieving a balanced and sustainable Green Metaverse.

Fuzzy logic embeds and encapsulates the relationships that can be inductively gleaned through the relationships among environmental, economic, and social parameters. The rule activation also suggests that environmental issues, which are complemented by innovation within technology, active participation in social issues, and responsible management of resources played important role in achieving a Green Metaverse. Meanwhile, Genetic Algorithms systematically optimize their results over generations with decreasing error and improving decision accuracy. These insights confirm that combining with AI techniques can enhance decision-making in sustainable governance for virtual ecosystems.

The findings of this study are realizable not only in the Green Metaverse but also in real-world situations such as smart cities, climate governance, and circular economy frameworks. The ability of fuzzy logic and genetic algorithms to address complex, uncertain, and multidimensional decision-making will render these approaches to be applicable in an array of sustainability-related domains. Eventually, the subsequent combination of environment-friendly technologies, localized governance systems, and AI-assisted policy making will be the capacities bringing the world to bear on sustainable environmental operations, significant reduction of carbon footprints, and judicious resource utilization. As technology is advanced, the Green Metaverse would become a playground for testing sustainability strategies in paradigm futures and feeding them into the actual and virtual policymaking realms. The triad of AI, sustainability, and governance promises to open a window of opportunity to deliver a resilient, adaptive, and eco-conscious digital and physical world.

References

1. Egliston, B., Carter, M., Clark, K.E.: Who will govern the metaverse? Examining governance initiatives for extended reality (XR) technologies. New Media Soc., 14614448231226172 (2024)
2. Uzun, M.M.: Metaverse governance. In: Metaverse: Technologies, Opportunities and Threats, pp. 231–244. Springer (2023)
3. Lnenicka, M., Rizun, N., Alexopoulos, C., Janssen, M.: Government in the metaverse: requirements and suitability for providing digital public services. Technol. Forecast. Soc. Chang. **203**, 123346 (2024)
4. Jobe, P.S., Yilmaz, M., Ilgın, H.E.: A roadmap for a metaverse-based digital governance: a case of the Gambia. PLoS ONE **19**(12), 0314388 (2024)
5. Rajguru, K., Brüggemann, P.: Sustainability meets metaverse: a conceptual framework of sustainable dimensions of the metaverse. J. Consum. Behav. **23**(5), 2720–2729 (2024)
6. De Giovanni, P.: Sustainability of the metaverse: a transition to industry 5.0. Sustainability **15**(7), 6079 (2023)
7. Tu, X., Silva, B.: Are we ready for the metaverse? Implications, legal landscape, and recommendations for responsible development. Digit. Soc. **4**(1), 9 (2025)
8. Hasan, R.: Governance and the metaverse: a systematic scoping review. Available at SSRN 4581731 (2023)
9. Effing, R.: Will the metaverse be out of control? Addressing the ethical and governance implications of a developing virtual society. Digit. Gov.: Res. Pract. **5**(3), 1–15 (2024)

10. Wan, J., Ho, A.T.-K.: Deciphering the paradigm shift of governance in the metaverse: a literature review. In: Proceedings of the 25th Annual International Conference on Digital Government Research, pp. 734–739 (2024)
11. Zhang, S., Lim, W.Y.B., Ng, W.C., Xiong, Z., Niyato, D., Shen, X.S., Miao, C.: Toward green metaverse networking: technologies, advancements, and future directions. IEEE Netw. **37**(5), 223–232 (2023)
12. Kshetri, N., Dwivedi, Y.K., Janssen, M.: Metaverse for advancing government: prospects, challenges and a research agenda. Gov. Inf. Q. **41**(2), 101931 (2024)
13. Maden, A., Yücenur, G.N.: Evaluation of sustainable metaverse characteristics using scenario-based fuzzy cognitive map. Comput. Hum. Behav. **152**, 108090 (2024)
14. Alam, T., Qamar, S., Dixit, A., Benaida, M.: Genetic algorithm: reviews, implementations, and applications. arXiv preprint arXiv:2007.12673 (2020)
15. Shah, D., Chatterjee, S.: An intelligent programmed genetic algorithm with advanced deterministic diversity creating operator using objective surface visualization. Evol. Intel. **13**(4), 705–723 (2020)

Consumer Behavior and Sustainability in Virtual Worlds

Dusari Sai Teja, Chevella Aravind Reddy, and Pritam Khan

Abstract Segmenting customers and predicting loyalty has become a powerful way for companies to work on customer retention and marketing optimization. This study goes through a hybrid model, which is a combination of clustering, fuzzy logic, and machine learning, in calculating customer loyalty scores with a focus on their behavior and sustainability in virtual worlds. Customer data are therefore preprocessed in terms of methodology: categorical feature encoding, numerical variable normalization, and treatment of missing values. In the customer segmentation phase, K-Means clustering is performed using purchasing patterns and behavioral characteristics. Next, the fuzzy logic system scores loyalty dynamically based on the behavior incidence and features specific to the segment, thus providing flexibility and interpretability over core threshold-based methods. To predict loyalty scores, the Random Forest Regressor is trained with features including purchase amount, purchase frequency, customer reviews, and transaction history. The model validation includes visual analysis of the cluster distributions, ranking of feature importance, and histograms of loyalty scores. The results prove that the combined methods of clustering, fuzzy logic, and machine learning enhance the accuracy and interpretability in customer segmentation and loyalty prediction. Data-driven methods, in this case, find a great applicability in intelligent consumer analytics with regard to virtual set-ups, where consumer behavior and sustainability are of immense significance. Businesses can leverage these findings to craft targeted marketing campaigns, stimulate customer dedication, and encourage habits of sustainable consumption.

Keywords Fuzzy logic · K-means clustering · Random forest regressor · Loyalty prediction

Dusari Sai Teja, Chevella Aravind Reddy are contributed equally to this work and are co-first authors.

D. S. Teja · C. A. Reddy · P. Khan (✉)
Department of Computational Intelligence, SRM Institute of Science and Technology, Kattankulathur, Tamil Nadu, India
e-mail: pritamk249@gmail.com

S. Kautish et al. (eds.), *Green Metaverse: Fuzzy Logic Approaches to Sustainable Virtual Worlds*, Information Systems Engineering and Management 74,
https://doi.org/10.1007/978-3-032-08183-4_10

1 Introduction

Digital commerce has evolved rapidly, and its fast evolution has greatly impacted the consumers buying behavior. In moving through this disruptive scenario, adoption of more advanced analytical mechanisms is key to understanding shopping patterns, predicting customer loyalty, and improving customer engagement. Typically, traditional segmentation models rely on demographic attributes in the segmentation process. These models do not hold up well in accommodating the dynamic qualities of consumer preferences. This limitation becomes a strong justification for the importance of data-driven methodologies that can bring forth underlying trends, accurately forecast customer behavior, and achieve lionization of the loyalty program. The proposed framework presented in this paper aims to develop and build new customer segments and loyalty prediction processes within the recently emerging framework of advanced machine-learning techniques, namely clustering algorithms, fuzzy modeling, and predictive modeling.

Hence, e-platforms absorb gigantic amounts of behavioral and transactional data, such as purchase frequency, spending habits, and product preferences.The multidimensionality of such datasets makes the extraction of insights computationally soo challenging that it requires very advanced analytical modeling. For example, this study adopts the methods namely K-Means clustering for customer segmentation, fuzzy logic for the dynamic assignment of loyalty score, and Random forest for predictive modeling. The hybridization brings segmentation accuracy validation, loyalty assessment refinement, and enhances the forecasting capabilities that lead to an integrated view of consumer behavior.

Customer segmentation is the key of business strategy planning, allowing companies to optimize engagement activities. However, the traditional models for market segmentation lack even a modicum of dynamism. Quite the opposite is true for machine-learning techniques that can dynamically assign customers into segments based on attributes in flux. Based on actual behavioral similarities, K-means clustering allows companies to develop targets for customer groups. In this way, this is an important advance in the more traditional segmentation methods with a value-added, flexible, expandable approach that may accurately and efficiently respond to changing customer behavior.

Assessing customer loyalty is an another key analytics impacting retention and profitability. Most traditional frameworks measuring loyalty rely on rigid tier-based systems and hence do not account for the varied nature of consumer interactions. This research proposes a fuzzy logic-based loyalty scoring system, which assesses customers across a wide range instead of multiple cutoff points. Such an approach, while clustering customers into high, medium, or low loyalty categories, will enable businesses to adopt retention tactics that fit the customer experience.

The introduced hybrid framework combines clustering, fuzzy logic, and predictive modeling to develop an exhaustive customer analytics solution. In contrast with the inherent nature of static segmentation methods, the proposed approach can be employed in businesses to detect changing in customer segments and their maturing shopping characteristics. Keeping that in mind, the fuzzy logic-based loyalty scoring mechanism delivers a refined and subtle view of customer engagement, which suggests that loyalty is fostered incrementally over time. Furthermore, predictive modeling captures changes in future loyalty, and businesses can modify marketing strategies according to such changes.

The research significantly contributes to this emerging domain of intelligent retail management and improvement of sustainability in virtual worlds, where the customer engagement strategies are data-driven-dependent. An unsupervised learning mechanism does the segmentation of customers, fuzzy logic takes care of the complex evaluation, and supervised learning deals with predictions, thus rendering a sufficiently scalable and accurate framework for consumer analytics. In this context, insights derived through machine learning can greatly enhance customer retention, increase profitability, and develop marketing strategies that are optimal both in physical and virtual commerce scenarios. The line of future research may explore developments in clustering algorithms, enhanced fuzzy systems, and deep learning methods to further refine loyalty prediction models while also contributing to sustain the digital marketplace.

1.1 Related Work

Recent explorations in sentiment analysis and cross-lingual challenges are provided in [1, 2]. The proposed approach of [1] is a rule-based fuzzy logic system to self-supervised sentiment classification of Hindi texts that brings three lexicons for dealing with common ambiguities in morphologically rich languages outperforming traditional supervised systems. Reference [2] analyzes fuzzy data from social media leveraging fuzzy neural networks (FNNs) for accurate marketing by integrating minor behavioral frequencies, content topics, and emotional states. In these contexts, it is said that fuzzy systems manage the well known linguistic complexities and can work even with noisy data. Reference [1] emphasizes cross-lingual sense disambiguation while [2] appears to be more inclined towards behavioral insights for targeted strategies.

Customer churn predictions can be found in [3–5]. For example, [3] shows some clustering (K-means, X-means) resulting in classification ensembles which achieved 94.7% accuracy across telecom datasets through cluster-driven feature engineering. Reference [4] employs SVM controlled by Coyote Optimization with Fuzzy Z-numbers as a basis to establish e-commerce churn decisions to capture the qualitative factors concerning the preferences of the customers. Reference[5] modifies Fuzzy C-Means to predict students' churn, based on a payment history and enrollment duration as proof of the discipline's versatility across various sectors. These models present hybrid methods, weighed between quantitative accuracy and qualitative insights, tailored for the telecom, e-commerce, and educational contexts.

References [6, 7] optimize customer behavior prediction. Reference [6] demonstrates the use of a hybrid model in which clustering is performed by fuzzy deep belief network while prediction is done using recurrent neural network, thus integrating butterfly optimization at the final stage to minimize the error. Stacking ensemble classification (KNNSGD) was achieved with an accuracy of 92.42% in [7]. In [8] a marketing-oriented algorithm consisting of a genetic algorithm designed to wrap around decision trees predicts with an accuracy of 90% from correlating demographics with decision-making variables. This collection of works demonstrates that it pays to combine feature engineering with ensemble or evolutionary approaches for greater robustness in the prediction for heterogeneous behavioral datasets.

In [9], a Mamdani-type fuzzy system is designed to assess the attractiveness of bank customers using CRM data based mainly on metrics such as wallet share. In [10], a fuzzy commitment matrix is adopted by a conventional FMEA application by eliminating uncertainty in data regarding customer groupings. In [11], fuzzy decision support systems (FDSS) were created to customize tourism products for older tourists in weighing the balance of accessibility and cultural preferences. E-Service Loyalty is described in [12] which connects quality of system and retention in fuzzy logic. These references highlight the extent of fuzzy systems audacity for making qualitative measures and actions in banking, tourism and e-commerce fields.

References [13, 14] clearly differ with respect to the analytical models; both models undertake dynamic multi-criteria analysis. According to [13] , the improvement in the incident resolution process involves setting a common set of RFID parameters (Recency, Frequency) with AHP-weighted linguistic evaluations for prioritization in real time of customer service requests. The work in [14] pools together RFM models with GIS-based data mining towards geo-demographic segmentation, which gives emphasis to visualization and clustering to create spatial marketing. In both cases, the frameworks lend themselves to fine context-aware strategies; for example, [13] takes the streamlining of a contact center, while [14] elaborates marketing through a spatial and behavioral integrated insights.

Adaptive Models are also key to the optimization techniques provided by [4, 6, 8]. Butterfly optimization method is applied in [6] for the deep learning parameter optimization; [4] tunes SVM for churn prediction using the Coyote Optimization Algorithm, while [7] applies genetic algorithms to increase the accuracy of decision tree classifiers, specifically in consumer classification. These three cases of studies have forwarded arguments in favor for the metaheuristic algorithms addressing the issues of local optima, overfitting, and enhancing generalizability in high-dimensional datasets. Optimizing the fuzzy and deep learning with adaptive and scalable solutions towards customer analytics justifies the development for adaptive and scalable solutions.

Comparative research in [3, 7, 15] is about methodologies for benchmarking in different domains where [3] compares state-of-the-art churn models with hybrid clustering-classification algorithms, [7] tests 11 classifier models and stacks set of classifiers on purchase behavior. 'Bank data have been transformed into feature engineering' in [15], which revealed how the structure of a dataset affects prediction accuracy. These papers make a domain-specific argument because of outside sectors, such as ensemble learning or feature transformation, may perform differently, which ultimately helps practitioners make contextually appropriate choices in tools, methodologies, and metrics for insights and retention strategies about customers.

In marketing and business intelligence application under [2, 11, 14] uses fuzzy neural networks to study social interactions in social media for targeted marketing to give an understanding of customer behaviors to improve their targeting. [11] creates packages of travel products individualized to the traveling experiences based on the likes and dislikes of elderly customers using fuzzy decision support systems. According to [14], spatial marketing is based on the integration of a GIS, that is geographic information systems, with customer segmentation through RFM clustering. This chapter presents the use of data to create wantable geography around consumers. These are examples of how fuzzy models assist in personalizing decision-making and strategizing relating to customers.

There is a sentiment-based decision-making approach in [1, 10, 12]. In [1], the work is based on sentiment classification for Hindi text using fuzzy logic, which still enables accurate working in morphologically complicated languages. [10] implements a fuzzy-based loyalty evaluation model to measure the level of satisfaction as a dynamic process which adds more conditions to the already existing retention strategies. On the other hand, [12] integrates fuzzy logic with e-service loyalty modeling to show how system quality affects the way customers decide on retention. All these contributions demonstrate the ability of fuzzy techniques to handle qualitative sentiment data in the field of consumer analytics.

The literature works reviewed, depict the increasing role of fuzzy logic and hybrid AI techniques in customer analytics, sentiment analysis, and decision-making. Whether dealing with customer churn prediction, behavioral modeling, loyalty assessment, or spatial marketing, these approaches are expected to improve predictive accuracy while managing uncertainty with respect to complex datasets. As fuzzy systems are increasingly combined with deep learning, optimization algorithms, and decision support models, intelligent customer analytics will continue to be revolutionized by these methods worldwide.

1.2 Contributions

The major contributions of this chapter are enumerated as follows:

1. K-Means clustering is combined with fuzzy logic in order to achieve a flexible customer loyalty assessment, enabling dynamic changes in shifting consumer behavior patterns.
2. The accuracy of predicting loyalty is enhanced by introducing the advantages of multiple decision trees through the Random Forest Regressor that captures the complex non-linear relationships among factors such as age, purchase quantity, review ratings, and purchase frequency.

3. Harnessing inputs through the Min-Max process and the label encoding to the categorical variables by converting them into numbers improves clustering accuracy and subsequently predictive model performance, while frequency mapping of acquisitions further fine-tunes the per-territory-dimension.
4. The results show graphical representations of feature importance, analyze scatter plots for relationships established, and also clustering distribution graphs that enable departments to derive insights about customer behavior. This will help in fine-tuning rewards and marketing strategies towards customer retention.

2 Customer Behavior Analysis Using Clustering, Fuzzy Logic and Predictive Modeling

Customer behavior analysis can be carried out leveraging techniques such as clustering, predictive modeling, and fuzzy logic to understand the patterns in consumer data. Similar buying patterns of customers can be identified by the clustering methods while regression-based prediction models can be used to anticipate the future purchase probability and churn risk. However, traditional models might not capture nuanced behavior as customer preferences tend to involve uncertainty. Therefore, fuzzy logic, which is a mathematical approach of dealing with imprecise and uncertain states to resemble people's thinking, can be incorporated to improve managerial decision making. The combination of these approaches will not only help in personalization of businesses and optimization of market strategies, but also ensure better customer engagement and experience.

2.1 Clustering Methods

In machine learning, clustering is a fundamental unsupervised learning technique applied for organizing data points depending on shared traits. Unlike classification, clustering methods find trends in labels free data. Clustering is a rather powerful tool for exploratory data analysis, consumer segmentation, anomaly detection, and recommendation systems. One of most often used clustering method is K-Means clustering. This approach aims to reduce the variance inside every one of the K designated clusters. Starting with the choice of first cluster centres, the procedure runs in iterative cycles. Each data point then is allocated to the closest centre, and the centres are recalculated in response to the revised assignments. The cycle keeps on until it reaches a steady condition. K-Means is effective in processing big data even if it depends on the choice of initial centres and requires a predefined number of clusters.

Hierarchical clustering is a divisive (top-down) or an agglomerative (bottom-up) way of establishing a hierarchy of clusters. Agglomerative clustering methodically combines each data point, starting with each as an independent cluster, producing a hierarchical representation also known as a dendrogram. Starting with a single cluster spanning all the data points, divisive clustering methodically divides it into ever smaller clusters via a recursive process. Since hierarchical clustering runs free of a predefined number of clusters, it is helpful in cases when the ideal number of groups is unknown. In the framework of big datasets, scalability can be constrained and computing expenses could be somewhat high. Increasingly sophisticated clustering is displayed by density-based spatial clustering of applications with noise (DBSCAN). Unlike K-Means and hierarchical clustering techniques, DBSCAN does not call for previous cluster specification. Rather, it views outliers as noise and finds dense data point clusters and then aggregate. Density- varying datasets or irregularly shaped clusters call especially for DBSCAN. Processing high-dimensional data might present difficulties. Therefore, it could need exact parameter tweaking for features such as the minimum points per cluster and neighbourhood radius. Gaussian Mixture Models (GMMs) are among alternative clustering methods that define data as a mix of Gaussian distributions, therefore offering a probabilistic frame-work. Unlike K-Means, which supposes rigid separation between groups, Gaussian Mixture Models (GMMs) provide adaptability and may reasonably depict overlapping clusters. Spectral clustering is another method using graph theory to find clusters in data by means of eigenvalues from similarity matrices. For datasets with intricate structures especially, this method is quite successful.

The type of data, the quantity of clusters, and computational limitations all influence the selection of the suitable clustering technique. K-Means is efficient and widely used; however, hierarchical clustering offers enhanced interpretability. DBSCAN demonstrates robustness against outliers, while Gaussian Mixture Models (GMMs) provide probabilistic flexibility. Clustering methods facilitate the identification of significant trends within large datasets, allowing organizations to optimize resource allocation, identify fraudulent activities, and gain insights from consumer demographics in practical applications.

2.2 *Fuzzy Logic Integration*

Fuzzy logic is a computer technique designed to manage uncertainty and imprecision, therefore enabling robots to reproduce human decision-making. While fuzzy logic allowed truth values to run between 0 and 1, standard binary logic regards inputs as either true (1) or false (0). In practical aspects, this makes data rather valuable when it is confusing, vague, or lacking precision. Fuzzy logic lets one admit partial facts, hence enabling more adaptable and human-like judgments. Fuzzy systems are frameworks designed to control uncertain or imprecise input by use of fuzzy logic. It comprises a collection of guidelines defining interactions between the intended results and the input factors. Unlike exact numerical value, fuzzy systems represent several types of data through textual language including "low", "medium", and "high". Fields such as artificial intelligence, control systems, robotics, medical diagnostics, and consumer behavior analysis find various applications for this method.

Three fundamental phases characterizing the functioning of a fuzzy system are fuzzification, inference, and defuzzification. Membership functions translate precise numerical inputs into fuzzy values in the fuzzification stage. These properties indicate the degree of membership of an input to a given fuzzy set. Under a system measuring customer loyalty, for example, someone with a purchase frequency of 15 times a month would fall between frequent buyer and moderate buyer categories rather than be categorically assigned either. To calculate the outcome, the system uses IFTHEN rules in the inference engine to appraise vague values. Regulations are design by experts or pulled from data patterns by the system. For example: A consumer who purchases products frequently as well as get strong review scores will earn a high loyalty rating. The inference process combines many conditions producing a meaningful result using logical operators such as AND, OR, and NOT.

In defuzzification, the vague output is transformed into a specific numeric value. Since the transformation of fuzzy outputs to numerical values during defuzzification is vital for decisions based on specific data points in the real world. Among the approaches used for defuzzification are the centroid technique, maximum membership principle, and weighted average computation. Defuzzied, 75 represents customer loyalty when the fuzzy model shows a loyalty score range of 60–90. Analyzing consumer behavior uses fuzzy logic quite a lot; it also plays a major role in medical decision support systems, as well as industrial automation and advanced self-driving vehicle technologies. Great at processing imprecise data, fuzzy logic becomes a potent tool in solving challenging issues that traditional mathematical models cannot address. Fuzzy logic provides more advanced and flexible decision making systems by including human like reasoning into computational models.

2.3 Proposed Hybrid K-Means-Fuzzy Logic and Predictive Modeling Approach Using Random Forest Regressor

The proposed approach builds a robust system for customer segmentation and loyalty forecasting into the green metaverse by combining fuzzy logic, cluster analysis, and virtual world simulations. Using machine learning, it groups customers according to buying behavior and predicts loyalty scores while analyzing customer interaction with the virtual world. The main components of data processing include clustering, fuzzy logic- based scoring and predictive modeling, together with lifestyle immersion experiences to smoothen engagement strategies. In this way, real and virtual behaviors are fused to provide dynamic segmentation, personalized reward systems, and optimization by artificial intelligence towards customer retention in the digital sustainable ecosystem.

The first stage of the process, data pre-processing, includes handling missing values, encoding categorical elements, and standardizing numerical ones. Using a Label Encoding technique, other classifying variables including gender, product bought, and payment mode is transformed into numerical data points. Furthermore, an already set mapping method changes the frequency of acquisitions into counts. Rising model clustering and predictive accuracy results from next Min-Max scaling ensures that every numerical trait falls within a standardized range.

K-Means clustering assists in customer segmentation once early measures have been taken. Based on similarity in purchasing behavior, K Means is an unsupervised learning technique that splits the data series into a set number of clusters. Experimental selection is used to choose three groups to best reflect various consumer categories, and the ideal number of clusters is obtained. Each group mirrors consumer with different buying habits: high, middle, low- that remain stable. Once grouping is done, consumers are moved to their assigned clusters, providing a systematic grasp of many forms of purchasing behavior. Figure 1 shows the workflow of the Hybrid K-Means Clustering with fuzzy system for loyalty score followed by a Random Forest Regressor for Predictive Modeling.

Customer loyalty is gauged using fuzzy logic. Fuzzy logic makes a more human-like assessment of loyalty possible since it varies with many factors including review scores, order frequency, and number of purchases. A set of imprecise rules provides loyalty scores based on the cluster membership values. Consumers belonging to clusters with less values have low loyalty scores, and are hence distinguished from

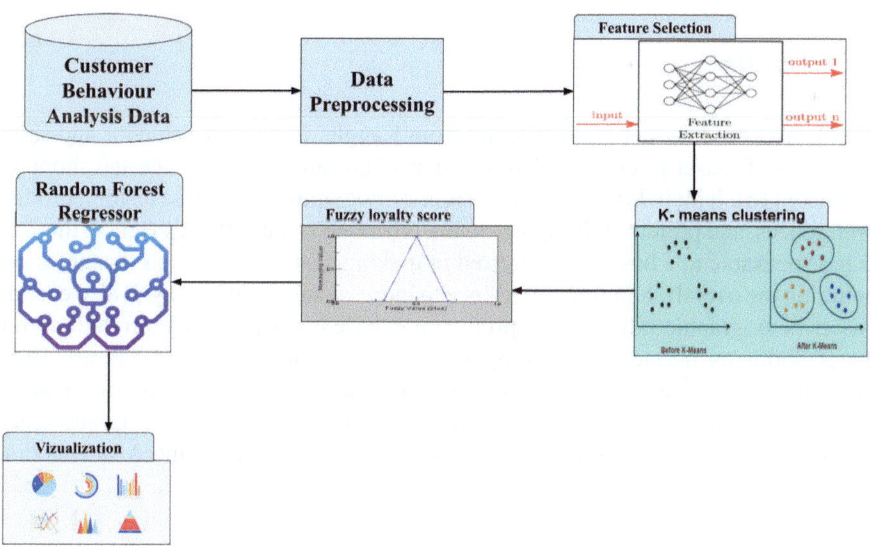

Fig. 1 Architecture of hybrid K-means-fuzzy logic and predictive modeling approach using random forest regressor

the ones belonging to clusters with higher values. This approach ensures an adaptable and accurate representation of consumer loyalty instead of relying on set boundaries.

Once loyalty ratings are given, a Random Forest Regressor is trained to forecast loyalty ratings based on important characteristics such as age, purchase volume, review rating, past purchases and buying frequency. Random Forest Regressor is an ensemble that trains several decision trees to get accurate predictions and avoid overfitting. Random forests are often used in regression tasks because they are good at fitting non-linear sets of data, can handle missing data, and do very well with high dimensional datasets. Training and testing are conducted to assess the model's predictive performance. In this chapter, the trained model is tested with a consumer behavior dataset to determine the customer loyalty prediction capability. Performance metrics such as Mean Squared Error (MSE), Mean Absolute Error (MAE), and R-Squared Score support the analysis of the model's performance.

The performance metrics are analyzed in terms of feature importance graphs, scatter plots, and cluster distribution plot. These graphics provide insights into customer segmentation, the power of fuzzy logic on customer loyalty scoring, and the top predictors of customer loyalty. The proposed approach offers a strong and data driven framework for anticipating and analyzing consumer behavior by means of cluster, fuzzy logic, and predictive model integration for companies looking to improve client retention and marketing strategies.

3 Results and Discussion

3.1 Dataset

The dataset used for our study is taken from Kaggle. Consumer behavior and shopping habits Dataset gives a detailed overview of consumer preferences and shopping behavior [16]. It includes information about demographics, purchase history, product preferences, and preferred shopping channels (online versus offline). Hence, this data is indispensable to a business interested in making customer-oriented strategies and bettering the overall shopping experience to increase sales and loyalty. It also encompasses demographic information, product purchase history, specific product preferences, shopping frequency, and shopping behavior (both online and offline). These data can thus allow analysts and researchers to study the consumer decision-making process in detail in order for businesses to devise relevant marketing strategies, improve their product offerings, and maintain better overall customerÂ satisfaction.

3.2 Results of the Hybrid K-Means-Fuzzy Logic and Predictive Modeling Using Random Forest Regressor

The horizontal bar chart in Fig. 2 display the different features and their importance in predicting customer loyalty. The importance score is on the x-axis, and features on the y-axis. By far, "Frequency of Purchases" is the most outstanding factor and blows away any other variable assessed. This means that customers who tend to have more frequent purchases are significantly more loyal. Next comes "Purchase Amount (USD)" indicating that spending behavior also matters a lot. Other features mentioned: "Previous Purchases", "Age", and "Review Rating" - do not help in prediction at all and thus, hint that past purchases and demographic factors do not have as much sway on loyalty when compared to frequency of purchase and value of purchase. By identifying these factors, businesses can better tailor loyalty schemes and marketing strategies, thus winning over their high-value customers. The bar chart in Fig. 3 demonstrates the customer segmentation done using the K-Means clustering algorithm in which three clusters are represented almost equally with nearly equal counts; this indicates a balanced and fair distribution of data points. Almost equal sizes of clusters indicate a well-behaved dataset without prevailing outliers or dominant clusters-the segmentation can be termed successful regarding analysis. Such a balance of distribution does ensure that all clusters will give worth while insights toward the identification of totally different customer segments. The graphical representation successfully depicts the clustering outcomes, and the characteristics of the clusters deserve special attention in order to derive more insights into customer behavior and preferences.

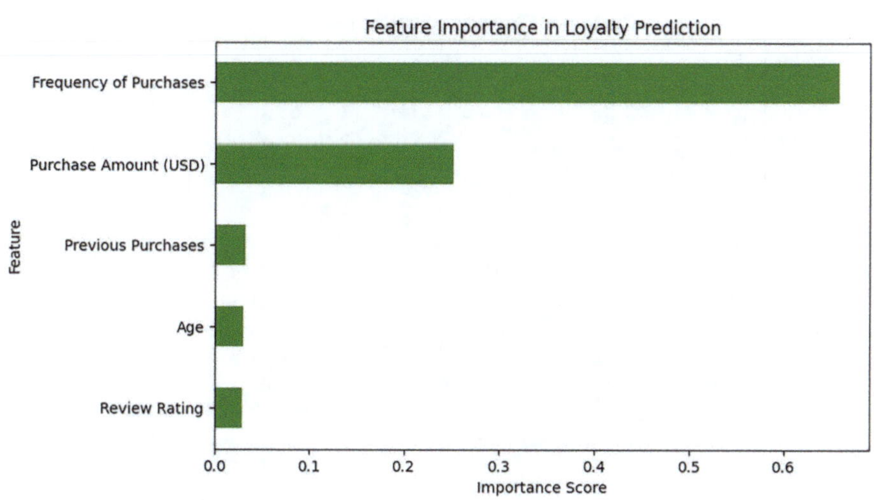

Fig. 2 Feature importance

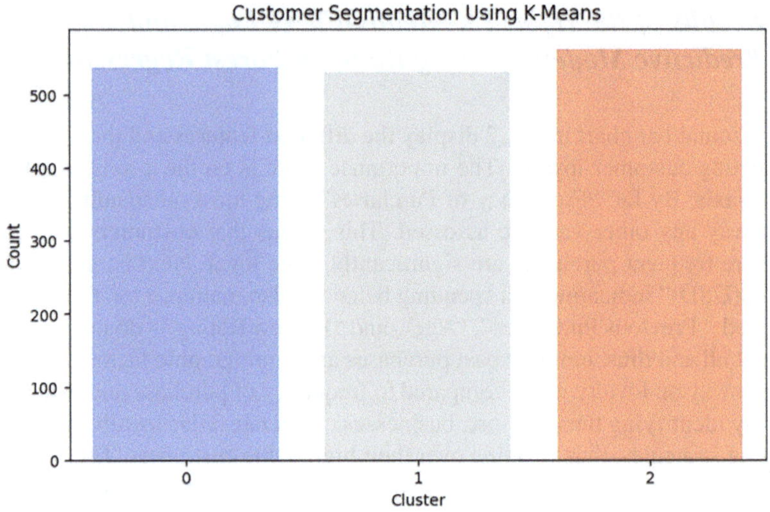

Fig. 3 Different clusters of customers

Figure 4 provides customer clustering through two scaled features in a scatterplot. The data point in their respective clusters, indicating the success of K-Means division into well-formed three clusters. Therefore, it seems that these clusters are distributed somewhat vertically-the lower part of the plot is primarily occupied by one cluster, while the upper region shows a combination of the other pattern of separation that

Fig. 4 Scatterplot of customer clustering

appears to be depicted by Feature 2. The colors even denote what is an "across-the-board" feature: the clusters against the gradient displayed in the legend signify the success of the divide and conquer strategy. This visualization therefore provides insights on how the features affect segmentation, which can assist in recognizing patterns and in data-driven decisions.

The loyalty score and sustainability scores are presented in the heatmap in Fig. 5 revealing a fairly strong positive relationship that is indicated through the correlation coefficient of 0.89. This indicates that customers who have high loyalty scores tend to have better sustainability scores, as these two scores may relate to the customers commitment to sustainable practices. The brilliant red diagonal shows how perfectly each variable is connected to itself, while the blue off-diagonal values emphasize just how much stronger the relationship is between the two metrics. Such a possibility can be a great insight for businesses that want to bring customers into a retention model based on sustainability-driven initiatives.

The Loyalty score distribution representing the frequencies of the customers' loyalty score results can be observed from Fig. 6. An overlaid density plot (in blue) gives the overall shape of the distribution with very little fluctuation and no sign of extreme skewness. There are peaks around lower and higher loyalty scores indicating that customers are clustered around certain score ranges as opposed to spread evenly along with it. Such an understanding would allow businesses to draw conclusions on customer loyalty patterns concerning high-loyalty versus lower engagement segments.

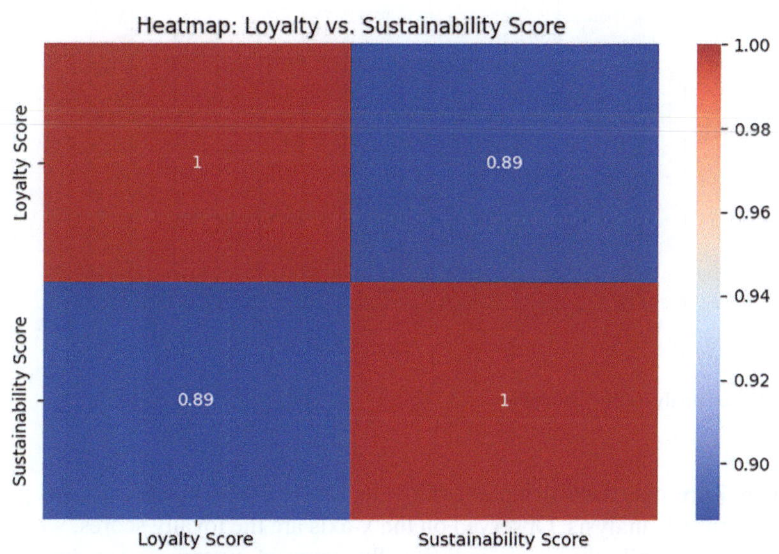

Fig. 5 Heatmap between loyality score and sustainability score

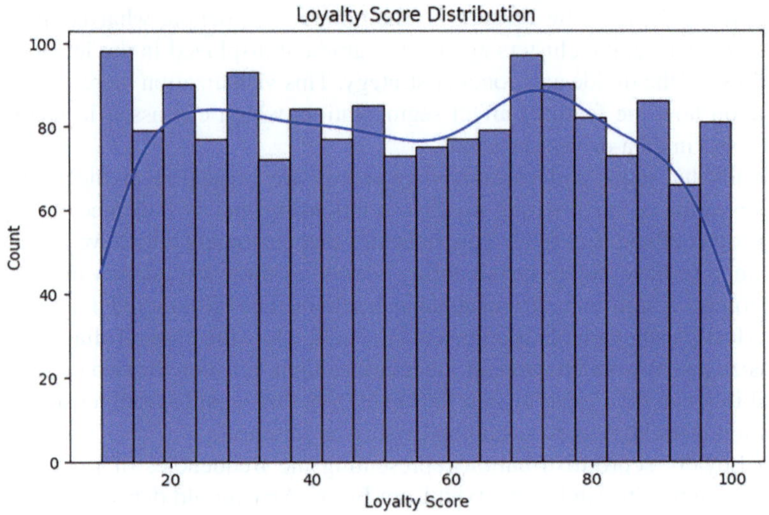

Fig. 6 Loyalty distribution curve

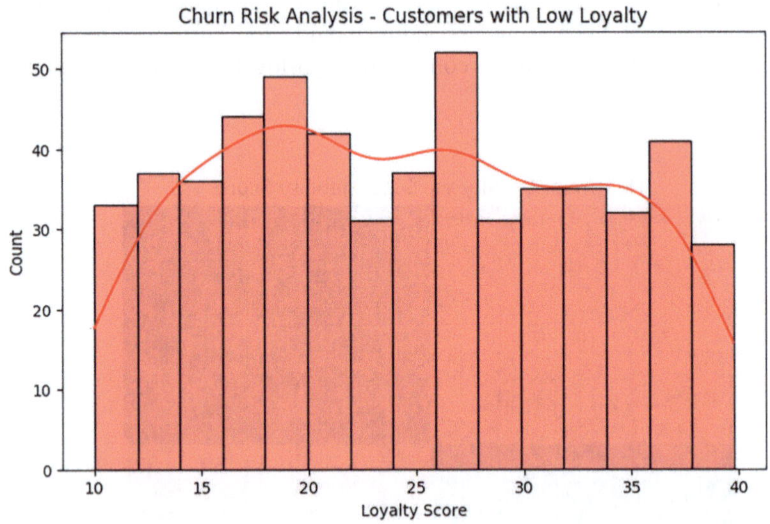

Fig. 7 Chrun analysis of customers

Figure 7 presents the distribution of loyalty scores of less loyal customers pertaining to churn risk analysis. Observed on the x-axis are the loyalty scores, which vary from 10 to 40, while the y-axis indicates the count of customers and the red KDE (Kernel Density Estimation) line smoothens the distribution of the data.

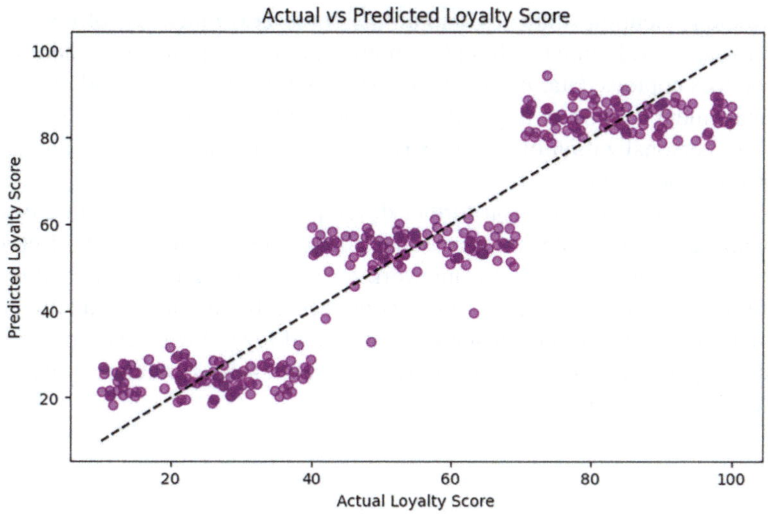

Fig. 8 Actual versus predicted loyality score

It is easy to compare between actual and predicted loyalty scores with scatter plot shown in Fig. 8. Actual scores are shown on the x-axis, predicted scores are shown on the y-axis. At the end, we can see the scattering data points exhibiting purple along the dashed diagonal line; this is the ideal condition where predicted values equal actual values. It's proximity towards that line tells us that the model indeed performs quite well in estimating the loyalty score as the deviation is small for most of the points. Some deviations are actually included in the middle range scores. It is a potential area for model tuning.

4 Conclusion and Future Scope

Timely segmentation of customers and forecasting their loyalty through clustering techniques, fuzzy logic, and machine learning applications will extend into virtual worlds and metaverse environments. The K-Means clustering examines customers purchasing patterns, thereby classifying them more advantageously. Fuzzy logic allows an assessment of the loyalty index more in keeping with the brochure's marketing intent by not forcing a strict binary classification scheme. Performance in predicted loyalty from the Random Forest Regressor is really improved by an excellent performance of Mean Absolute Error: 7.73; Mean Squared Error: 82.30; R^2 Score: 0.87. By this, it acquires further insight into data driven marketing through visual analysis like cluster distribution display, feature importance ranking, and loyalty histogram. Such a framework allows companies to personalize engagement, improve their retention strategies, and ultimately leverage their use of loyalty forecasting toward developing the digital economy in those environments where simulation produces copious data. AI-driven analytics will facilitate this kind of personalized engagement defined by targeted promotions, optimization of rewards within the game, and personalization of user experience to ensure maximum satisfaction and lifetime customer value.

Future research will focus on further developing best practices with clustering, developing fuzzy logic rules to increase interpretability, and integrating deep learning refinements into the current framework for superior segmentation and loyalty-prediction techniques. As industry players increasingly migrate into the immersive digital landscape, this approach shows how AI and fuzzy logic serve to strengthen customer engagement, loyalty, and long-term value across traditional and virtual ecosystems alike.

References

1. Chauhan S, Shet JP, Beram SM, Jagota V, Dighriri M, Ahmad MW, Hossain MS, Rizwan A (2023) Rule based fuzzy computing approach on self-supervised sentiment polarity classification with word sense disambiguation in machine translation for Hindi language. ACM Trans Asian Low-Resour Lang Inf Process 22(5):1–21
2. Luo B, Luo R (2024) Application and empirical analysis of fuzzy neural networks in mining social media users' behavioral characteristics and formulating accurate online marketing strategies. Int J Comput Intell Syst (1):273
3. Bilal SF, Almazroi AA, Bashir S, Khan FH, Almazroi AA (2022) An ensemble based approach using a combination of clustering and classification algorithms to enhance customer churn prediction in telecom industry. PeerJ Comput Sci 8:854
4. Hu S, Chen A (2023) Data-driven customer retention strategies in e-commerce: a fuzzy z-number approach. IEEE Access
5. Supangat S, Bin Saringat M, Kusnanto G, Andrianto A (2021) Churn prediction on higher education data with fuzzy logic algorithm. SISFORMA 8(1):22–29
6. Altameem AA, Hafez AM (2022) Behavior analysis using enhanced fuzzy clustering and deep learning. Electronics 11(19):3172
7. Chaubey G, Gavhane PR, Bisen D, Arjaria SK (2023) Customer purchasing behavior prediction using machine learning classification techniques. J Ambient Intell Humaniz Comput 14(12):16133–16157
8. Gkikas DC, Theodoridis PK, Beligiannis GN (2022) Enhanced marketing decision making for consumer behaviour classification using binary decision trees and a genetic algorithm wrapper. In: Informatics, vol 9. MDPI, p 45
9. Bojanowska AB, Kulisz M (2023) Using fuzzy logic to make decisions based on the data from customer relationship management systems. Adv Sci Technol Res J17(5)
10. Geramian A, Abraham A (2021) Customer classification: a mamdani fuzzy inference system standpoint for modifying the failure mode and effect analysis based three dimensional approach. Expert Syst Appl 186:115753
11. Lu X, Zhu J (2024) Design and development of middle-aged and elderly tourism products based on fuzzy decision support system. Int J Comput Intell Syst 17(1):115
12. Guo L-X, Lin C-C, Huang P-F, Jhou S-Y, Chen S-C, Tsai F-S (2021) Fuzzy logic analysis for key factors for customer loyalty in e-shopping environment. Front Psychol 12:742699
13. Marín Díaz G, Carrasco González RA (2023) Fuzzy logic and decision making applied to customer service optimization. Axioms 12(5):448

14. Ernawati E, Baharin S, Kasmin F (2021) A review of data mining methods in rfm-based customer segmentation. J Phys: Conf Ser 1869:012085
15. Abedin MZ, Hajek P, Sharif T, Satu MS, Khan MI (2023) Modelling bank customer behaviour using feature engineering and classification techniques. Res Int Bus Financ 65:101913
16. Aslam S (2024) Consumer behavior and shopping habits dataset. Accessed 29 Mar 2025. https://www.kaggle.com/datasets/zeesolver/consumer-behavior-and-shopping-habits-dataset

Fuzzy Logic-Based Framework for Risk Management and Economic Optimization in Sustainable Virtual Economies

Anup Kumar Thander and Shantanu Chakravarty

Abstract The blockchain ecosystem operates in rapidly changing perspectives based on market volatility, investor sentiments, regulatory ambiguity, and transaction fees, so risk management and economic optimization becomes inevitable. This fuzzy logic framework implementation is based on a Mamdani Fuzzy Inference System (FIS) in MATLAB that integrates risk evaluation with economic decision optimization for Blockchain-based economies. Our proposed FIS model uses fuzzy membership functions and a rule-based inference system to process these four important input parameters and provide output indicating risk assessment. Factors such as market volatility, investor sentiment, regulatory uncertainty, and transaction fees create challenges and opportunities for market participants, shaping the behavior of buyers and sellers and influencing investment strategies. Through fuzzy framework, it gives a strong way to deal with the uncertainties occurring in the blockchain operations and also to take best decisions. FIS toolbox in MATLAB will enhance the interpretability of results. Contour and surface graphs show how risks vary along different input conditions while the scatter diagram reveals trend in economic behavior. A coloured gradient of the risk scores, both visually and dynamically, is shown by the heatmap, and some insight into the fuzzy sets describing input–output relationships is provided by the membership function plots. These visualizations will enable financial decision makers to effectively interpret the patterns of risks and adjust economic strategies appropriately. These results validate the model's ability to spot high-risk conditions, optimize transaction strategies, and improve economic stability within blockchain networks. Also, this adaptive data-driven methodology provides investors, policymakers, and developers with potent sources of risk mitigation and strategic decision-making in decentralized financial ecosystems.

A. K. Thander (✉) · S. Chakravarty
Department of Applied Science and Humanities, Guru Nanak Institute of Technology, Kolkata, India
e-mail: anup.thander@gnit.ac.in

S. Chakravarty
e-mail: shantanu.chakraborty@gnit.ac.in

S. Kautish et al. (eds.), *Green Metaverse: Fuzzy Logic Approaches to Sustainable Virtual Worlds*, Information Systems Engineering and Management 74,
https://doi.org/10.1007/978-3-032-08183-4_11

Keywords Economic optimization · Blockchain · Mamdani FIS · Fuzzy logic · Risk management · Market volatility · MATLAB

1 Introduction

Blockchain technology [1, 2] drives a very fast-paced manner of integration of economy and finance, provides transparency and security, as well as decentralization, but it also comes with substantial risk and uncertainty. In turn, market volatilities drive sentiment and hence, with regulatory uncertainties and transaction costs, the role of the stable and efficient level of blockchain-based economies plays out. It is important to come up with effective risk management and economic optimization strategies to make the best of these digital ecosystems and reduce financial vulnerabilities to a minimum possible level.

Fuzzy logic, particularly the Mamdani Fuzzy Inference System (FIS) [1, 3], offers an effective basic structure that encourages fundamental scaffolding to engage the complexities and ambiguities inherent in blockchain societies. In contrast to traditional mathematical models that primarily require precise input, fuzzy logic will allow decisions to be made through approximate reasoning, which is rightly adapted to the volatile and stochastic environment of markets in blockchain technologies. Fuzzy rules and linguistic variables from Mamdani FIS yield a sophisticated estimate of risk levels and economic efficiencies thereby providing more relevant and resilient financial strategies [2].

The study presented utilizes MATLAB's fuzzy Mamdani FIS framework to assess practices to optimize the economy of blockchain ecosystems. The system construes the focal points that lead to the identification of risk assessment and determination of the best economic strategies, such as ratios of market volatility, investor sentiment, regulatory uncertainty, and transaction fees. Several graphical aids-surface plots, heat plot, contour plots, bar and scatter plots, and membership function visualizations-make their contribution towards the interpretability and meaningfulness of results for various stakeholders. Through the application of fuzzy logic principles and techniques, the study hopes to lay the ground for developing intelligent decision-support systems for economies operated on blockchain technology. This framework is meant to assist investors, policymakers, and financial analysts in reducing risk management strategies, improving economic stability, and higher strategic decision-making in the decentralized digital economy. Future research directions will involve transferring the proposed model into a fully operational model, introducing other economic factors [4–8] for consideration, and integrating real-time data from blockchain to generate more predictive accuracy and applicability.

This paper is arranged as follows: Sect. 2 deals with the Literature Review. Section 3 explains in detail Fuzzy Inference Systems for Blockchain Investment Risk Assessment in MATLAB. The results are presented and discussed in Sect. 4. Concluding thoughts are given in Sect. 5, while Sect. 6 attends to future directions.

2 Literature Review

Extensive research has been conducted in this research area, focusing on fuzzy inference system to enhance our understanding of financial risk assessment and economic optimization in the blockchain ecosystem. A MATLAB-based fuzzy inference system was developed by Jamshidi et al. [3] to assess financial ergodic risk under an uncertain situational gambler. Alternative groups of researchers investigated fuzzy logic within the aspects of a blockchain process to improve risk management, adding to economic sustainability. Kaur et al. [9] have studied finance risk assessment in decentralized finance with fuzzy AHP. Another example of applying fuzzy cognitive maps in risk assessment came from Zhang et al. [10] in the area of blockchain innovation in education. In their research, Anita et al. [11] have given a concept based on fuzzy logic engine application for providing trust in IoT blockchain network development. Uncertainties related to real-world risk assessment were considered within a model developed by Pokoradi [12]. A methodology for organizational risk management, based on fuzzy logic, was introduced by Zlateva et al. [13]. Implementation was done by Chinna Rani et al. [14] for enhancing blockchain mining efficiency using fuzzy logic. Ali et al. [15] developed a blockchain-based fuzzy trust management framework (BFT-IoMT) to detect Sybil nodes in medical data exchange, while Gardas et al. [4] designed a fuzzy-driven node selection methodology for blockchain-enabled edge IoT. Atlam et al. [5] combine fuzzy logic knowledge and expert judgment in a risk-based access control model while developing a reputation evaluation system for blockchain-based IoT access control. Decision-making can be considerably improved in permitting processes.

Also numerous studies dedicated to Fuzzy Inference Systems have been undertaken after the MATLAB computer environment was developed for risk management and economic optimization in the blockchain ecosystem. Alqbaishi et al. [6] put forth a fuzzy logic model for decision-making support in blockchain systems through the definitions of linguistic terms for both input and output variables. Nicolas et al. [16] explored a Mamdani FIS aimed to let management control decision-making for an SMEs accountant by giving insights regarding economic optimization. Hsieh et al. [17] designed a software project risk assessment model based on fuzzy logic adaptable towards blockchain risk management. John et al. [18] proposed a trust evaluation model for cloud service providers with an assessment of the QoS characteristics relevant to the blockchain framework of trust. Bamakan et al. [19] introduced Di-Ali converted to Di-ANFIS within the context of service supply chain optimization, while combining blockchain, IoT, and big data with an Adaptive Neuro-Fuzzy Inference System (ANFIS). Alkhdour et al. [20] combined fuzzy logic and blockchain to enhance the techniques of authentication and key management in healthcare, ensuring secure and efficient data management.

3 Fuzzy Inference System for Blockchain Investment Risk Assessment

The FIS implemented in MATLAB [21, 22] is supposed to evaluate the investment risk characteristic of the specific blockchain markets considering the main financial and regulatory factors taking place during the investment period. The model is based on FIS in the Mamdani type so that the system uses fuzzy logic to manage the uncertainties that result from the investment in blockchain. The model comprises four main input variables: Market Volatility, Investor Sentiment, Regulatory Uncertainty, and Fees; each input is defined by Gaussian membership functions so that transitions between levels (for instance Low, Medium, and High) can be performed smoothly. The variable of output, Investment Risk, is determined by Gaussian membership functions into which risks are categorized into five parts: Very Low, Low, Medium, High, and Very High. This system describes a sharply defined rule base characterized by nine rules that connect input conditions with output risk levels. These rules depict situations of concrete investment reality, for example, high volatility and negative investor sentiment combining into a very high-risk assessment. For blockchain investors, a MATLAB-based system like this will be of immense visuals-only assistance since, within a very structured and interpretable methodology, they can have a proper assessment of the risks. With this system, therefore, they will be capable of looking at various factors at the same time so that an informed investment decision can be made within a high degree of uncertainty. Eventually, it will need some more financial parameters and better rule sets in combination with real market data to have predictive more accuracy (see Table 1).

This table puts in order the important factors of the Fuzzy Inference System (FIS) [20, 23, 24] that has been drawn regarding the investment risk in blockchain. The main input parameters have been described, Market Volatility, Investor Sentiment, Regulatory Uncertainty, and Transaction Fees, having a specific variation from 0 to 100. These inputs are associated with linguistic terms (Low, Medium, High, etc.) via Gaussian membership functions as to distinguish the degree of risk smoothly. Very Low to Very High is used for the output variable—Investment Risk, therefore making it possible to categorize risks properly within a range of different market conditions. Each input has its individual effects in totality on the risk at stake. For

Table 1 Parameters for blockchain investment risk assessment

Parameter	Range	Membership functions	Categories
Market volatility	0–100	Gaussian	Low, medium, high
Investor sentiment	0–100	Gaussian	Negative, neutral, positive
Regulatory uncertainty	0–100	Gaussian	Stable, uncertain, risky
Transaction fees	0–100	Gaussian	Low, moderate, high
Investment risk (Output)	0–100	Gaussian	Very low, low, medium, high, very high

example, high market volatility and a negative investor sentiment typically point toward higher investment risk, whereas stable regulations and low transaction fees might indicate lower risk. These fuzzy rules [25–27] are here to amalgamate the inputs and generate output risk scores that use the bisector defuzzification method, which in itself provides a balanced risk assessment. The parametric structure helps investors and analysts form de facto judgments based on an informed assessment, underpinned by a flexible, rule-based model that can incorporate the dynamics of any market [28–30].

4 Results and Discussion

Various plots are generated to visualize the behavior of the FIS in system analysis. The membership function plots present a classification of both inputs and outputs into different fuzzy sets, reflecting a smooth transition between levels of risk. The rule viewer provides a graphical representation of how different input conditions influence investment risk interactively. The surface plots, meanwhile, depict the three-dimensional relationship between pairs of various input variables and output, revealing how factors such as regulatory uncertainty and transaction fees interact to influence levels of risk. A contour plot might be applied to map the level of intensity in terms of susceptibility to risk versus variability in both market conditions and investor sentiment hence making a very simple interpretation for investors into how changes in those set of conditions add to risk. A scatter plot would enable interested parties to view the relationship between return volatility on the asset and its investment risk, thus showing patterns and the existence of certain anomalies. The histogram shows how risk is distributed across scenarios and gives one the least common risks bound by extremes. A heatmap simply shows these values in colors.

Figures 1, 2, 3, 4, and 5 are fuzzy classifications of the key financial factors which directly influence investment risk. Figure 1 segregates market volatility as low, medium, or high using Gaussian membership functions in a smooth transition that outlines financial uncertainty. Figure 2 describes the investor sentiment as negative, neutral, or positive, with the latter agenda moving towards decision-making when selling panic is induced by negative sentiment and the positive sentiment makes a push for growth. Figure 3 explains financial stability as secure, uncertain, or risky, thereby emphasizing regulations currently molded around the decision-making of investors during pivotal selling moments. Figure 4 depicts transaction fees as low, moderate, or high-where they influence profitability and trading volumes. Figure 5 describes investment risk as very low, low, medium, high, and very high. It explains that in some cases, market conditions dynamically shape certain risk levels. This graphically depicts an alternative perspective of financial decays that facilitates nuanced valuations of financial uncertainties and investment decision processes undertaken.

Fuzzy inference systems used for investment risk assessment have been completely visualized in Figs. 6, 7, 8, 9, 10, 11, 12, 13 and 14, showing interaction between factors when decision-making is influenced. Figure 6 shows the Rule

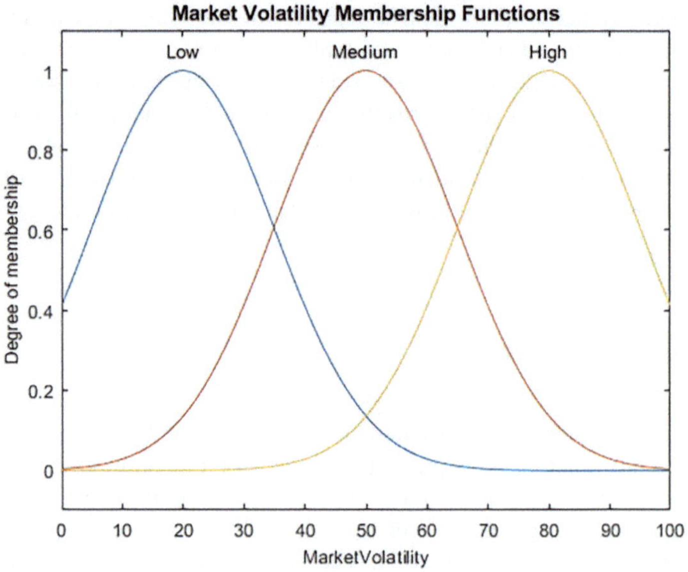

Fig. 1 Visual representation of membership functions including high, medium, and low market volatility levels

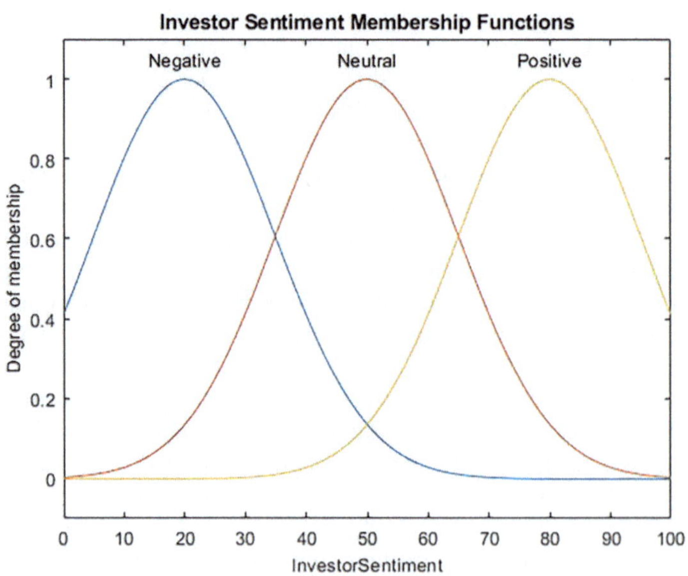

Fig. 2 Visual representation of membership functions including negative, neutral and positive investor sentiment levels

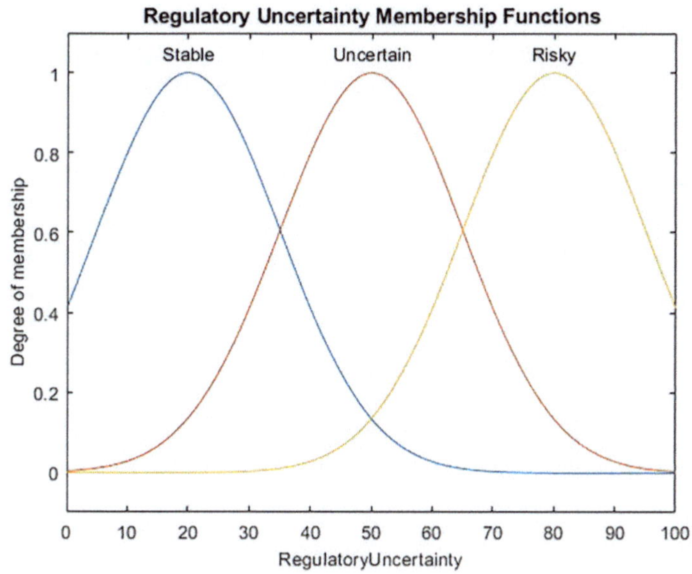

Fig. 3 Graphical representation of membership functions including regularity uncertainty levels

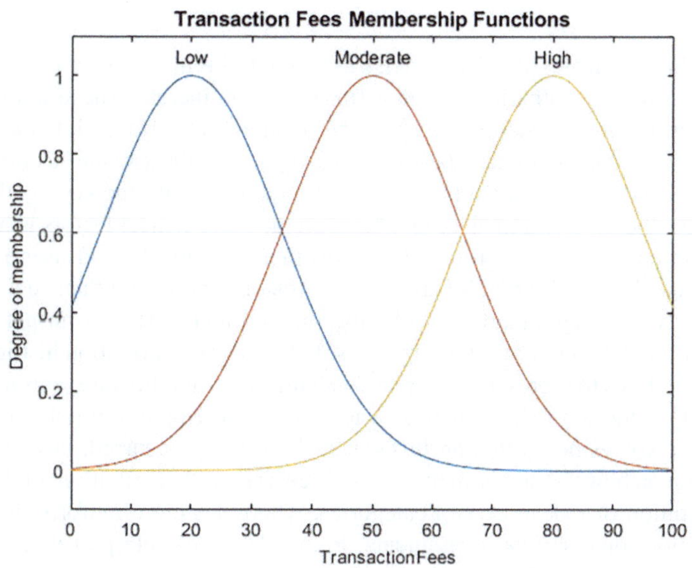

Fig. 4 Visual representation of membership functions including low, moderate and high transaction fees

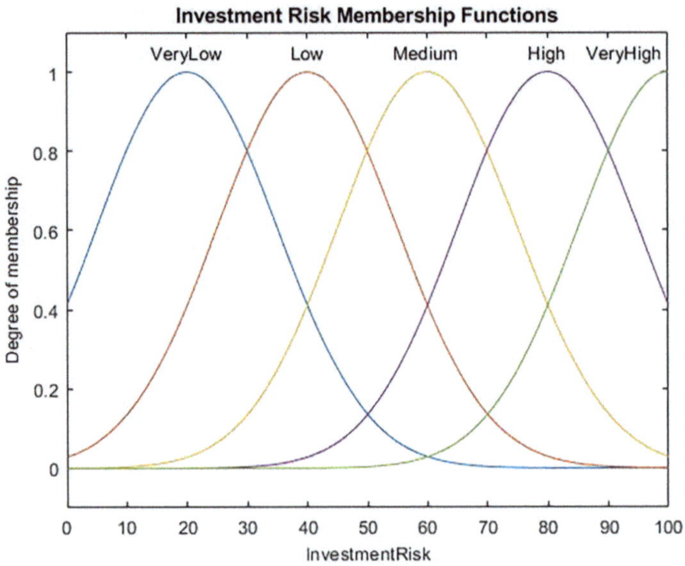

Fig. 5 Graphical representation of membership functions including very low, low, medium, high and very high investment risk levels

Viewer, which indicates the effect on risk of using different combinations of input so that analysts can test scenarios and refine fuzzy rules for increased accuracy in prediction further. Figures 7, 8, 9 and 10 proceed to show 3D surface plots that provide information on the interaction between two input variables while keeping others constant. These 3-dimensional plots are beneficial to identify the key trends in risk, which allow the investors to place emphasis on the more critical factors in achieving greater accuracy in financial forecasting. Together with this, Fig. 11 uses a heatmap showing visually how Market Volatility and Investor Sentiment influence investment risk, identifying risky situations with color-graded areas. This is further complemented by the 2D contour plot in Fig. 12, which is a risk distribution histogram that uses contour lines to represent where small changes in volatility and sentiment cause significant changes in risk. Statistically assessing a distribution of risk, in Fig. 13 a histogram displays how often an Investment Risk value occurred, providing some information on how risk assessments skew. Such a skewed distribution might suggest a concentration of risk in certain categories, whereas a more uniform distribution suggests more diverse risk assessments. Figure 14 is a scatterplot that shows the distribution of Market Volatility against Investment Risk, revealing patterns, correlations, and possible thresholds where risk dramatically amplifies. Combined with all these, the presentational visuals deepen insight to create more informed strategies for investments, risk management, and financial decision-making.

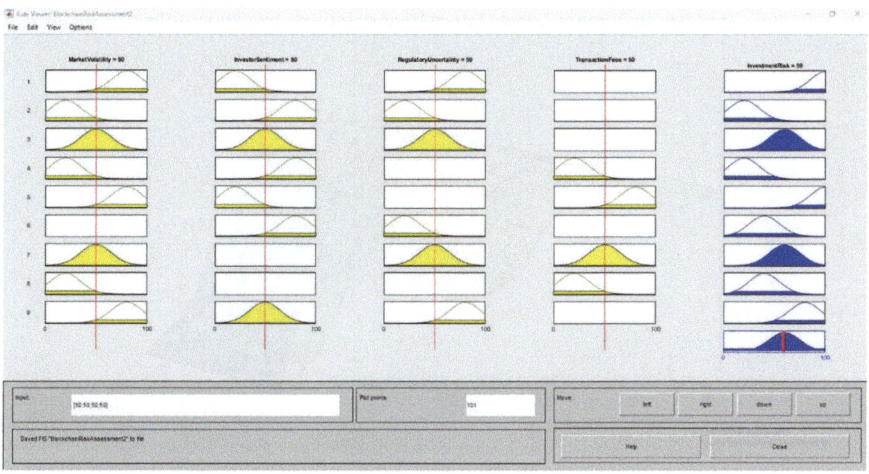

Fig. 6 Graphical representation of application of fuzzy rules for particular input values

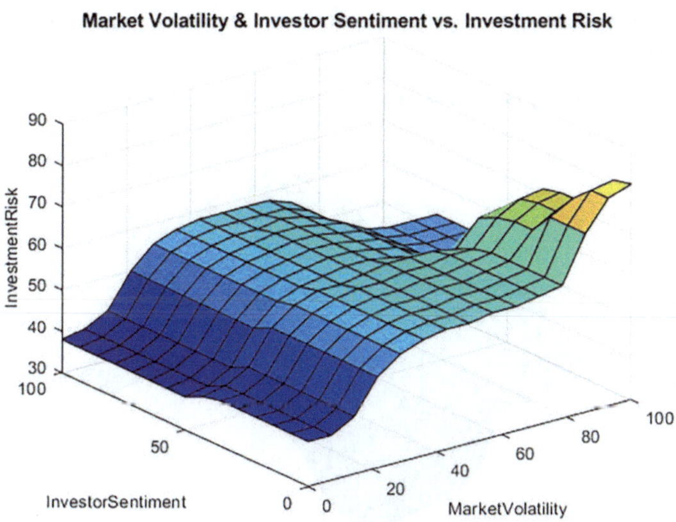

Fig. 7 Three dimensional surface plot for market volatility and investor sentiment versus investment risk

Therefore, FIS in MATLAB acts strong as a computational engine that coordinates uncertainty management in operations on the blockchain and in investment risk assessment. It brings in market volatility, investor sentiment, regulatory uncertainty, and even transaction fees, hence enabling systematic risk assessment while influencing market players' decision making. This behavioral change in buyers and sellers comes from an adaptive approach that changes dynamically based on changing conditions in financial markets. This is through enhancing investment strategy due

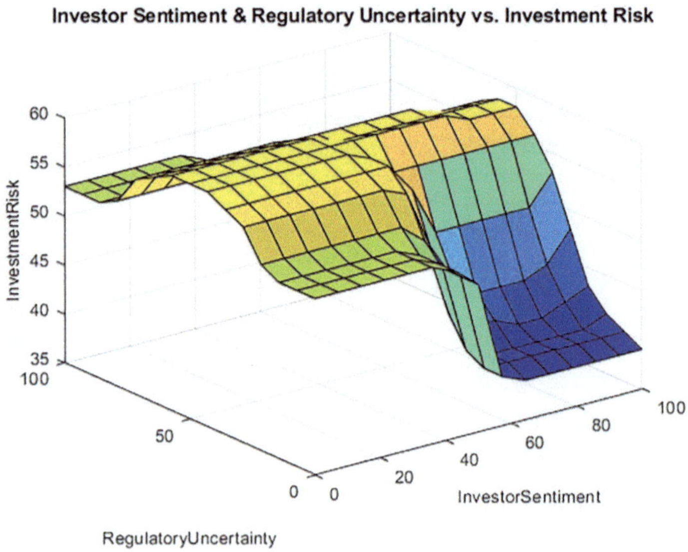

Fig. 8 Three dimensional surface plot for investor sentiment and regularity uncertainty versus investment risk

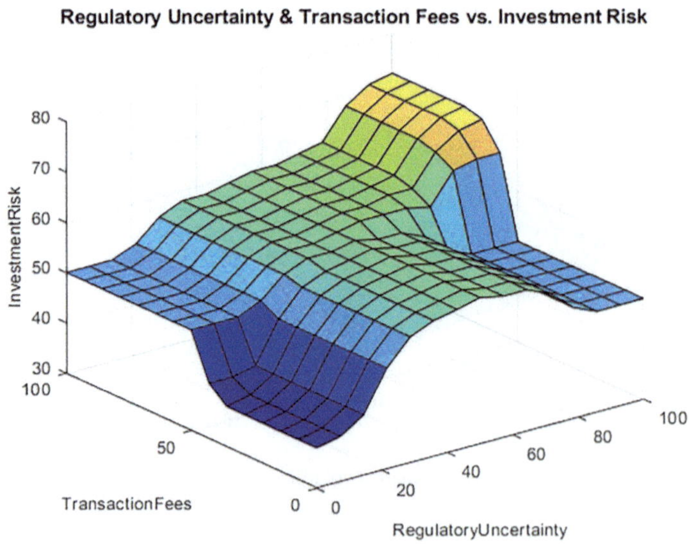

Fig. 9 Three dimensional surface plot for regularity uncertainty and transaction fees versus investment risk

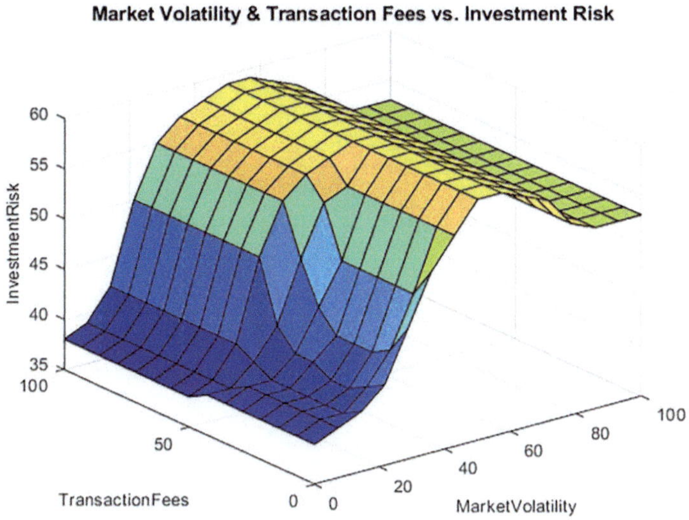

Fig. 10 Three dimensional surface plot for market volatility and transaction fees versus investment risk

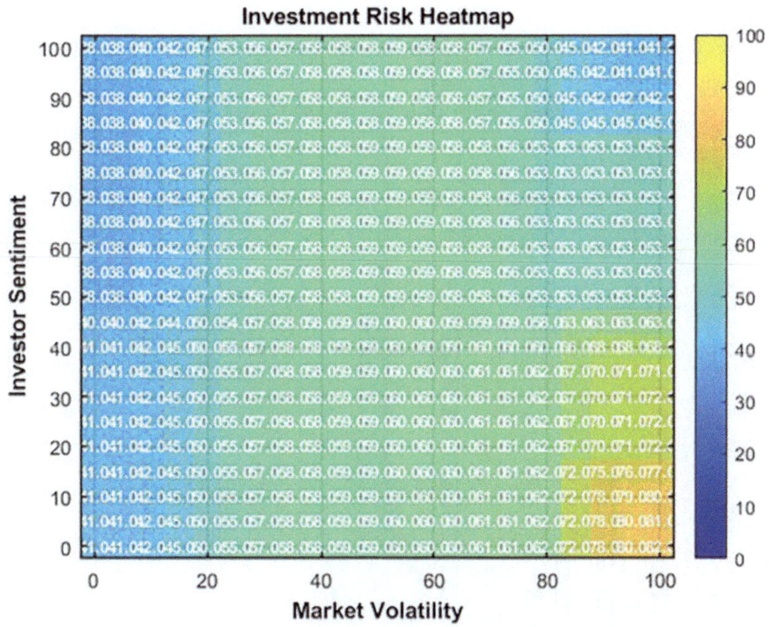

Fig. 11 Heatmap visualization of risk assessment based on fuzzy logic evaluation

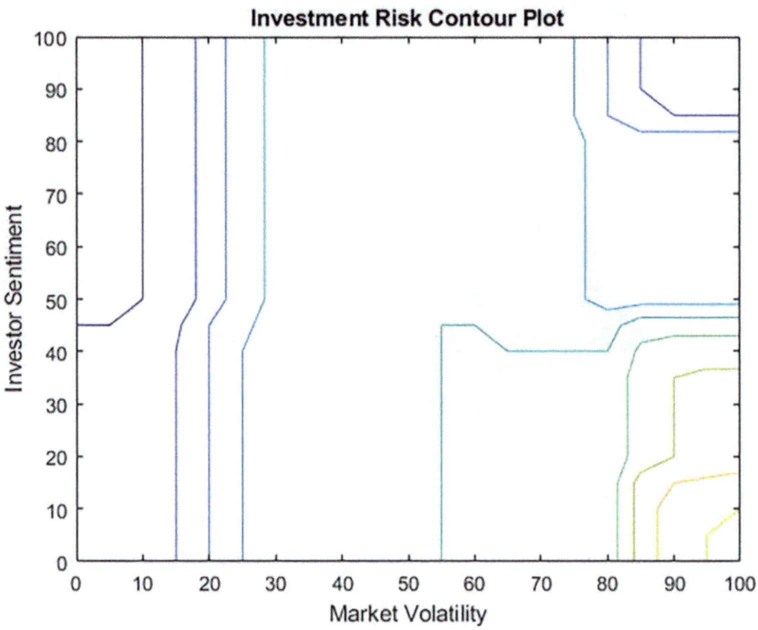

Fig. 12 Two dimensional contour plot for risk assessment

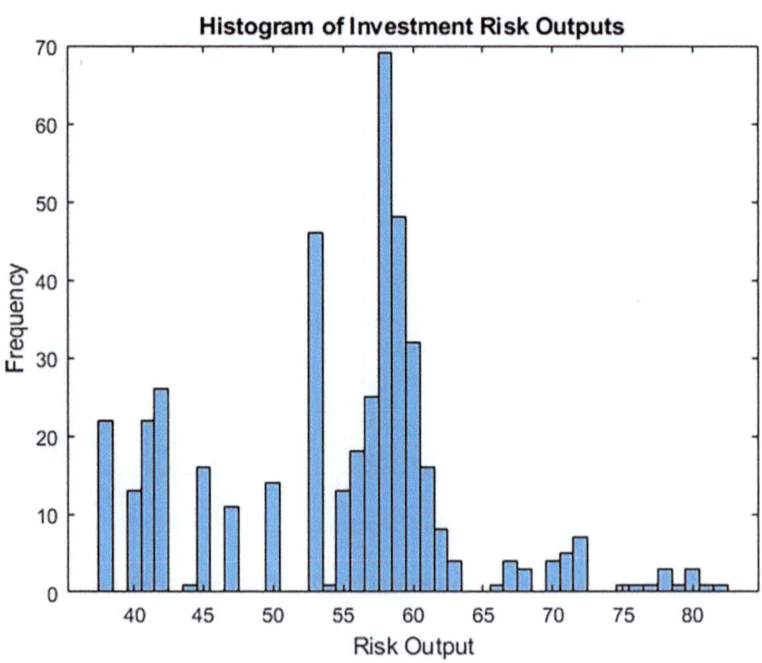

Fig. 13 Histogram representation of risk output distribution in fuzzy risk assessment

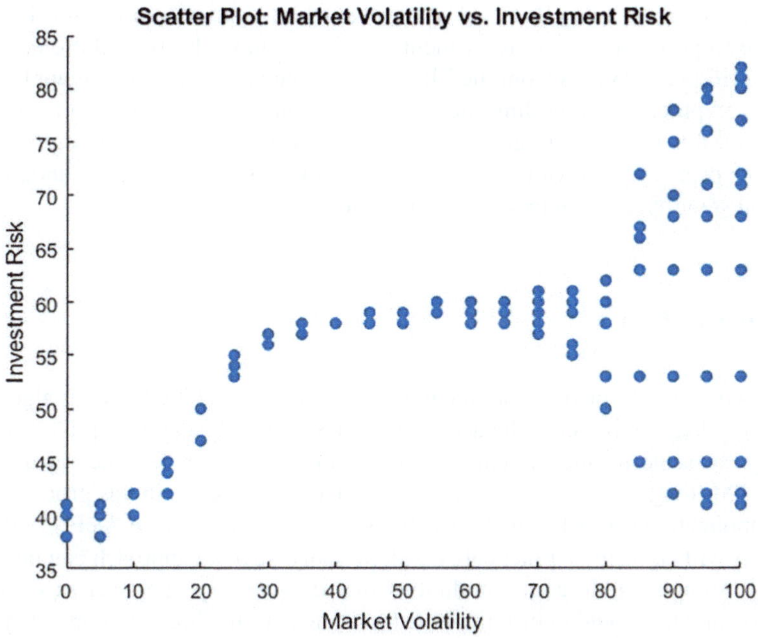

Fig. 14 Scatter plot depicting the relationship between investment amount and associated risk levels

to the structured, data-driven approach for risk management under conditions of uncertainty in markets.

5 Conclusion

This study shows how Mamdani fuzzy inference system (FIS) is effective in risk management and economic optimization in blockchain ecosystems. The proposed fuzzy logic framework incorporates inductive parameters such as market volatility, investor sentiment, regulatory uncertainty, and transaction fees. Thus, it offers a very practical decision-making framework with respect to the uncertainties presented in decentralized financial systems. Additionally, statistical visualizations conducted in MATLAB, such as surface and contour plots, bar and scatter plots, a heatmap, and visualizing membership functions, reveal insights regarding risk patterns. These visualization tools convey to stakeholders how various economic factors influence blockchain-based financial environments in an intuitive and data-driven manner. Then numerical simulation results provide evidence of the FIS-based method advantages by appropriately revealing high risk events. This minimizes gains/lives in alleviating risks through enhanced optimal investment strategies, transaction policies, or regulatory compliance measures. Developing such flexible FIS-based assessments continues relies on fuzzy logic as a fair evaluation approach in design. It is also valuable to perform revised assessments on a continuous basis, making necessary changes. Representations from the FIS can translate across blockchain applications such as cryptocurrency trading, decentralized finance (DeFi), or smart contracts. Therefore, this research suggests assessing risk profiles using fuzzy logic demonstrates its primacy in previous risk valuation work while providing a foundation for financial stability and enhancing outcomes in a blockchain network.

6 Future Plan

Future work should involve the integration of several machine learning algorithms with fuzzy logic to increase the accuracy of risk predictions and adaptive decision-making. To scale up the systems by optimizing fuzzy rule sets, neural networks like LSTM (long short term memory) is a decisive application in this process. Real-world application in blockchain transactions, decentralized finance (DeFi), and smart contracts will confirm its practicality. Additionally, such an approach can be implemented through multi-agent simulations that dynamically characterize economic behavior, and this could in turn improve financial stability. Market-adaptive systems should be explored for even more dynamic economic optimization within blockchain ecosystems, thus guaranteeing stronger and smarter risk management strategies.

References

1. Janková, Z., Dostál, P.: Type-2 fuzzy expert system approach for decision-making of financial assets and investing under different uncertainty. Math. Probl. Eng. **2021**(1), 3839071 (2021)
2. Bayram, M., Akat, M.: Market-neutral trading with fuzzy inference, a new method for the pairs trading strategy. Eng. Econ. (2019)
3. Jamshidi, A., Yazdani-Chamzini, A., Yakhchali, S.H., Khaleghi, S.: Developing a new fuzzy inference system for pipeline risk assessment. J. Loss Prev. Process Ind. **26**(1), 197–208 (2013)
4. Gardas, B.B., Heidari, A., Navimipour, N.J., Unal, M.: A fuzzy-based method for objects selection in blockchain-enabled edge-IoT platforms using a hybrid multi-criteria decision-making model. Appl. Sci. **12**(17), 8906 (2022)
5. Atlam, H.F., Walters, R.J., Wills, G.B., Daniel, J.: Fuzzy logic with expert judgment to implement an adaptive risk-based access control model for IoT. Mobile Netw. Appl. **26**(6), 2545–2557 (2021)
6. Alqbaishi, A.A., Alaa E.S.A.: Reputation evaluation using fuzzy logic for Blockchain-Based access control in an IoT environment. IEEE Access (2024)
7. Thander, A.K., Paul, S., Maitra, P.: An improved Shamanskii method for finding zeros of linear and nonlinear equations. Appl. Math. Sci. **6**(86), 4277–4281 (2012)
8. Abdullayev, T., Rahib, I., Niyar, U.: Improving decision-making in blockchain-based systems using fuzzy logic. In: International Conference on Theory and Applications of Fuzzy Systems and Soft Computing. Springer Nature Switzerland, Cham (2023)
9. Kaur, S., Singh, S., Gupta, S., Wats, S.: Risk analysis in decentralized finance (DeFi): a fuzzy-AHP approach. Risk Manage. **25**(2), 13 (2023)

10. Zhang, L., Hu, H., Zhang, D.: A credit risk assessment model based on SVM for small and medium enterprises in supply chain finance. Financ. Innov. **1**, 1–21 (2015)
11. Anita, N., Vijayalakshmi, M., Mercy Shalinie, S.: A lightweight scalable and secure blockchain based IoT using fuzzy logic. Wireless Person. Commun. **125**(3), 2129–2146 (2022)
12. Pokoradi, L.: Fuzzy Logic-Based Risk Assessment (2002).
13. Zlateva, P., Velev, D., Raeva, L.: A fuzzy logic method for assessment of risk management capability. Int. J. Innov. Manag. Technol. **6**(4), 260 (2015)
14. Chinna Rani, K., Fathima, L., Anuradha, M.P.: Fuzzy-Enhanced optimization algorithm for blockchain mining. In: International Conference on Futuristic Communication and Network Technologies. Singapore, Springer Nature Singapore (2021)
15. Ali, S.E., Tariq, N., Khan, F.A., Ashraf, M., Abdul, W., Saleem, K.: Bft-iomt: a blockchain-based trust mechanism to mitigate sybil attack using fuzzy logic in the internet of medical things. Sensors **23**(9), 4265 (2023)
16. Nicolas, C., Müller, J., Arroyo-Cañada, F.-J.: A fuzzy inference system for management control tools. Mathematics **9**(17), 2145 (2021)
17. Hsieh, M.-Y., Hsu, Y.-C., Lin, C.-T.: Risk assessment in new software development projects at the front end: a fuzzy logic approach. J. Ambient. Intell. Humaniz. Comput. **9**, 295–305 (2018)
18. John, J., John Singh, K.: Trust value evaluation of cloud service providers using fuzzy inference based analytical process. Scientific Rep. **14**(1), 18028 (2024)
19. Bamakan, S.M., Hosseini, N.F., ZareRavasan, A.: Di-ANFIS: an integrated blockchain–IoT–big data-enabled framework for evaluating service supply chain performance. J. Comput. Design Eng. **8**(2), 676–690 (2021)
20. Alkhdour, T.A.Y.S.E.E.R., Almaiah, M.A., Ali, A.I.T.I.Z.A.Z., Lutfi, A.B.D.A.L.W.A.L.I., Alrawad, M.A.H.M.A.O.D., Tin, T.T.: Revolutionizing healthcare: unleashing blockchain brilliance through fuzzy logic authentication. J. Theor. Appl. Inf. Technol. **102**(4), 29 (2024)
21. Bhattacharyya, S., Thander, A.K.: Conventional education & online education in engineering: a case study. J. Phys. Conferen. Series **1797**(1), 012069. IOP Publishing (2021)
22. Thander, A.K., Mandal, G.: Optical waveguide analysis using alternative direction implicit (ADI) method in combination with successive over-relaxation (SOR) algorithm. J. Opt. **53**(1), 475–481 (2024)
23. Kickert, W.J.M., Mamdani, E.H.: Analysis of a fuzzy logic controller. In: Readings in Fuzzy Sets for Intelligent Systems. Morgan Kaufmann, pp. 290–297 (1993)
24. Li, H.X., Gatland, H.B.: A new methodology for designing a fuzzy logic controller. IEEE Trans. Syst. Man Cybern. **25**(3), 505–512 (1995)
25. Lee, C.C.: Fuzzy logic in control systems: fuzzy logic controller. IEEE Trans. Syst. Man Cybernet. **20**(2), 404–418 (1990)
26. Asemi, A., Asemi, A., Ko, A.: Adaptive neuro-fuzzy inference system for customizing invest ment type based on the potential investors' demographics and feedback. J. Big Data **10**(1), 87 (2023)
27. Mandal, G., Thander, A.K., Dey, S.: Numerical comparison of two different single step iterative methods in complex plane for finding basins of attraction. In: AIP Conference Proceedings. AIP Publishing, vol. 2876 (2023)
28. Nedumparambil, E., Bhandari, A.K.: Risk factors, uncertainty, and investment decision: evidence from mutual fund flows from India. Indian Econ. Rev. **57**(2), 349–372 (2022)
29. Thander, A.K., Bhowmik, D.: Artificial neural network with the Levenberg-Marquardt algorithm for numerical solution of two-dimension Poisson's equation. BOHR Int. J. Smart Comput. Inform. Technol. **4**(1), 55–61 (2023)
30. Li, L., Muwafak, B.M.: Adoption of deep learning Markov model combined with copula function in portfolio risk measurement. Appl. Math. Nonlinear Sci. **7**(1), 901–916 (2021)

Effectiveness of Education in Metaverse Environment: Issues and Opportunities

Baritule Prince Saue

Abstract The paper analysed the digitisation of education as the use of modern information technology into educational methodologies to improve teaching and learning experiences. It involves the use of digital tools, resources, and platforms to enhance communication, teaching, learning, and assessment. The Metaverse is a groundbreaking instrument for the digitisation of education, enabling interactions among learners inside a virtual setting. The Metaverse is an immersive digital environment where users interact with their own avatars and those of others. The paper conceptualises the metaverse as having certain characteristics, including the Internet of Things, digital reality, and human–computer interaction, among others. It later classified metaverse technologies into four unique categories: virtual and augmented reality, Internet of Things (IoT), Artificial Intelligence (AI), and digital twin. The study noted that the implications of the metaverse are many, particularly in its gradual erosion of the distinctions between virtual and real-life contexts. The metaverse has impacted education via several methods, including visualisation, the deterrence of academic dishonesty, immersive experiences, and customisation, among others. The advent of the metaverse in education has presented serious issues, notably substantial disruptions to ontological and epistemological underpinnings. Nevertheless, the paper demonstrated that the integration of the metaverse into the educational system offers various opportunities, such as the establishment of an immersive learning environment that engages all senses, the enhancement of the teaching process constrained solely by our imagination, and the facilitation of personalised education and collaborative work options, among others.

Keywords Education · Metaverse · Issues and opportunities

B. P. Saue (✉)
Department of Metalwork Technology Education, Federal College of Education (Technical) Akoka, Lagos State, Nigeria
e-mail: baritule.saue@fcetakoka.edu.ng

S. Kautish et al. (eds.), *Green Metaverse: Fuzzy Logic Approaches to Sustainable Virtual Worlds*, Information Systems Engineering and Management 74,
https://doi.org/10.1007/978-3-032-08183-4_12

1 Introduction

There has been a significant increase in the development and demand for technological mobile applications, cloud technologies integrated into these applications, and information-media-communication tools, since internet use and dependence have become essential to contemporary life. As novel electronic communication and information modalities emerge, conventional classrooms are being supplanted by virtual environments, prompting an evolution in instructional methodologies and pedagogical frameworks. Mobile gadgets, encompassing laptops, cellphones, smartwatches, and analogous things, have become essential in daily life and significantly influence interactions with technology. Consequently, these improvements make internet use essential [1]. This underscores the essential need of understanding how innovative tech-enhanced learning methods that prioritise student needs may transform educational dynamics. As shown by [2, 3], technology-enhanced learning is increasingly prevalent in contemporary classrooms, stemming from a 21st-century educational paradigm that amalgamates conventional teaching techniques with digital resources.

Moreover, students' perceptions of successful learning and instructional requirements evolve as they navigate a globally diverse digital world that offers adaptive, personalised information [4]. Consequently, there is an increasing need to provide students more engaging and immersive learning experiences that facilitate active participation rather than passive observation [5]. The adjustments consist of two primary components. Teachers and students are forming a novel kind of interaction, to begin with. Educators already possess an unprecedented array of tools to design classes, engage with students, and collect data more precisely, facilitated by online learning environments and communication platforms. The Metaverse educational framework can promptly communicate student requirements to educators, while big data technologies may enhance students' understanding and retention. Moreover, students will actively exhibit interest in interacting with teachers in the virtual environment, while professors will adopt a less formal demeanour compared to traditional instructional models. This, however, underscored the need of integrating digital tools and information communication technology (ICT) into the educational sphere, such as the metaverse, according to [6].

2 Metaverse

The fusion of "meta" with "universe" gives rise to the concept of the metaverse. If there is a Turkish phrase for this concept, it may be "beyond the physical universe," however no such word really exists [7]. Apart from facilitating the formation of virtual communities, the metaverse expands beyond the domains of commerce and entertainment. The metaverse seems, at first look, to be a great platform for students to hone their social media skills and apply what they've learnt in the real world to a virtual setting. With its vast digital expansion in cyberspace and its three-dimensional virtual environment where users converse via avatars, Metaverse represents the future generation of the internet from the viewpoint of [8]. While virtual reality (VR) allowed people to engage with computer-generated environments, the idea of a metaverse is still in its infancy in academic circles. It paves the way for the creation of avatars, or digital personalities, that stand in for our actual selves online. A metaverse is an all-encompassing virtual environment where users may communicate via their avatars and those of other users, as stated in [9]. The author asserts that the metaverse allows users to experience many realities simultaneously.

Also, as stated in [10], the metaverse is a kind of virtual reality that enables real-world-like social structures and activities. the eleventh One may create a frictionless transition between the real and virtual worlds by using these digital identities to enter the metaverse and engage in activities that are analogous to those in the actual world. As per [11], the metaverse, a large network of interconnected digital universes, is rapidly transforming the technological landscape. As described in [12], it is a hybrid of augmented, virtual, mixed, and artificial intelligence technologies that operates on the internet and provides a hyper virtual reality experience. One definition of the metaverse is an alternative reality where digital and physical elements coexist; other characteristics of the metaverse include its multi-user nature, permanence, and continuity [13]. One of the several ways in which the continuous growth of the metaverse may impact people's everyday lives is by promoting sustainable education techniques in Nigeria. You might say it's a bridge between the actual and virtual worlds.

3 Basic Features of Metaverse

Metavers is a strategy to promote human progress in every area. For everyone, it's a whole new world. Metaverse differentiating traits include high degrees of human–computer interaction, decentralisation, quick improvements in Internet connection protocols and hardware platforms, and a combination of virtuality and reality.

- **Decentralization originates** in the blockchain technology. By doing away with the conventional social operation mechanism, it allows for the equal and collaborative development of the Metaverse by all users, rather than its exclusive production by a select few.
- **The Internet of Things (IoT):** Which means that everything is online. It combines virtual and real worlds, allowing users to seamlessly transition between the two at any given moment. A user's sense of self (who am I?) and their social connections in two realms (how do I relate to them?) will usher in a novel way of living in a Metaverse.
- **Human–computer interaction** is the basic function of the Metaverse. How useful the Metaverse is to people is directly related to how well it performs, which in turn defines the limits of human perception.

- **Digital reality** will provide an immersive experience for users. Mixed reality, augmented reality, virtual reality, and 360° video are its components.
- **The great economic value** of Metaverse in education is far greater than that of traditional education. In the Metaverse, everyone is capable of creating various digital products. Besides, transactions among people will be more flexible and convenient (e.g., virtual to virtual and virtual to realistic).

Petrigna and Musumeci [14], divided metaverse technology into four groups: VR/AR, IoT, AI, and data twins.

i. **Augmented and Virtual Reality**: Technology like virtual reality (VR) and augmented reality (AR) is leading the way in developing metaverse immersion experiences [15]. By simulating a real-life setting on a computer, users of virtual reality (VR) headgear and gloves may experience a whole new universe. Conversely, augmented reality (AR) improves users' perceptions and interactions with their environments by superimposing digital material on top of the actual world [16]. When used in tandem, AR and VR provide the metaverse the spatial and sensory elements that make it come alive and invite people to engage with it in ways that are eerily similar to the real thing.
ii. **Internet of Things (IoT)** In order to link physical edge devices in the actual world to the metaverse, technology acts as a bridge between the two realms [17]. Everything from household appliances to cars may now connect to the internet and exchange data with one another thanks to the Internet of Things. With the help of the Internet of Things (IoT), the metaverse may become an even more natural extension of the actual world via the interchange and interaction of data in real time [18]. With this innovation, people may manipulate actual objects in a virtual setting or transfer data from the real world to the metaverse to see how it all fits together. By capturing and digitising real-world locations, things, and people, this technology faithfully transports them to the metaverse.
iii. **Artificial intelligence (AI)** technology powers the metaverse's reactivity and intelligence. The metaverse relies on AI for a number of features, such as the generation of dynamic and adaptable landscapes and the creation of life-like virtual characters that users may interact with. According to [19], virtual teaching assistants, language processing for students in various parts of the world, and evaluation of learning outcomes are all examples of how artificial intelligence (AI) is a crucial technology for improving Metaverse education. An important challenge for educational AI systems is developing learner-specific virtual teaching assistants; removing language barriers in communication; and developing fair methods for assessing students' progress towards learning goals.
iv. **Digital Twin**: The primary function of a digital twin in the Metaverse, as stated in [20], is the real-time digitisation of physical items into the virtual environment. Digital twins must meet stringent standards for accuracy and real-time performance in order to participate in some high-precision studies conducted as part of Metaverse schooling. This necessitates the capability for digital twins to detect and rectify mistakes [21].

In this regard, the novel concept of the metaverse, a hybrid of several technologies, may play a crucial role in enhancing and developing society's ability to provide education over the long term.

Consequently, education's goal should be to provide students with the cognitive and physical tools they'll need to become contributing members of society [22]. Experts in the field of educational psychology of teaching and learning agree that a number of factors contribute to high-quality education, including clear objectives, a conducive physical and social learning environment, intrinsic motivation on both the part of teachers and students, good classroom management, subject expertise, enthusiasm from teachers, and students' actual progress towards those objectives. Citizens and the nation alike hold education in the highest regard [23]. In the twenty-first century, we must move towards a stronger society if we are to address the complicated educational problems that exist today [24]. From a social perspective as well as one concerned with the spatial quality and its impact on students' development, the school's physical space is crucial to learning and education. This is because schools serve as the first model that shapes kids' personalities [24]. Educators must strengthen their own pedagogical knowledge, beliefs, and practices before implementing them in the classroom if they want their students to achieve success [25].

The metaverse has just come to light as a revolutionary technological development with exciting potential uses in the field of education. By constructing a virtual realm that provides users with a plethora of new chances and an improved learning environment, metaverse technology has the potential to improve education [26]. Effective pedagogical approaches are those that increase students' ability to retain and apply course content in a noticeable way [27]. The following table contrasts and compares traditional classes with online and metaverse courses according to certain criteria.

S/N.	Parameters	Traditional education	Online education	Metaverse education
1.	Location school	Location school	School, home	School, home
2.	Equipment	Book, pen, blackboard	Computer, mobile phone, tablet	Brain-computer interface, wearable d
3.	Teaching form	One-to-many	One-to-many, one-to-one	One-to-many, one-to-one
4.	Educator	Teacher	Knowledge sharer	Knowledge sharer
5.	Educated	Student	Learner	Learner
6.	Teaching content	Social and natural science	Interest, social and natural science	Customization
7.	Teaching purpose	Personal training	Personal training, enrich lives	All-round education
8.	Technology support	None	Web 2.0	Web 3.0

Source [28]

4 Implication of Metaverse on Education Effectiveness

There is a gradual blurring of lines between our reality and the metaverse. The power of virtual reality lies in the lifelike experience it can provide via varied user interactions [29]. Educators and lawmakers must put forth the effort for virtual learning opportunities to materialise; otherwise, Metaverse education initiatives will only transfer students to other locations rather than improve their academic performance. Imagine a world where advanced technology and artificial intelligence make everything conceivable, and where learning can go on forever, regardless of location or time [30]. By providing a setting that is both interactive and immersive, the Metaverse universe can meet the virtual and real-world needs of teachers, which is the most noticeable advantage it has for education. A wide variety of educational institutions are now investigating and testing out various uses of the metaverse in the classroom [31].

Thanks to advancements in Extended Reality (XR) technology, students are able to engage with virtual animals and acquire valuable knowledge in these settings [32]. Virtual learning environments (VLEs) are also important for building online learning communities, which in turn encourages students to collaborate, boosts their analytical skills, and makes interactive resources more accessible. Researchers explore potential applications of the metaverse across disciplines by means of cutting-edge virtual reality and augmented reality technologies that provide immersive educational experiences [33]. In addition to enhancing both conventional and online learning environments, they empower students to take an active role in shaping their own educational trajectories.

5 Impact of Metaverse on Education

Over the course of many centuries, the educational system has changed to include new information and technology. Developments within the metaverse include the miniaturisation of technical equipment and native plug-ins, a focus on the positive impacts of virtual tools on learning from elementary school groups to college students, and other noteworthy achievements [9]. But the most critical thing to consider is how future school systems will integrate technology into the all-around skill, sense, and knowledge development of its kids. Multimodal support for the visual, auditory, and kinaesthetic senses is required to enable effective learning in the metaverse [34].

Metaverse has positively impacted a changes in education in the under listed ways:

i **Visualization**: A microscopical picture of molecules or biological cells, for example, is something that students would have a hard time seeing in the physical world, but thanks to digital technology, the Metaverse makes it possible. On top of that, it may concretise theoretical concepts like relativity (Einstein's theory) by simulating ideal circumstances in physics [35] ii **Preventing Academic Misconduct**: Metaverse uses blockchain technology to root out instances of academic dishonesty. The use of this function to copyright protection allows for easy tracing and supervision of academic works' publishing, distribution, and dissemination [36]. Furthermore, smart contracts are only functional upon fulfilment of the contractual duties.

iii **Low Learning Costs and Risks**: It is common practice to require students to do experiments in certain courses. As part of Metaverse schooling, however, digitisation makes it possible to mimic all of these trials. Reducing resource utilisation is possible with this. The operational risk of a learner will also be reduced if they practise in environments with high levels of risk, such as with chemical ingredients that are flammable or explosive or in air crash simulators.

iv **Promoting Communication**: The present state of online classrooms suffers from ineffective communication and engagement as a result of geographical distance [37]. Students are unable to pay attention in class, and instructors are unable to pace their responses (such as body language and facial expressions) to maximise learning. Metaverse enables educators to construct digital meeting spaces. Students also have the option to design their own study spaces where they may go to get some work done, hang out with friends, or do all three. All of these things work together to make classroom interactions (and friendships among students) more positive.

v **Immersive Interactive Experience**: As a result of metaverse education, which removes the limitations of web-based instruction in Web 2.0, students will have a more enjoyable experience in a teaching class and learn more effectively in a real-world context (including observation and work).

vi **Unrestricted Time and Space**: Using metaverse education does not include any time limitation. Assuming digital representations of the topics are accessible, students may achieve this objective via the use of the Metaverse. Nevertheless, it removes restrictions based on location.

vii **Personalization**: Using digital twin generators or simulators to create their own avatars could boost students' confidence and enthusiasm in their academics. Furthermore, the educational system may utilise the data to develop regulatory-compliant lesson plans and instructional materials if users provide their permission.

6 Issues of Metaverse in Education

Some disadvantages do exist in the metaverse. Due of their ability to remain hidden behind the averter, these invisible beings may cause an increase in cybercrime. A distributed and infinitely free system with a mountain of data and easily accessible personal information can give rise to an illicit atmosphere [38]. Metaverse may further shake things up by threatening established theories of knowledge's ontological and epistemological underpinnings [39]. A lack of central authority or a feeling of groundlessness in this context might lead to social breakdowns, as described in [39], while concerns with virtuality and identity confusion can cause political uneasiness.

Other issues of Metaverse in education are shown as follows.

1. **Inclusiveness**: The original goal was to increase participation in the Metaverse and education. Make sure that as many people as can fit into the virtual space have their requirements met. Though challenging, it is filled with love.
2. **Privacy and Security**: People will undoubtedly spend even more time than before on the internet once the Metaverse comes online. To fully understand their learners' mental processes and behavioural patterns, businesses in the Metaverse will constantly collect massive amounts of personally identifiable information from them. Obtaining this data will definitely be a first. Under the framework of these three aspects of Metaverse education, we anticipate the following scenarios:

    i ***Confidentiality***: In order to achieve better educational effects, Metaverse education needs to collect identity information or personal emotion tracking [40] the sensitive personal information should be protected in multiple ways. It should be emphasized that the collection of information is not mandatory. If users choose to keep their information at the expense of a better experience, this should be considered reasonable and permissible.

    ii ***Integrity***: All information, especially about digital twins, on the educational system cannot be maliciously tampered with, and thus the system should strictly review the uploaded files to avoid penetration (e.g., Trojan horse programs, computer viruses). Otherwise, changing the information may cause damage to the digital twin. In this regard, combining the Metaverse with a blockchain is a direction worth exploring.

    iii ***Availability***: Users are able to enjoy services from the educational system of the Metaverse at any time and any place. Some systems of the Metaverse education can stand up to cyberattacks like DoS and DDoS.

3. **Addiction**: An unhealthy fixation with cyberspace may lead to a variety of medical, social, and mental diseases known as "cyber-syndrome" [41]. The more immersed a user is in their connection, the more likely they are to suffer from this disorder. By contrast, the XR experience amplifies the visual and auditory stimuli, which may motivate students to work more harder than they do now. Immersive gaming technologies seem to be crucial to Metaverse education at least for the foreseeable future.
4. **Governance Challenges**: Community governance should take precedence over other areas of the Metaverse due to the fact that learners' moral standards differ. So, to prevent moral concerns (including slang, insults, bullying, and shaming), it is crucial to promote and enhance communal standards of behaviour [42]. Given the size of the Metaverse cosmos compared to Web 2.0, another critical problem to address is the cost of monitoring.
5. **Technical Implementation**: To build the Metaverse, immersive interactive technologies like VR, AR, and MR are essential. Extended reality (XR) is the collective name for the aforementioned three technologies.

7 Opportunities for Exploring Meataverse in Education

The educational opportunities that Metaverse environment will provide according to [39, 43, 44], are portrayed below:

i. To provide education and learning opportunities anytime and anywhere by exceeding the limits of time and space.
ii. To create a rich learning environment that appeals to all senses
iii. To expand the teaching process as much as our imagination
iv. To offer individualized education and group work options
v. To provide student participation motivation.
vi. It helps to eliminate physical inequalities in education.
vii. Conditions such as race, gender and physical disability are features that will remain behind avatars in the Metaverse education world
viii. Metaverse offers zero risk and very low cost options, especially for high-risk areas such as military training / exercises.

Other opportunities in that could be found in exploring metaverse are;

i. Investigating and Achieving More Accessible and Inclusive Metaverse Environments: 0-5-10-15-20-25 Online persona administration Online marketplace Education and Training Information architecture Problems with the Copyright and IP Regulation and Ethics Procedures Annual Conference on Enterprise Information Systems (ICEIS) 2024 Platforms that are accessible to users with diverse abilities, backgrounds, and degrees of technical access are the goal of 492 development initiatives.
ii. Ethical Design for Better User Experience: Projects that prioritised protecting users' privacy and security while also creating a welcoming environment for them to interact with one another in the metaverse.
iii. Implementing Agile and Scalable creation Processes: Studies or initiatives that study integrating agile approaches for the dynamic creation of metaverse platforms, ensuring they are scalable and can grow with user demands.
iv. Regulatory and Governance Frameworks: Create thorough regulatory and governance frameworks to meet metaverse interoperability requirements, data protection, and privacy.
v. Research into enhancing VR/AR hardware, computing power, and storage solutions: these areas are crucial for meeting the demanding requirements of metaverse settings.
vi. Cybersecurity Enhancement: Initiatives to fortify metaverse platforms, shield users from any cyber dangers, and provide a secure virtual environment for communication.
vii. Efforts to Develop and Implement Standards for Interoperability: This section focusses on the development and implementation of standards for interoperability, which aims to facilitate the use of diverse metaverse platforms by ensuring consistent experiences across different virtual environments.

viii. Programs to Close the Digital Divide: These programs aim to remove barriers to participation in the metaverse by levelling the playing field between those who have and those who do not have easy access to digital technology.

8 Conclusion

The metaverse is a state-of-the-art platform for many businesses, including governance, education, and commerce, because to its distinctive features and strong technological underpinnings. There are several areas (human, technological, and procedural) where the Metaverse may need significant advancements as it is still in its early stages as a field. If our goal is to create virtual worlds that are more realistic, approachable, and practical, then it is technologically imperative that we upgrade the infrastructure. Resolving current restrictions and laying the groundwork for a more immersive and inclusive metaverse are both possibilities.

References

1. Akkoyunlu, B., Yilmaz, M.: Information literacy levels of teacher candidates, frequency of internet use and purposes of internet use. Euras. J. Educ. Res. **3**, 154–196 (2007)
2. Colomo-Magaña, E., Cívico-Ariza, A., Ruiz-Palmero, J., Sánchez-Rivas, E.: Problematic use of ICTS in trainee teachers during COVID-19: a sex-based analysis. Contemp. Educ. Technol. **13**(4), 314 (2021)
3. Januszewski, A., Molenda, M.: (2013). Educational Technology: A Definition with Commentary. Routledge
4. Anastasiadis, T., Lampropoulos, G., Siakas, K.: Digital game-based learning and serious games in education. Int. J. Adv. Sci. Res. Eng. **4**(12), 139–144 (2018)
5. López-Belmonte, J., Moreno-Guerrero, A.-J., López-Núñez, J.-A., Pozo-Sánchez, S.: Scientific production of flipped learning and flipped classroom in Web of Science. Texto Livre: Linguagem E Tecnologia **14**(1), 1–26 (2021)
6. Daniel, S.J.: Education and the COVID-19 pandemic. Prospects **49**, 91–96 (2020)
7. Çelik, R.: What is metaverse? conceptual evaluation and overview Balk. Near East. J. Soc. Sci. **8**, 67–74 (2022)
8. Lee, J.Y.: A study on metaverse hype for sustainable growth. Int. J. Adv. Smart Converg. **10**(3), 72–80 (2021)
9. Suh, W., Ahn, S.: Utilizing the metaverse for learner-centered constructivist education in the post-pandemic era: an analysis of elementary school students. J. Intell. **10**(1) (2022). https://doi.org/10.3390/jintelligence10010017
10. Lee, H., Woo, D., Yu, S.: Virtual reality metaverse system supplementing remote education methods: based on aircraft maintenance simulation. Appl. Sci. **12**, 2667 (2022)
11. Cheong, B.C.: Avatars in the metaverse: potential legal issues and remedies. Int. Cybersecur. Law Rev. **3**(2), 32–49 (2022)
12. Zhao, Y., Pino, A.M., Sánchez, M.C.: Digital competence in higher education research: a systematic literature review. Comput. Educ. **168**, 104–212 (2021)
13. Mystakidis, S., Christopoulos, A., Pellas, N.: A systematic mapping review of augmented reality applications to support STEM learning in higher education. Educ. Inf. Technol. **27**, 1883–1927 (2022)

14. Petrigna, L., Musumeci, G.: The Metaverse: A new challenge for the healthcare system: a scoping review. J. Funct. Morphol. Kinesiol. **7**, 63 (2022)
15. Dincelli, E., Yayla, A.: Immersive virtual reality in the age of the Metaverse: a hybrid-narrative review based on the technology affordance perspective. J. Strateg. Inform. Syst. **31**, 101717 (2022)
16. Wang, H., Ning, H., Lin, Y., Wang, W., Dhelim, S., Farha, F., Ding, J., Daneshmand, M.: Survey on the metaverse: the state-of-the-art, technologies, applications, and challenges. IEEE Int. Things J. **10**, 14671–14688 (2023)
17. Aloqaily, M., Bouachir, O., Karray, F., Al Ridhawi, I., El Saddik, A.: Integrating digital twin and advanced intelligent technologies to realize the metaverse. IEEE Consum. Electron Magn. **12**, 47–55 (2022)
18. Padmanaban, S., Nasab, M.A., Hatami, M., Milani, O.H., Dashtaki, M.A., Nasab, M.A., Zand, M.: The Impact of the Internet of Things in the smart city from the point of view of energy consumption optimization. In: Biomass and Solar-Powered Sustainable Digital Cities, Hoboken, NJ, USA, pp. 81–122 (2024)
19. Merabet, G.H., Essaaidi, M., Haddou, M.B., Qolomany, B., Qadir, J., Anan, M., Al-Fuqaha, A., Abid, M.R., Benhaddou, D.: Intelligent building control systems for thermal comfort and energy-efficiency: a systematic review of artificial intelligence-assisted techniques. Renew. Sustain. Energy Rev. **144**, 110969–111007 (2022)
20. Jones, D., Snider, C., Nassehi, A., Yon, J., Hicks, B.: Characterising the digital twin: a systematic literature review. CIRP J. Manuf. Sci. Technol. **29**, 36–52 (2020)
21. Gadekallu, T.R., Huynh, T.W., Yenduri, W., Ranaweera, G., Pham, P.V., da-Costa, D.B., Liyanage, M.: Blockchain for the metaverse: a review. arXiv preprint arXiv:2203. 09738:1–17 (2022)
22. Boden, K.K., Zepeda, C.D., Nokes-Malach, T.J.: Achievement goals and conceptual learning: an examination of teacher talk. J. Educ. Psychol. **112**(6), 1221 (2020)
23. Şentürk, C., Baş, G.: An overview of learning and teaching from the past to the present: new learning and teaching paradigms in the 21st century. In: Paradigm Shifts in 21st Century Teaching and Learning. IGI Global, pp. 1–19 (2020)
24. Zin, D.M.M., Mohamed, S., Kashim, M.I.A.M., Jamsari, E.A., Kamaruzaman, A.F., Rahman, Z. A.: Teachers' knowledge and practice in implementing the thematic approach In Pre-School. Int. J. Civil Eng. Technol. (2019)
25. Thomas, S.M.: School and teacher value-added performance and the relationship with teacher professional development in mainland China. In: International Perspectives in Educational Effectiveness Research. Springer, Cham, pp. 209–229 (2020)
26. Gonzalez-Moreno, M., Andrade-Pino, P., Monfort-Vinuesa, C., Piñas-Mesa, A., Rincon, E.: Improving humanization through metaverse-related technologies: a systematic review. Electronics **12**, 1727 (2023)
27. Westbrook, J.: Pedagogy, curriculum, teaching practices and teacher education in developing countries. In: *Education Rigorous Literature Review* (2013)
28. Hong, L., Shicheng, W., Wensheng, G., Jiahui, C., Han-Chieh, C.: Metaverse in education: Vision, opportunities, and challenges (2022). https://www.researchgate.net/publication/365820950
29. Lee, S.G., Trimi, S., Byun, W.K., Kang, M.: Innovation and imitation effects in metaverse service adoption. Serv. Bus. **5**(2), 155–172 (2022)
30. Park, S., Kim, S.: Identifying world types to deliver gameful experiences for sustainable learning in the metaverse. Sustainability **14**(3) (2022). https://doi.org/10.3390/su14031361
31. Mehta, M., Pancholi, G., Saxena, A.: Metaverse changing realm of the business world: a bibliometric snapshot. J. Manag. Develop. **42**, 373–387 (2023)
32. Lampropoulos, G., Keramopoulos, E., Diamantaras, K., Evangelidis, G.: Augmented reality and gamification in education: a systematic literature review of research, applications, and empirical studies. Appl. Sci. **12**(13), 6809 (2022)
33. Hajjami, O., Park, S.: Using the metaverse in training: Lessons from real cases. Euras. J. Train. Develop. **48**, 555–575 (2023)

34. Lee, H., Hwang, Y.: Technology-enhanced education through VR-Making and Metaverse-linking to foster teacher readiness and sustainable learning. Sustainability **14**(8), 4786 (2022)
35. Georgiou, Y., Tsivitanidou, O., Ioannou, A.: Learning experience design with immersive virtual reality in physics education. Educ. Tech. Res. Dev. **69**(6), 3051–3080 (2021)
36. Sharples, M., Domingue, J.: The blockchain and kudos: A distributed system for educational record, reputation and reward. In: European Conference on Technology Enhanced Learning. Springer (2016)
37. Kanematsu, H., Fukumura, Y., Barry, D.M., Sohn, S.Y., Taguchi, R.: Multilingual discussion in metaverse among students from the USA, Korea and Japan. In: The International Conference on Knowledge-Based and Intelligent Information and Engineering Systems. Springer (2010)
38. Kye, B., Han, N., Kim, E., Park, Y., Jo, S.: Educational applications of Metaverse: Possibilities and limitations. J. Educ. Eval. Health Profes. **18**, 1–13 (2021)
39. Akpınar, B., Akyıldız, T.Y.: Metaversal Öğretim: Fırsatlar ve Tehditler. 4. Küresel Pandemiler ve Multidisipliner Covid-19 Çalışmaları Uluslararası Sempozyumu. Ankara (2022)
40. Miller, M.R., Herrera, F., Jun, H., Landay, J.A., Bailenson, J.N.: Personal identifiability of user tracking data during observation of 360degree vr video. Sci. Rep. **10**(1), 1–10 (2020)
41. Ning, H., Dhelim, S., Bouras, M.A., Khelloufi, A., Ullah, A.: Cybersyndrome and its formation, classification, recovery and prevention. IEEE Access **6**, 35501–35511 (2018)
42. Inceoglu, M.M., Ciloglugil, B.: Use of metaverse in education. In: The International Conference on Computational Science and Its Applications. Springer (2022)
43. Duan, H., Li, J., Fan, S., Lin, Z., Wu, X., Cai, W.: Metaverse for social good: a university campus prototype. In: 29th ACM International Conference on Multimedia (ss. 153–161) (2021, Ekim)
44. Kalkan, N.: Metaverse evreninde sporun bugünü ve geleceğine yönelik bir derleme. Ulusal Spor Bilimleri Dergisi **5**(2), 163–174 (2021)

Environmental Challenges and Opportunities in the Metaverse

Govind Singh, Mahesh Mishra, and Ira Rana

Abstract The Metaverse, an expansive virtual ecosystem integrating augmented and virtual realities, presents both significant environmental challenges and opportunities for sustainability. This paper explores the ecological impacts of virtual worlds, including high energy consumption, carbon footprints from data centers and devices, and e-waste generation. It examines sustainable practices such as adopting renewable energy sources and energy-efficient technologies to mitigate these effects. Technological innovations, including blockchain and NFTs for carbon offsetting, AI-driven resource optimization, and low-energy computing, are highlighted as tools to enhance environmental responsibility. Social and economic implications, such as the digital divide and inclusivity issues, are discussed alongside regulatory frameworks for global standards and compliance. Case studies of green virtual events and best practices demonstrate practical applications, while emerging trends in green technology point to future directions. Drawing on interdisciplinary literature, the study calls for collaborative efforts among governments, industries, and users to build a sustainable Metaverse that minimizes real-world harm and leverages digital platforms for environmental awareness and innovation.

Keywords Metaverse · Environmental sustainability · Carbon footprint · Renewable energy

G. Singh (✉)
Department of Commerce & Finance, Quantum University, Roorkee, Uttarakhand, India
e-mail: govind.qsb@quantumeducation.in

M. Mishra
Department of Commerce, GLA University, Uttar Pradesh, Noida, India
e-mail: mahesh.mishra@lloydbusinessschool.edu.in

I. Rana
Department of Law, Quantum University, Roorkee, Uttarakhand, India
e-mail: ira.law@quantumeducation.in

S. Kautish et al. (eds.), *Green Metaverse: Fuzzy Logic Approaches to Sustainable Virtual Worlds*, Information Systems Engineering and Management 74,
https://doi.org/10.1007/978-3-032-08183-4_13

1 Introduction to the Metaverse

The concept of the Metaverse, a collective virtual space created by the convergence of virtually enhanced physical reality (VEPR) and persistent virtual reality (PVR), is not a new one. The word "Metaverse" itself was first popularized in a science fiction novel [1]. However, it wasn't until the past few years that it began to appeal to the general public and gained significant attention. As of the end of March 2022, Meta has shifted its strategic focus to the Metaverse, and recently Google began modeling its search engine around the Metaverse as well [2]. Nevertheless, the Metaverse doesn't pertain strictly to the Metaverse businesses being promoted by Meta and Google. Instead, it is an expanding or possible accumulated universe of the entirety of the "virtual worlds" domain. This domain includes video games, 3D chat rooms, AR spaces, VR spaces, web-based worlds, and more.

Calls for papers on the Metaverse have been circulating across various disciplines after 2021. These calls have originated from disciplines as varied as communication studies, virtual environment design, philosophy, computer science, political science, statistics, womxns studies, "gardenria", as well as tourism and hospitality. This speaks to a rapid growth of both positive and critical investment in the Metaverse occurring across the academy and industry. In particular, driven by the Meta piggyback, Google response, various scientific endeavors, and cultural shifts associated with COVID-19, the popularity and coverage of the term Metaverse have skyrocketed. In fact, the number of publications related to the Metaverse in 2022 directories has already quintupled that from all of 2021. So what accounts for this recent explosion of interest regarding the Metaverse that is so prevalent today? There are plenty of likely contributing factors involving advancements likely in technology, cultural conditions related to the pandemic and labor dynamics in tech, and economic impacts of technological investment. Beyond those factors, the Metaverse discourse serves as an interesting case study for research considerations within the progression and circulation of academic tropes.

1.1 Definition and Evolution

Metaverse is defined as a multi-layered, parallel digital universe that is a combination of various AR and VR applications, 2D application, and other sorts of digital platforms. In the digital plane, this universe imitates the true domain and provides engagement to users. For companies as well as users, the Metaverse ecosystem has created fresh possibilities. It paved the way for users to interact in new ways through their virtual selves, and thus a new environment that facilitates this engagement can cultivate a broad variety of financial returns for companies. In the Metaverse, compatible virtual experiences can be developed by users of gaming, social media, and shopping users, which will result in synergistic interaction. If people have established networks, digital cultures evolve so rapidly that they arise spontaneously. VR enables consumers to enter simulated 3D conditions where they can connect with their situations to performers or additional users. By turns, AR creates an interaction between the real, actual, physical world and the objects existing in it and the virtual world. AR uses numerous multi-senses, particularly vision and audition. The AR world has both digital and genuine elements, with digital material overlaid on genuine objects. Starting with visual notifications, the subsequent step is to inject interactive holograms into the real environment, which makes the production accessible to thousands of users nearly instantly. The introduction of Messenger's initial piece of AR knowledge was made by Facebook in 2018. On the other hand, Metaverse League, a virtual kickball game, worked as an online tournament simulator in the Roblox. Broadcasting sports is a market price approximately $204 billion. 62% of the sport industry's overall income comes from marketing uses [3].

2 Environmental Impact of Virtual Worlds

Virtual spaces have opened the door to tremendous socialization and creativity, fostering communities, businesses, and multifaceted applications. Concurrently, the overwhelming growth and spread of digital environments have raised a critical inquiry: what is their ecological impact? Server operations and data centers hosting the ever-expanding infrastructures that support virtual worlds are linked to energy demand with tangible consequences for the earth [4]. There are already concerns regarding the oversized carbon oligopsony produced to power these digital metaverses. But the engagement with virtual settings also requires an array of technological devices that follow energy-intensive processes in their manufacture. Such electronic gadgets encompass consoles, computers, VR headsets, and smartphones but also components like batteries, controllers, and peripherals. Discerning the complex stages of their life-cycle—as they are obtained, utilized, and discarded—unveils diversified paths of environmental repercussions. After all, the downgraded, hazardous, or unmanaged disposal of e-items produces a forward-propagating vortex of multiply negative impacts on environmental sustainability in a promise rarely aligned against the touting virtual enhancements. At a time when numerous companies and artists are preparing the commercialization and broadcasting of products and events in digital realms, it is essential to have a comprehensive interpretation of the duality characterizing the sustainability of the Metaverse. The design and establishment of virtual worlds are therefore cast as the innovative takeoff for a pertinent and conscientious conversation on the environmental problems trailed by the formidable expansion and devotion to digital settings. Ultimately, more mindful engagement backed by sustainable practices and projects in these realms—and stipulated from the industries commoditized by these artificial spaces—is persuasively contended as a crucial kuntute to the still incipient dialog enfolded by the phenomenal extent of the digitally assisted landscapes. On a quest to bring homeroom to the enclosure of the virtual cosmos with this pressing discussion, a decentralized endeavor is formulated involving routine analysts, digital worker consulting, and predestined emporium organizers in a gradual stroke of arguments taking upside in unraveling a mainstreamed virtual arrangement which is progressively built and engaged. The up thorough study of said materials is regarded as instrumental in developing an understanding of the crucial environmental issues in the increasingly complex and widespread urban landscapes enabled by digital technology.

2.1 Energy Consumption and Carbon Footprint

Understanding the disparities and congruity between different worlds is important because such comprehension may influence design decisions matching human requirements with computational capabilities. As digital platforms encompass a growing portion of human activity, there is a growing literature intent on leveraging digital traces to gain insights regarding how people interact with others and with information. This interest has found places in many disciplines, and studies have looked into social networks, search engines, e-commerce websites, mobile phones, sensors, social media, just to name a few. Insights are gained into how the world works, how it might be designed to work better, and the societal implications of these design and predictions [5]. Some of these studies begin by exploring how digital footprints associate with off-line phenomena including research on electoral outcomes, international development, and epidemiology. Digital twins promise to be detailed 3-dimensional models of the physical universe, continually updated with live sensor data. Analysis of digital twins is expected to monitor the health of the system, prescribe preventative maintenance, and allow for new designs or operation strategies in a safe virtual environment.

While such human-real world models have attracted growing attention, there has been relatively little work on the increasingly detailed digital worlds that people create and inhabit. With computational environments- like ambient displays, digital signage, smart homes, smart cities, e-textiles, mobile phones, augmented/virtual reality, and so on- play an increasing role in everyday lives, there is an opportunity to study digital footprints in a different context with different questions. [the Plush fabric in-home light-display] can make inferences regarding activities and moods within the situated, sensed reality. They segment the literature, discussing, first, how physical, social, and psychological attributes associate with different data modalities, and, second, considering ecological validity and ethical issues.

3 Sustainable Practices in Virtual Environments

As the Metaverse becomes more pervasive in our lives, it is crucial to consider its environmental consequences. Despite the virtual nature of these digital spaces, they have tangible effects on the planet when sustained by energy, materials, and devices in the real world [4]. Given the rapid growth of the Metaverse, fostering its sustainability is imperative to avoid exacerbating ecological issues. This delineates a set of recommendations to create and maintain the Metaverse in a sustainable manner, further empowered by multiple examples of innovative and sustainable practices. The Metaverse can serve as a platform for innovative solutions to pressing ecological problems. Virtual spaces can be used to visualize and understand complex environmental phenomena, changing the underlying approach to awareness, intervention, and mitigation. It involves not only integrating sustainability into the Metaverse but also utilizing the Metaverse for sustainability, cultivating a two-way relationship with a dual aim: to limit the negative footprint of digital environments on the real world and to leverage digital environments to promote awareness of sustainability issues.

Energy is the leading aspect of sustainability here, as the electricity needed to power online servers and devices is drawn from fossil fuels for the most part, which drives up carbon emissions. Therefore, it is imperative to adopt energy-efficient server technologies while maintaining high performance and effectiveness. The Metaverse should strive to be a sustainable digital ecosystem, with a reduced online resource consumption of water, food, packaging, and devices. This furthermore looks at the efforts of companies and organizations to integrate sustainability principles into the Metaverse. Through a combination of technological, economic, and educational methods, businesses foster online practices with more environmental and social responsibility.

3.1 Renewable Energy Sources

In the next decade, the Metaverse Industry Consortium is estimated to demand 1000-folds more computing power and push data usage by 20 times above 5G trends, which is beyond the scale of the 7 nm process. Sustainable development of networks, data centers, and devices is essential for the survival and push of the Metaverse. Accordingly, to the best of solely knowledge, the first effort to optimize energy efficiencies of all network components is made for realizing the Metaverse, and an analytical framework is provided in a technology-based approach for evaluating energy efficiency. Efforts should be driven towards a Green Metaverse in a timely manner [1]. As virtual worlds rapidly evolve, players and companies have a unique opportunity to influence how these worlds are powered. From the very beginning, vast digital landscapes populated by human avatars were run on servers in massive data centers. These servers consume a lot of energy and by 2007, virtual worlds, video games, and digital entertainment as a whole were already responsible for approximately 2% of global carbon emissions. However, virtual entertainment production was responsible for much fewer emissions than its digital representation. In 2022, most virtual entertainment productions are digital-based, from video editing to set designs and advertising campaigns for in-game promotions. Networks of computers and servers in data centers represent the physical basis for virtual representation, and they have a sizable carbon footprint. The Metaverse is heading toward mainstream adoption on a massive scale. It is predicted that now is the time to ponder how the virtual worlds are run and what path should be taken. However, some organizations have already started to see virtual lands and avatars run entirely on renewable energy. They have decided to go for a fairly aggressive timeline while still facing challenges and working to overcome them.

4 Technological Innovations for Environmental Sustainability

Environmental sustainability emerged as an increasingly pressing issue in digital spaces, which not only constitute physical areas but also virtually rendered landscapes that have an impact on the real world as well. Given the rise of the Metaverse as the next generation of the Internet, the special issue uses the opportunity to shed light on the environmental challenges and opportunities that come with the Metaverse, the entirely digital delegates of the real world, composed of collection of 3D virtual worlds, augmented reality, and digital artifacts. To do so, a series of essays by professional scholars are disseminated on various topics from different perspectives. Technological innovations have been at the forefront of fostering environmental sustainability in these digital landscapes, while providing perspectives across a variety of disciplines. These contemporary technologies not only harness the cutting-edge to minimize their environmental impact, they also optimize user experiences while promoting more socially attuned mitigative practices. Artificial intelligence (AI) and data analytics are highlighted tools that may be utilized to optimize the use of energy and allocation of resources in the digital infrastructures that already proliferate the globe. It is argued that virtual reality (VR) simulations can provide another novel landscape for environmental education and enhancing the awareness of the environmental challenges that face current and future generations. It is suggested that VR simulations can be cultivated to provide experiential learnings of these challenges, thereby promoting behavioral changes in both the physical and digital worlds. Low-energy computing is developed and discussed that aims to significantly reduce the power consumption of digital infrastructures, which may be disseminated across an emergently digital world. Smart grids are suggested as an emergent technological solution that may be utilized to monitor and manage the energy use of such landscapes in real-time [4]. By showcasing exemplary successful innovations that have been taken, the conversation is nurtured to continue the inter-, multi-, and trans- disciplinary approach needed to foster environmental responsibilities across the physical and digitally rendered landscapes. It is paramount that with the further proliferation of the Metaverse and the inevitably continued merging of the physical and digital, the necessity of collectively integrating the cutting-edge with the environmental responsibility is not only recognized, but also acted upon.

4.1 Blockchain and NFTs for Carbon Offsetting

The Metaverse represents a combination of physical reality and digital space, holding both promises and challenges for environmental sustainability. Digital currencies and the Metaverse have received substantial attention from the cryptocurrency market, as well as significant financial investment by high-profile investors. However, the environmental effects of this form of decentralised finance are still being explored. Blockchain technology offers the potential to enable the transparent determination of custody, rights ownership, and environmental integrity of forest carbon offset arrangements. Experimental analysis has shown that the deployment of a smart ledger is able to improve the accuracy, verifiability, and coordination of project-level forest carbon offsets. Using smart ledger on blockchain to register, issue, and transfer carbon credit NFTs allows for a transparent history of each credit to be maintained.

Blockchain has the potential to substantially enhance the transparency and accountability of forest carbon offsetting. This will require the development of standardised and extensible digital assets based on blockchains, as well as online applications that can read and write this data, and integrate it with off-chain databases. Every financial transaction made by a company generates a carbon footprint. By utilizing NFTs, a new digital asset can be created on blockchain with a unique badge, image, and/or metadata related to an environmental project. Each time this asset is transferred (including sales, gifts, or leases) a percentage of the linked project's operations is offset, and a linked site displays the current sales market value. Similar NFT car models have been used to fundraise for and promote the restoration of an extensive wetland.

The interest in cryptorforests underscores the need for rapid action on measurement, monitoring, and verification to counteract the preponderance of inadequate or counterproductive blockchain-based projects. To date, concern has primarily been on the failure of projects rated as "red" to obtain ongoing emissions reductions [6]. Nevertheless, projects producing substantial positive environmental and social outcomes have also been rated as "red." An emerging body of scientific literature is also refuting the basic premise of blockchain forestry projects: that they involve interventions which would not otherwise have taken place [7].

5 Social and Economic Implications

1. **Introduction**

Recent discourse around the Metaverse has sparked numerous conversations about what the digital trend entails, how it might unfold, and its potential implications. A growing field of research aims to project the shape of a digital society parallel to the tangible reality we know today [2]. Rooted in science fiction and long embracing various virtual manifestations, the notion of the Metaverse has taken on a more practical imperative with advancing technology and shifting norms. As this cyberspace contender becomes more tangible, it is critical to consider the advantages and pitfalls it entails, and the role it can play in shaping the broader world.

2. **Social and Economic Implications**

The Metaverse encourages introspection about the kind of virtual world that will arise, and how it will run. While there is potential for greater connection and global understanding, the paradigm shift to a virtual society often encourages unequal privilege. The so-called "digital divide" underlines that unequal access to technology can deepen prevailing social divisions. It is concerning if only some can benefit from the Metaverse's capabilities or flourish in a world with such technologies. Regarding inclusivity, the challenge is to provide an equivalent virtual space as the real world. Without action, there is a risk of fragmented lifetimes existing in parallel virtual realities rather than a democratic digital utopia. A more collaborative effort is needed to bridge this gap and ensure a more equitable virtual realm. The Metaverse challenges a broad range of economic norms and the nature of work. In some places, gig economies flourish, where workers can take on irregular temp jobs with limited rights, are preoccupied with current social security and lack of compensation. There are concerns about how digital equivalents operate. Furthermore, notions of possession become more abstract when commodities exist only in virtual spaces. It is also easy to engage in illegal activity in a digital society, be it pirating rare skins, laundering digital money, or tweaking real-world revenue streams. Conversely, the Metaverse also opens up areas for employment. Previously, many jobs and industries were developed within virtual environments. New professions may arise, such as digital architects, Cyberspace security specialists, or even philosophers dedicated to digital ethics. New markets can also emerge within virtual realities. For example, a burgeoning sphere of financial tools helps bank tokens inside the game, converting assets from the game into real currency and back again.

5.1 Digital Divide and Inclusivity

With the developments in the digital realm in recent years, there is no better organic medium for this paradigm shift than the Metaverse. The Metaverse has its own environmental challenges and opportunities, and while exploring its current form from the viewpoints of digital divide and inclusivity, virtual world operators, telecommunication companies and governments need to act together right now to develop a resilient Metaverse which actively prevents social disadvantages. As personal computers and internet networks have become popularized, the digital divide has been partially bridged. However, there are many people who do not experience the Metaverse because of their environment and social status. This phenomenon has now been classified as the third digital divide, which is still a problem in both developed and developing countries.

One difficulty in experience stems from incapabilities of internet communication, which are caused by using low powered personal computers, lack of machine availability and the high cost of internet connection. The telecommunication companies should make an effort to provide these services at more affordable prices, such as imposing a provider fee or expanding the service area. Another reason for this internet divide is the lack of service providers, local information and a shortage of broadband contents. Therefore, the government agencies should strengthen their financial aid towards content creation and local projects. Meanwhile, the municipalities and public organizations as a neutral entity should focus on user education and the notification of free service area. It is crucial to connect their domestic actions and collaborate with other global cities to form a global metropolis, which promotes an advanced and sustainable Metaverse. Cultural organizations or educational institutions should take part to serve as an administrator and mediator between the nations to sustain dialogue and academic debate. Looking at the current situation with isolation, exclusion, traced the distancing and declining privacy of physical space, the Metaverse should be an inclusive place where everybody can go. To do so, cooperation between the public and the private sector is essential. It is necessary for the government to provide guidance for the Metaverse companies to adopt the most suitable security and privacy technology. At the same time, technology manufacturers should design a form of technology where security and privacy considerations have been prioritized. Given the new nature of the Metaverse, the issue is particularly important for the government to maintain it as a socially safe environment.

6 Regulatory Frameworks and Policies

The Metaverse represents a novel digital landscape that includes a three-dimensional, interactive world containing users generated immersive experiences that extend many aspects of commerce, culture, and society. It is not a single entity or platform but is underpinned by emerging enabling technologies including virtual reality (VR), augmented reality (AR), spatial computing, haptic display, etc. The development of the Metaverse has brought many opportunities and challenges. The opportunities include: (1) access to timely information, (2) internet of value, (3) digital asset trade, (4) crossing the boundaries of space and time, (5) breaking through the limitations of reality in entertainment services, (6) multi-terminal services, (7) new lifestyle, (8) exploration of new business models, (9) business collaboration, (10) city governance, (11) public safety, (12) grand strategic value, and (13) a new market.

As digital representations of the physical world, emerging metaverse implementations offer measurable features and pose new opportunities for investigating the physics of public outdoor spaces. This study utilizes data acquired in an open-source multiplayer online digital environment simulating the real-world city of Berlin to quantify pedestrian movement patterns and social behaviors. Integral to this work is a summarization of the unique utility and challenges faced in the analysis of large-scale movement data sets distilled from metaverse environments—dynamically evolving digital spaces where throngs of virtual avatars navigate at man-made scales but display alien, bounded behaviors differing from both fully artificial systems and the real-world. Further empirical observations include the interplay of metaverse-evoked factors with socioeconomic predictors. In particular, it is found that crowd behavior is driven by individual, homogenous, and macroscopic forces comparable to those shaping actual human movement in streets and public squares. Intuitively created protection bubbles to regard virtual movement as a safety measure by assessing the difference in rates of social interactions and the distribution of speeds. Coercive social measures serve as a useful model for both increasing and decreasing physical social isolation to impact avatars' contact rates. Miyoshi data, a variant of Shen model implementing crowd persuasion, attains high compliance rates and a controlled increase in the likelihood of social connections. By repurposing across dataset applications, metaverse flows further help approximate the effects of secular changes by tweaking model parameters such as exogenous noise levels and population compositions. Despite their simplicity and reliance on digitality-specific structures, these strategies reveal qualitative improvements to user proximity and can gestate more parties to come into contact with each other—with lessons on behavior manipulation extending to the broader domains of event management and public health [1].

6.1 Global Standards and Compliance

The Metaverse is challenging and reshaping the real world, including through the creation of new environments akin to the imposition of the treaty ports in China, which generated governance difficulties given their extra-territorial nature. Unlike a landbound building, however, the affordances of digital environments are fully manageable by code, allowing for privacy auto-enforcement and ensuring accountability in the physical world even where compliance support is not readily available. Addressing digital regulatory capture may therefore better enable a similar embracing of progress and free trade beyond the artificial chasm between the real and digital worlds [1].

International treaties, regional coordination, and an eventual development of universally recognized and binding rules are expected to address many of the cross-jurisdiction concerns, however they are outstanding matters requiring a considerably more granular consideration. The meteoric rise of the Metaverse within a year has caught many by surprise, raising the need for a very learns and adaptable policy approach. The dictates of speed should not stop the policy and the political discussion, both of which are vital to adhere to the principles of openness, transparency, fairness, and inclusivity. The unilateral and sudden changes in digital public policy-makers are indeed disruptive and often have a global fallout solidarity with a multiplicity of other information society developments provides a decent foundation for co-operation that should be deployed on a wider scale.

International collaboration amongst like-minded countries and civic bodies on the basics of what constitutes acceptable digital behavior can lay a good foundation for global agreements. Companies with a view to expand the scope and uptake of a digital operation could appreciate that cross-border compliance is a trust investment that guarantees that the digital hosting environment is compliant with best practices. Leading by example and sharing experience and lessons on efforts to promote digital safety and accountability is essential; it is expected that once the visitor of the paths to pursue and the landscape model to search for, the adaptation of policies and regulations will become easier in a rapidly evolving digital technology and information environment. Success would also rely on opening the lead-making process to create public trust in a Metaverse governance and to construct a resilient and prosperous infrastructure that takes advantage of the new frontier and at the same time mitigates its intrinsic sustainability and innovation challenges.

7 Case Studies and Best Practices

Impactful initiatives promoting sustainability in the Metaverse have been put into place by several organizations that can inspire both practitioners and researchers to adopt more sustainable practices. This section reflects on a variety of case studies and best practices within the Metaverse environment, considering the challenges observed and the outcomes realized. Instances analysis from a plethora of sectors within the virtual environment will clarify sustainable challenges faced by the Metaverse. It involves case studies from video games, virtual reality musical experiences, blockchain, music virtual events, and virtual exhibitions. Moreover, various best practices will be shared as evidence of the belief that best practices should be exchanged and shared not just to be replicated but to inspire others with novel ideas or innovative realizations.

Implementing successful sustainable virtual practices can supply meaningful experiences to users while maintaining high engagement. In video games, initiatives such as more engaging storytelling and the implementation of ecological details can result in a positive outcome where players are given the option to confront environmental challenges. Likewise, a music virtual event with environmentally friendly blockchain enablement can be more informative about its carbon impacts through exposure to Eco-Chats. However, failures can also come out of adopting specific detailed concepts into virtual and sustainable practices. Due to financial reasons and time constraints, executing and exploring fully the greenhouse gas calculation part had to be discarded, which then led to an overall effect-ridden carbon footprint. It is emphasized that collaboration and innovation can evoke significant and long-lasting effects on sustainable challenges posed by the Metaverse [4].

7.1 Green Virtual Events

The Metaverse is an online virtual world that resides alongside the physical world, providing users access to experiences and interactions that are persistent, in real-time, and diverse [4]. It is a collective online virtual space with a collective system of interoperable telecommunication activities that are immersive in nature or Metaspring-branded instances connected. Embedded in the virtual world is metaverse city. "Metaverse in Metaverse City is not simply an online computer game, chat room or social network that can be played or used on a computer, a games console or smart phone, but is a virtual place or space that is a whole different 3D symbolic representation. From avatar representations, users can engage in creative endeavors, build trust and facilitate the identity integration of themselves and their participants." As artificial intelligence continues to evolve, realism in 3D content has become part of our everyday lives. This creates an opportunity for individuals to work and play virtually. I sense people will reside in the metaverse as an avatar inhabiting the local land within the metascape to surf the capital space for both virtual and web-physical experiences. In talk shows, interviews, computer games and concerts the metaverse landscape hosts farm, urban or any imaginable purchase avatar-based ventures. In two commercial settings, this division shows metaverse, a conference room, store, lecture theatre or a login version. In order for physical commerce to occur on the virtual platform, they set closed-off representations of real world venues. Towns, cities and metropolis scenes offer inclusive venues for conference. A richness of sites in the metaverse. These representations of world cities as metaverses are applied to the global usage of development in cities to meet amorphous political issues. Other examples demonstrate the chance for creative trade exchanges and human resource growth to be expanded. In doing so, easier trade flows and livelihood growth increase the richness and well-being of all desired video games' 3D metaverses, though seven buildings model towns and cities. Treatment on the model depends on similarities to cities worldwide reliant on how metaverse models are derived. When virtual metaverse cities are developed, what conditions should planning economists expect to be considered?

8 Conclusion and Future Directions

The Metaverse—a dynamic digital environment that blends augmented and virtual realities, connecting users in real time through persistent online spaces. The Metaverse enables new forms of digital life, breaking down the boundaries between the online and offline worlds. It creates boundless opportunities for businesses, individuals, and communities over the internet. Owing to fantastic visions depicted in science fiction over the past decades, the Metaverse projected in the near future was eagerly expected by the public. However, with the expansion of the scale and the increasing demand for the virtual world, the sustainability of the Metaverse was put into question. The energy footprint, e-waste, and transmission bandwidth were underestimated concerns as the Metaverse matured, but they attracted widespread attention from academia and industry when digital products showing the irreparable environmental harm as well as the rapid growth of the digital universe emerged.

In this paper, a comprehensive review of the distinctive features, status quo, and environmental impacts of the Metaverse was conducted, identifying the research gaps as well as key trends on exploring digital technologies contributing to enhancing environmental sustainability in the Metaverse. Furthermore, this work accentuates the sustainable perspective and calls for a common effort of governments, enterprises, and the public to create a sustainable Metaverse, which will inspire more stakeholders to walk hand in hand on the way of building a desirable virtual world.

8.1 Emerging Trends in Green Technology

The Metaverse continues to make substantial progress toward materialization as established industry players are strongly committed with substantial investments and other technical industries are taking the opportunity to substantiate their offerings. Recently, a diverse set of emerging trends in green technology are reshaping the path toward future Metaverse by considering the concurrent optimization of communication, physical computing and human elements that would define the spatial Internet (Zhang et al. 2022). Substantial reductions in carbon emissions and resource optimization can be enabled in digital environments through the utilization of supercomputers and low-power and renewable energy sources.

Advancements in a variety of established communication technologies, mostly with respect to their energy efficiency, are described. New technologies will be employed across the life cycle of all products that make up the digital environment, most notably sustainable and biodegradable materials. Green technology is also expected to have a huge impact on how devices allowing access to the Metaverse are manufactured. Artificial intelligence (AI) can also provide valuable tools to optimize resource usage and energy expenditures. Consumer demand can drive substantial supply chain improvements and industry innovation. An analysis of how recent developments in the green technology field have established a new foundation for the Metaverse is presented. Given ongoing and unsolved challenges, the importance of sustained investment in research and development over the long term is underlined. A future is envisaged in which the development and deployment of new technology is marked with stewardship of the environment and sustainability at its core.

References

1. Zhang, S., Yang Bryan Lim, W., Chong Ng, W., Xiong, Z., Niyato, D., Sherman Shen, X., Miao, C.: Towards green metaverse networking technologies, advancements and future directions (2022)
2. Lee, L.H., Braud, T., Zhou, P., Wang, L., Xu, D., Lin, Z., Kumar, A., Bermejo, C., Hui, P.: All one needs to know about metaverse: a complete survey on technological singularity, virtual ecosystem, and research agenda (2021)
3. Chen, Z., Gan, W., Sun, J., Wu, J., Yu, P.S.: Open metaverse: issues, evolution, and future (2023)
4. Pellegrino, A., Wang, R., Stasi, A.: Exploring the intersection of sustainable consumption and the metaverse: a review of current literature and future research directions (2023). ncbi.nlm.nih.gov
5. Liu, F., Pei, Q., Chen, S., Yuan, Y., Wang, L., Muhlhauser, M.: When the metaverse meets carbon neutrality: ongoing efforts and directions (2023)
6. Howson, P., Oakes, S., Baynham-Herd, Z., Swords, J.: Cryptocarbon: the promises and pitfalls of forest protection on a blockchain (2019)
7. Kotsialou, G., Kuralbayeva, K., Laing, T.: Forest carbon offsets over a smart ledger (2021). osf.io

Smart Grids and Virtual Power Plants: Enabling the Future of Decentralized Energy Systems

M. Yuvarani, R. Saranya, V. Karthi, T. Govindraj, and T. Vinitha

Abstract With the development of Smart Grids and Virtual Power Plants (VPPs), energy systems are developing towards decentralized, intelligent, and sustainable structures. With the goal optimize energy generation, consumption and distribution and improve efficiency, dependability, and resilience, smart grids make use of use of advanced connectivity, automation, and data analytics. For the purpose of to facilitate grid balancing and smooth coordination, VPPs serve up as digital a platform that integrate Distributed Energy Resources (DERs), includes solar panels, wind turbines, and energy storage devices. By enabling peer-to-peer continuous tracking, demand response, trade of energy, and these developments collaborate to enhance decentralized energy systems. By employing predictive analytics and machine learning they make possible the integration of renewable energy, enhance grid stability and reduce dependency on fossil fuels. They can also open up transformative potential by achieving energy sustainability and decarbonization. Therefore, this chapter discusses key technologies, benefits, and challenges associated with the adoption of these systems in the near future and the important role that these will play in enabling flexible and intelligent energy infrastructures.

M. Yuvarani (✉) · R. Saranya · V. Karthi · T. Vinitha
Department of EEE, Builders Engineering College, Tiruppur, India
e-mail: my.eee@builderscollege.edu.in

R. Saranya
e-mail: rs.eee@builderscollege.edu.in

V. Karthi
e-mail: vk.eee@builderscollege.edu.in

T. Vinitha
e-mail: tv.eee@builderscollege.edu.in

T. Govindraj
Department of EEE, Velalar College of Engineering and Technology, Erode, India

S. Kautish et al. (eds.), *Green Metaverse: Fuzzy Logic Approaches to Sustainable Virtual Worlds*, Information Systems Engineering and Management 74,
https://doi.org/10.1007/978-3-032-08183-4_14

1 Introduction to Virtual Power Plants (VPPs) and Smart Grids

Because they provide incorporate renewable grid stability, energy sources, improve energy efficiency, and lower carbon emissions, smart grids and VPPs have become very important components of modern energy systems. Smart Grids and VPPs make use of decentralized power generation, resulting in real-time energy stability, improved energy security, and demand management, when compared with conventional centralized grids with long transmission lines and giant power plants.

VPPs, however, are cloud computing applications specialized in the distributed energy resource management (DERM) of products such as battery storage, wind turbines, solar panels, and demand response equipment as a coordinated, flexible control generation entity. Smart Grids and VPPs are developed on the basis of the growing need for sustainable alternatives to energy, reflecting a shift from centralized power systems to more robust and adaptable infrastructures [1].

By integrating distributed energy production techniques, new technologies for communication and automated systems. Virtual Power Plants (VPPs) and clever grids point to the peak of energy management growth. A smart grid is a strong electrical grid which employs digital knowledge to predict, track, and get better electricity flow, ensuring effectiveness and reliability.

2 Key Technologies and Components in Smart Grids

For demand response and real-time monitoring for most favorable energy management in grids, Advanced Metering Infrastructure (AMI) allows dual utility-consumer notifications. Micro grids, wind turbines, and Solar panels are instances of distributed energy resources (DERs) which tender grid elasticity by decentralizing production while lessening addiction on conventional power plants. For the reason of to preserve the grid, Energy Storage Systems (ESS) such as pumped hydro store, batteries, flywheels, and overload energy and bring it when insist is high [2]. Monitoring systems based on conversation and the IoT makes fault discovery, real-time information compilation, and predictive preservation possible, ensuring grid operation and minimizing downgrade. When these technologies work together, a more vigorous, sustainable and brighter power grid that incorporates renewable energy sources and may get better energy efficiency is possible.

3 Virtual Power Plants: Architecture and Operation

A virtual power plant unites power-generating, controllable devices that are connected, decentralized and flexible- among other things. This includes battery storage, wind turbines, solar panels, and combined heat and power systems [3]. The primary purpose of virtual power plants is to optimally embed disseminated energy resources into the grid, thereby coordinating their operation to give the illusion that they function as a large power plant.

3.1 Structure of Virtual Power Plants (VPPs)

The main components of a virtual power plant consist of

a. **Distributed Energy Resources**: These are the specific power-generating units that form the virtual power plant, namely:
   - **Renewables**: Solar PVs, wind, small hydro systems, etc.
   - **Energy storage**: Batteries, pumped hydro storage, etc.
   - **Combined heat and power systems**: Combined generation of heat and electricity for efficiency.

b. **Centralized Control System**: This is the brain of the VPP-similar to a master control room in an energy management system [4]. It is a software platform that monitors and controls the operations of the DERs, ideally optimizing their dispatch and output to suit the instantaneous requirements of the grid. Some of the basic functionalities of control system are:
c. **Monitoring and Data Analytics**: Managing data from all of the distributed units, covering collection and analysis.
d. **Forecasting**: Predicting power generation (in the holder of renewable energy) or consumption (in the case of demand response).
e. **Optimization Algorithms**: Deriving the optimal dispatch of scattered energy storage and energy resources to help with balancing supply and demand efficiently, avoiding costs, stabilizing generating future energy supply without jeopardizing cost.
f. **Communication infrastructure**: Through the appropriate communication systems, real-time data regarding DERs may pass between each other, the control system, and the grid so that the system keeps the units organized and able to respond in unison to changes in grid conditions with immediate effect [5].
g. **Market interface**: In general terms, a virtual power plant interacts with power markets to sell energy or ancillary services, which may include frequency regulation, voltage control, or spinning reserves. They can be involved in various segments, such as real-time markets and day-ahead, balancing services, or even capacity markets (Fig. 1).

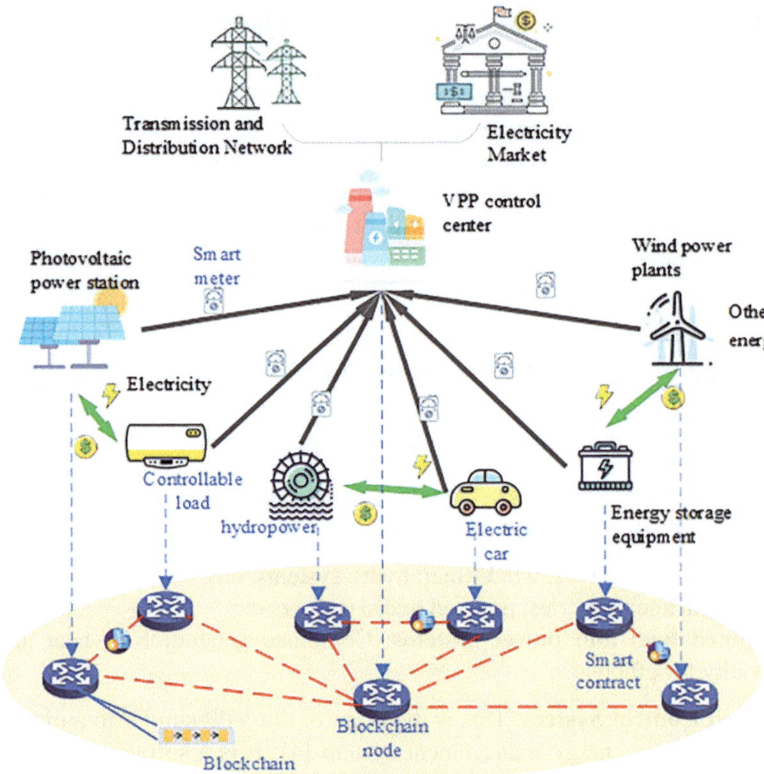

Fig. 1 Structure of virtual power plants—Demand side management

3.2 The Working Principles of VPP

The working of VPP is understand by studying how it integrates its resources into grid operations while optimizing that very operation.

a. **Aggregation of Resources**:

- The VPP aggregates energy from many distributed resources, normally generating power for many smaller generators such as rooftop solar, small wind farms, batteries, and flexible loads, into one virtual unit: it creates the same effect as that of a big-centralized generating station while keeping the decentralized structure [6]. The resources of a VPP can be located anywhere inside the grid area.

b. **Coordination and Control**:

- The VPP control continuously observes the operational performance and availability of the DER.

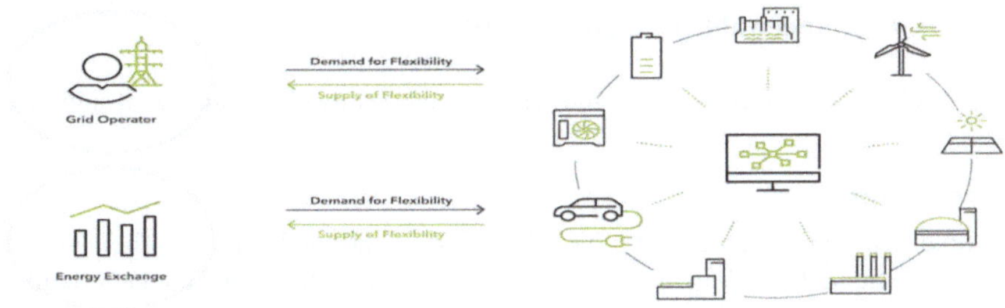

Fig. 2 Energy exchange in VPPs

- It modifies the total output of the distributed units to meet the grid demand on a near real-time basis. It uses forecast models (e.g., weather forecasting affecting solar or wind) and demand forecasts in order to optimize energy production and consumption.

c. **Dynamic Response to Grid Signals**

The VPP is highly responsive to signals from the grid operator [1]. These signals may include:

- Frequency regulation; such as, adjusting generation or consumption to sustain balance between demand and supply.
- Voltage control; that is, providing reactive power to stabilize voltage levels.

d. **Energy Trading:**

- It is also possible for the VPP to initiate trading on the energy market, selling power at times of high prices, buying power at low prices. To optimize its position in the market, the VPP communicates with the demand-side resources in categorize to generate, store, or consume electricity on demand based on grid needs and market prices [7] (Fig. 2).

3.3 Integration of Renewable Energy Sources

A virtual power plant (VPP) is a set of energy sources disseminated over decentralized energy resources (DER), including wind turbines, battery storage, solar panels, and system responses to insist [8]. Such resources are brought together via a digital platform for coordinated operation, monitoring, and optimization. The combination of renewable energies such as solar and wind into Virtual Power Plants (VPP) is another approach to effectively address the challenges specifically in relation to decentralized energy production; it has a vital role in improving grid constancy.

3.3.1 Use of Renewable Energies in VPPs

But in VPPs, the baseline renewable energy sources available are wind and solar energy.

Solar Energy

a. Solar Energy resources are typically placed at various locations. Their yield is intermittent, depending on the hours of the day, the climate, and the geographical position. VPP software, it can aggregate energy production from multiple solar installations, optimize, and channel the effectively generated energy.
b. Solar energy is thereby compatible with VPP because it can be harvested at different scales, including residential, commercial, and industrial [9].

Wind Energy

- Wind farms play a major part in renewable energies. As with solar energy, wind energy is intermittent and dependent on location.
- VPP can manage variability in wind power by predicting wind conditions and adjusting other resources, together with battery storage to take advantage of high periods of wind production.

3.3.2 Combination of VPPs with Renewable Energy Sources:

The effective addition of VPPs with renewable energy sources is accomplished by analysing the following data:

a. **Real-Time Data and Forecasting**:
    - Advanced forecasting techniques allow VPPs to predict renewable production (solar and wind) in a more refined manner, taking into consideration meteorological patterns and historical data.
    - Real-time monitoring of energy output assures that the energy mix remains stable while optimizing dispatch.
b. **Storage of Energy**:
    - Energy storage is an absolute must for VPPs with solar and wind. Energy in excess of demand can be stored in batteries for later discharge, assisting in minimizing the level of variances in contributing periods.
    - The battery storage ensures grid stability through regulation of frequency.
c. **Demand Response**:
    - Demand side management is done through levels of load control and adjustment to consumption by connected consumers of VPPs and returning it based on levels of active generation.
    - This flexibility can smooth out supply and demand and therefore eliminate recourse to dispatching the less eco-friendly sources during low-renewable production times.

d. **Backup Supply**:

- VPPs can provide the grid with frequency regulation, voltage support, and load balancing. Bringing in renewable energy sources will create less dependence of fossil fuel and reduce carbon emissions.
- VPP shackling will bring in more stability and reliability in bulk support of grid executing fluctuating renewable energy generation.

In Virtual Power Plants, integrating solar and wind energy open a way toward increasing energy system reliability and sustainability [10]. By using state-of-the-art technologies such as energy storage, forecasting and demand response-management of renewable energies along with the optimal utilization of decentralized resources allows them to trigger the green revolution toward resilient power grid development.

3.4 Role of Aggregation and Optimization Algorithms in VPP

Aggregation and optimization algorithms are greatly responsible for how efficiently a Virtual Power Plant (VPP) operates. They take ownership of the efficient management of a varied mix of distributed energy resources (DERs) under ideal coordination between demand-side assets, storage, and generation. Let's examine closer the functions of optimisation and aggregation in VPPs.

3.4.1 Role of Aggregation in VPPs

Usually, aggregation is the process of aggregating several distributed, decentralised energy resources—such as wind, battery storage, solar, demand response units—to form a single virtual entity [11]. This aggregated entity can then be controlled and dispatched as if it were a single, large power plant. The benefits of Aggregation is as follows:

a. **Operational Ease**:

- Instead of managing each individual distributed energy resource (DER) separately, aggregation allows for the coordinated control, monitoring, and operation of multiple assets in an efficient manner.
- The aggregation of resources increases system flexibility enabling superior integration of renewable energy sources [12].

b. **Wider Market Participation**:

- While the individual resources may not be independently viable to enter into energy markets themselves, aggregated DERs stand to turn profitable from entering into those markets.
- By pooling resources, VPPs can meet the size and dispatch requirements for market participation, increasing revenue potential and stability.

c. **Improved Risk Management**:

- Aggregation brings smoothness to variability and intermittency of individual assets [13]. If one particular wind turbine, for example, is generating a lesser amount of energy due to certain weather conditions, the other turbines or solar panels in the aggregation can work toward compensating that, thus stabilizing overall energy output.

d. **Better Load Balancing**:

- It facilitates effective balancing of supply and demand across different sectors (residential, commercial, and industrial) to avert congestion on the grid. Thus then, optimal load-balancing operations are easily possible.

3.4.2 Role of Optimization Algorithms in VPPs

Optimization algorithms are used to resolve the finest way to dispatch and utilize the aggregated resources in a VPP [14]. Their goal focuses on the optimization of the VPP operation by minimizing costs, maximizing efficiency, ensuring stability and balance of the power grid.

a. **Areas of Optimization in VPPs**

- Energy Generation Optimization
- Energy Storage Optimization
- Demand Response Optimization
- Profit Maximization
- Grid Stability and Reliability

b. **Types of Optimization Algorithms Used in VPPs**

- Predictive Algorithms
- Real-Time Optimization Algorithms
- Mathematical Programming
- Machine Learning and AI Algorithms
- Stochastic Optimization.

Aggregation and optimization algorithms are the foundation of Virtual Power Plants. Aggregation integrates diverse energy resources into one entity, while optimization algorithms maximize efficiency, cost-effectiveness, and grid stability in their operation [15]. Together, they allow VPPs to balance supply and demand, make additional successful use of renewable energy sources, and support reliable grid functioning; altogether are very important in the evolution to a resilient energy future and sustainable.

4 Benefits and Challenges of Decentralized Energy Systems

Decentralized energy resources (DERs), are small-scale electricity generation facilities, energy storage space devices, and demand-side management devices that are situated at or near the point of use [16]. Among the systems are microgrid configurations, small-scale hydroelectric plants, solar photovoltaic systems, wind systems, battery storage facilities, combined heat and power (CHP) systems, While presenting special difficulties, decentralisation of power generation and control offers a spectrum of advantages at once [17].

4.1 *Energy Efficiency Improvements*

The efficiency of DERs is improved by the following ways:

a. **Reduced Transmission Losses**:

In conventional energy systems, electricity is being transported over long distances from power plants to consumers. This results in considerable energy loss due to resistance in wires. Decentralized energy systems generate power locally where it is being consumed, which decreases such losses and improves the efficiency of the system.

b. **Local Generation**:

Decentralized systems such as rooftop solar photovoltaic panels or mini windmills are able to generate power where it is being consumed. This means that the generated power can be utilized instantly at the point of generation or the surrounding area, reducing the need to transport electricity over long distances and conserving energy from being wasted.

c. **Improved Demand Matching**:

Distributed systems are able to match local energy requirements better. For example, solar power is best when individuals consume more energy during the day, and wind power can be utilized during peak usage hours. Such matching reduces the need for backup power from large plants to match peak usage, improving the overall efficiency of the system [6].

4.2 *Reliability Improvements*

Grid Outage Resistance

Decentralized systems, especially microgrids, can function autonomously if the central grid fails or has a power outage. This makes communities immune to natural disasters or grid failures. Microgrids can supply backup power locally and restore service faster than waiting for central grid maintenance.

Distributed Redundancy

Many little power producing plants provide backup choices for the energy system. Should one generator fail, others can still provide electricity, so reducing the possibility of total service loss.

Enhanced Response Times

Distributed systems bring power generation closer to the consumers so they may react faster to changes in demand or supply [18]. When generating is low, for example, battery storage systems can offer instantaneous power, so stabilizing the grid.

Enhanced Grid Stability

By providing extra services like frequency control, voltage management, and load balancing, including distributed energy resources to the grid keeps it stable. These services particularly help in cases when the grid depends more on variable renewable energy sources like solar and wind.

From more energy efficiency to more dependability, more environmental sustainability, and financial savings, decentralized energy systems have many benefits. By means of distributed power generation and storage, we can create more sustainable energy systems with less reliance on centralized fossil-fuel power plants, so fostering security. To fully realize such systems, however, intermittent grid connection, regulatory hurdles, and grid connection must all be overcome [1]. Decentralized energy can be at the leading edge of the future of the global energy system with suitable technologies, policies, and investments.

4.3 Grid Stability, Scalability, and Resilience in Decentralized Energy Systems

When discussing distributed energy systems, resilience, scalability, and grid stability are especially important ideas since these systems are meant to function in ways either complementing or replacing conventional centralized power grids. New dynamics into the generation, distribution, and consumption of electricity are brought by decentralized energy systems including distributed generation (e.g., wind turbines, solar panel, micro grids), energy storage (e.g., batteries), and demand-side management.

4.3.1 Grid Stability in Decentralized Energy Systems

Grid stability is the capacity of the electricity grid to sustain a level of power even in the course of supply or demand fluctuations. Since consumption and production are not concentrated in one location, decentralized energy systems have a more complicated stability. Issues with grid stability are.

- Intermittency of Renewable Energy Sources
- Bidirectional Power Flow
- Voltage and Frequency Regulation.

These challenges can be overcome by the following solutions to maintain stability in the system.

- Energy storage space systems
- Highly developed Grid Management Technologies
- Demand Response
- Grid-Forming Inverters.

4.3.2 Scalability in Decentralized Energy Systems

Scalability in systems is their capacity to expand and grow as demand increases or technology develops free from appreciable performance or efficiency loss.

- Particularly in cases where the grid was not initially designed to control distributed energy; major changes can be demanded from energy sources including solar panels, wind turbines, and storage facilities into current grid architecture.
- Different interconnection policies for different technologies can make seamless integration of new systems into the grid difficult, so restricting scalability. By means of advanced optimization tools, one can balance producing, storage, and demand while coordinating and controlling these increasing resources.
- Decentralized systems can solve these using modular design, which can be added just without disturbing the system overall.
- Modern control systems and software platforms using artificial intelligence and machine learning can efficiently manage and maximize huge distributed energy resource (DER) systems.
- Microgrids offer a scalable approach to enable localized energy distribution that can expand with demand without endangering grid stability.
- Blockchain technology can also enable peer-to–peer energy trading and help secure and transparent transactions, so enabling efficient integration of new resources. Scalability is then further promoted.

4.3.3 Resilience in Decentralized Energy Systems

By decentralized energy systems, we simply mean that they may analyze how the resilience of a system can deals with disruptions, whether due to natural disasters, technical failures, or cyber attacks [19].

- They share a number of weaknesses and challenges with respect to the toughness of the decentralized energy systems. One is the vulnerability of small-scale distributed generation units to local disruptions, such as extreme weather events or accidents, which could lead to outages, assuming there is no storage or energy backup.

- Being decentralized and interconnected, their level of exposure to risk increases due to the cyber security threat where an attacker can exploit the communication network to undermine grid security [20].
- The coordination of such responses would be quite painstaking, such as those from extreme weather or grid failures without a centralized system that will make the various distributed assets act in a coordinated fashion.
- Possible measures involve appropriate solutions like microgrids which can operate in an islanding mode during grid failures, allowing the system to continue to run while waiting for the grid to resume normal operation.
- They provide backup power during disruptions, batteries should be cold, power supply continuity even when the active supply is out, should have backup via redundant paths. Restoration of service can be carried out within seconds, and rapid grid restoration tools can restore power following the repair of the fault [21].
- Cybersecurity is prime in protecting decentralized systems against cyber-attacks, providing safe and secure operation of the grid with absolute assurances of compliance with work toward stability guarantees.

4.4 Cyber Security Risks and Data Privacy Concerns in Decentralized Energy Systems

Key problems that must be resolved if we are to guarantee the safe and effective operation of distributed energy systems are cyber-attacks and data privacy issues. Based on digital technologies and communications for monitoring, control, and coordination, decentralised energy systems including distributed energy resources (DERs), smart meters, microgrids, energy storage, and demand-side management [22] are Cyberattacks and data space to yourself invasions rise as these systems develop and get more linked, so endangering system integrity, user data, and public safety.

4.4.1 Cybersecurity Risks in Decentralized Energy Systems

- The vulnerabilities in communication networks, particularly in smart grids and DERs, create windows of opportunity for cyber attacks such as data manipulation or unauthorized control. This can result in operational failure, incorrect energy supply, or mass outages if cyber attacks compromise critical infra structure [23].
- Malware, ransom ware, and Denial-of-Service (DoS) attacks complicate the security situation further by targeting communication channels or crashing system operations altogether.

- Moreover, the decentralized nature of these power systems, while ensuring efficiency, also provides windows of opportunity for cyber attacks to take advantage of vulnerabilities in remote monitoring and control systems.
- Unauthorized access to smart grids or DERs can lead to manipulation of energy flows, system failure, or hazardous conditions. A successful cyber attacks on such systems, particularly in the event of poor security at endpoints or network segments, can severely destabilize grid stability and public safety [24].
- While there is greater automation and a greater role for IoT in managing energy, securing decentralized energy systems against cyber attacks is important to avoid large-scale damage and ensuring reliable service.

4.4.2 Data Privacy Concerns in Decentralized Energy Systems

The data generated from smart meters, patterns of energy consumption, dispersed energy resources, and consumer behaviour supply decentralized energy systems with an ample amount of data [25].

- Optimising its energy management, performance, and decision-making by consumers and utility operators depends on this knowledge. Still, it begs quite grave privacy questions.
- Smart meters and Internet of Things (IoT) devices record minute information on energy consumption. Should such data be hacked, personal behaviour patterns could be exposed, so raising the possibility of privacy invasions including identity theft or burglary.
- Private Consumer Information: Personal data including names, addresses, and payment details linked to energy consumption could be revealed during a security breach. From this one can follow financial fraud and invasions of personal rights.
- Data Flow Among Many Entities: Among many stakeholders, decentralized energy systems include utilities and outside service providers who access consumer data [26]. Inappropriate data sharing or management calls great concern about invasions of privacy.\Third Party Analysis: Data for predictive analytics or customer segmentation could lead to commercial exploitation devoid of sufficient privacy protections, so aggravating privacy issues. Data breach risks: Decentralised systems let private consumer data on personal energy consumption to be exposed. Thanks for this exposure, cybercrime has opportunities to participate in fraud or other illegal activities.
- In absence of data encryption When unencrypted data is passed between devices—say, from smart meters to cloud services—it becomes susceptible to interception, so compromising privacy and security policies.
- Variability in rules on privacy: The always changing terrain of privacy rules—including GDPR—can lead to legal uncertainty for consumers and utilities about their responsibilities within distributed energy systems [27].

- Consumers have to be suitably informed and consent to the ways of data collecting and distribution. Lack of openness or clarity in these consent processes increases likelihood of consumer rights being violated.
- Giving consumer data to Third Parties without suitable permission could lead to Unjustified Monitoring or Personal Information Abuse.

4.4.3 Mitigating Cybersecurity Risks and Data Privacy Concerns

Cybersecurity Measures

- Encryption and Secure Communication
- Multi-Factor Authentication (MFA)
- Regular Software Updates and Patches
- Intrusion Detection Systems (IDS).

Data Privacy Protections

- Data Anonymization and Minimization
- User Consent and Transparency
- Access Control and Auditing
- Compliance with Data Privacy Laws.

Regarding grid resilience, sustainability, and flexibility, the much acknowledged advantages of distributed energy systems present more cybersecurity and data privacy risks. Strong cybersecurity systems and encryption mechanisms together with user consent mechanisms and rigorous adherence even to local rules safeguarding privacy are needed to protect these systems and their users.

5 Market and Regulatory Structure for Smart Grids and VPPs

- Smart grids improve dependability, efficiency, and resilience of power distribution by means of their integration of advanced communication, automation, and real-time monitoring technologies. This helps the market and legal environment for Smart Grids and Virtual Power Plants (VPPs) fit the path towards distributed energy systems, renewable integration, and digitalisation. By combining distributed energy resources (DERs) including wind, batteries, solar and demand response systems, VPVs increase electricity generation and consumption in the meantime.
- These changes guarantee smooth involvement of all the participants, including manufacturers, aggregators, and utilities; they also imply exact rules and market systems that guarantee grid flexibility and unbiased competition.
- The spread of Smart Grids and VPPs will rely much on guidelines directing data management, pricing policies, and grid connectivity.

- Incentives and regulatory assistance have been established in a number of countries to encourage demand-side management, dynamic pricing, and grid upgrading [4].
- Market structures are changing for Smart Grids and VPPs, moving more towards flexible and competitive models, where new entrants can be well included in energy markets. Aggregators and independent power producers increasingly balance supplies and demands via real-time energy trading and grid services [5]. Blockchain, AI, and IoT enhance transparency and efficiency within such networks.

In the future, the standard rules of policy need to be well defined in such a way that it will encourage innovation, provide grid stability, fair pricing, and consumer protection. The clearly defined market and regulatory framework is crucial to realize the potential of Smart Grids and VPPs for evolution towards a resilient, sustainable, and well-organized energy future.

5.1 Policy and Standards for Decentralized Systems in Smart Grids and VPPs

- Distributed power systems, such as Smart Grids and Virtual Power Plants (VPPs), require specific rules and regulations to ensure their effective operation, collaboration, and adaptability.
- Governments and regulatory bodies all around the world have put in place mechanisms that support distributed energy resources (DERs), such as wind, battery storage, photovoltaic solar electricity, and demand side planning frameworks [8]. Key policies focus on grid interconnection standards, dynamic pricing models, and incentives for renewable energy adoption. Regulations like FERC Order 2222 in the U.S. and the EU's Clean Energy Package will now allow DERs to interact with energy markets, thereby allowing decentralized energy solutions to remain fair and economically viable.
- Standards are essential for distributed systems to be suitable, safe, and dependable [9]. Institutions like the Institute of Electrical and Electronics Engineers (IEEE) and the International Electro technical Commission (IEC) have created technical strategy on cybersecurity, electrical grid stability, smart metering, and networking standards.
- In particular, IEEE 2030.5, also known as SEP 2.0, permits DERs and providers to work cooperatively securely, while IEC 61850 provides a structure for intermediate automation. By putting the above recommendations into execution, VPPs and Smart Grids can be easily incorporated, avoiding grid breakdowns and maximizing efficiency in operations.

Future policymakers will concentrate on enhanced cybersecurity precautions for decentralized networks, blockchain-powered cooperative energy trading, and powered by AI grid management. In order to test cutting-edge technologies, accelerate the approval process, and strengthen market rules, governments must establish unambiguous regulatory sandboxes.

5.2 Demand Response and Pricing Mechanisms in Smart Grids and VPPs

- Demand response (DR) and Dynamic pricing are very important components of the market and regulatory frame for smart grids and virtual power plants (VPPs) [28] For the reason of to maintain a successful supply–demand balance, demand response technologies permit customers to modify their electrical utilization in response to promotions, actual time grid conditions.
- The regulators and utilities adopt incentive based and price based demand response approaches to motivate users to shift or reduce their electricity during peak insist periods [29]. This flexibility enhances grid reliability, lowers stresses on infrastructure, and supports the incorporation of renewable resources such as wind and solar.
- To support pricing in Smart Grids, cost-effective energy distribution and VPPs depends on real-time energy demand and supply. Time-of-Use (ToU) pricing and Critical Peak Pricing (CPP) and Real-Time Pricing (RTP) are frequently utilized methodologies that tender customers price signals that optimize their consumption of energy. Blockchain-enabled peer-to-peer (P2P) energy trading models allow producers to purchase and advertise excess power openly in decentralized markets, promoting consumer involvement and energy democratization [30]. Peer-to-peer (P2P) power trading models, supported by block chain technology, enable prosumers to buy and sell excess energy directly within decentralized markets, further encouraging consumer participation and democratization of energy.

However, what also has to come in handy for these mechanisms would be the establishment of transparent market rules on reliable communication protocols and fair models of compensation in case something appears overly impractical. In summary, smart sensors and advanced metering infrastructure provide automated DR solutions and real-time pricing modifications. With the purpose to support cost savings, grid stability, and improved consumer engagement in the energy transition, regulatory frameworks will require to make confident that demand response and pricing mechanisms are adaptable translucent and easily accessible as energy markets change.

5.3 *Role of Government Incentives and Subsidies in Smart Grids and VPPs*

Incentives and subsidies by the government will accelerate the Smart Grids and Virtual Power Plants adoption by offering financial support and policy-driven motivation for the stakeholders. It motivates utilities, businesses, and consumers to invest in renewable energy technologies, power storage systems, and sophisticated grid infrastructure. To motivate the integration of distributed energy resources (DERs) and improve grid modernization, governments across the world enact policies which involve feed-in tariffs (FiTs), tax credits, subsidies, and rebates [31]. By contributing vital resources for the deployment of smart grids, measures like the the US's Investment Tax Credit (ITC) and EU's Green Deal promote the general availability and profitability of sustainable energy solutions.

Subsidies are also provided on demand-side participation and energy efficiency plans. Subsidies encourage consumers towards DR schemes, time-of-use pricing plans and P2P energy trading. Governments provide assistance to the installing of smart metering infrastructure, battery storage solutions and AI-driven energy management solutions so as to make grids more flexible and reliable. Additionally, integrators and companies accept compensation for achieving targets on pollution reduction, energy efficiency, and grid stability by means of performance-based incentives (PBIs) intended by regulating bodies. By humanizing the incorporation of the storage, consumption and production of renewable energy sources, these actions afterward assurance a bendable and dependable energy ecosystem.

Policymakers must gradually move from direct subsidies to self-sustaining, competitive energy markets and establish an appropriate equilibrium among incentives and market-based mechanisms in order achieve long-term sustainability. By introducing rules and regulations which facilitate the testing of innovative smart grid technology and business models, governments can further tackle hurdles to regulation, accelerate the permitting process, and promote innovation [32]. Public–private cooperation is supported, renewable energy transitions have been accelerated, and smart grid and VPP technologies have been rendered easier to access, cheaper, and cost-effective overall with a designed effectively economic framework (Fig. 3).

Here is a diagram of the Market and Regulatory Framework for Smart Grids and VPPs. This is important in depicting government policies, pricing mechanisms, incentives, consumer participation, grid modernization, and renewable integration among other elements of how they connect to each other.

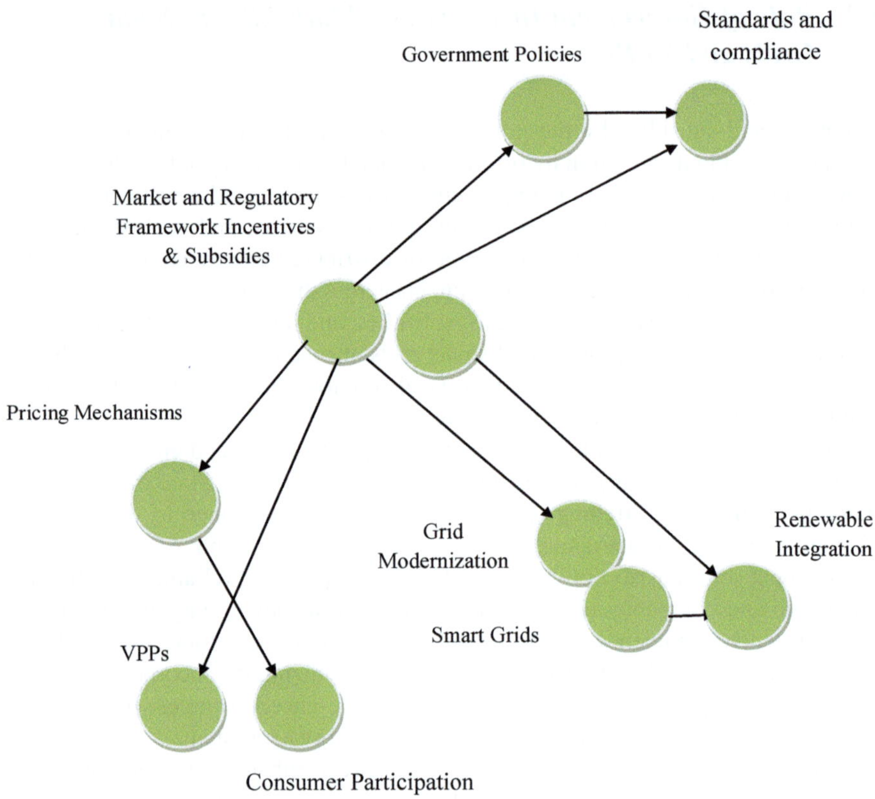

Fig. 3 Market and regulatory framework for smart grids and VPPs

6 Future Trends and Innovations in Decentralized Energy Systems

- In order to increase grid resilience, efficiency, and sustainability, decentralized energy systems continuously change through improvements in technology and reforms to legislation. Real-time energy management, maintenance predictions, and computerized responses to demand from the incorporation of artificial intelligence (AI) and machine learning are a few of the primary topics that will guide the future. Block chain technology is also rising as a game-changer, which enables secure and transparent peer-to-peer (P2P) energy trading.
- Producers can sell and buy excess electricity without intermediaries. They complement the development of Virtual Power Plants (VPPs) and micro grids [33].They optimize scattered sources of solar power, wind energy, battery storage, which together create flexibility in a far more decentralized approach to power grids.

- The main other trend refers to the upsurge importance of energy storage solutions, of course, not limited to some new advanced versions of batteries-like solid-state ones or lithium-sulfur based ones, bringing benefits in increased energy retention levels and grid stabilization.
- V2G integration and hydrogen-based storage systems are also varying with electric cars performance as mobile sources of energy. Decentralized systems present predictive analytics, smart metering, and real-time energy tracking through enhanced connections made probable by IoT and 5G [34]. In terms of consumer organization on energy consumption, pricing, and the degree of sustainability gained, this supports the creation of another consumers-centric energy marketplace.

Looking forward, policy and regulatory frameworks will be a vital enabler for decentralized energy innovations. Governments will introduce market-based incentives, regulatory sandboxes, and standardization protocols to help accelerate the smart grid solutions and DERs adoption. The power industry will see a transformation as it moves toward localized energy markets, energy-as-a-service models, and community-driven energy projects. A future in which energy is fully interconnected, resilient, and sustainable and provides power to the consumer with global energy security on the horizon lies ahead of the decentralized energy system.

7 AI and Machine Learning for Predictive Analytics and Grid Management

7.1 Predictive Analytics and Grid Management by Machine Learning and AI

- ML and AI are transforming predictive analytics and grid management within the framework of decentralized energy systems [35]. According to these utilities, advancements, VPPs, and microgrids be able to at the present successfully control energy delivery and operation all the way through forecasting, automation, and real-time data analysis.
- Predictive analytics with AI can anticipate renewable energy generation patterns, power demand fluctuations, and equipment failures, enabling grid operators to take proactive measures in preventing outages and enhancing grid reliability.
- By using the large datasets of smart meters, IoT sensors, and weather forecasts, machine learning models go through continuous improvements in accuracy that enable grid operations to be efficient and balanced with energy.
- In grid management, AI-based algorithms for energy optimization actively participate in real-time adjustments of power flows and balances between supply and demand. AI-enabled autonomous energy trading platforms are able to establish

real-time pricing of electricity, peer-to-peer (P2P) transactions, and response to demand from the grid through reduced reliance on centralized grid infrastructure.

- In addition, artificial intelligence has enhanced extrapolative maintenance and fault recognition, allowing businesses to distinguish grid deviations, expect evils and assemble revival when main interruptions appear.
- These AI-driven substances' victorious integration of disseminated electricity sources, like solar, battery storage and wind will permit their greatest utilization with centrally forbidden systems [36]. Future directions in AI and ML will shape smart grid automation further, towards self-healing grids that would seek power flow reconfiguration upon faults or failures.
- Reinforcement learning algorithms will facilitate adaptive grid management where AI adapts to real-time energy market conditions and behaviors of the users for optimization of operations [37]. With the broad growth of decentralized energy systems, regulatory frameworks need to evolve into accommodative formats about the decision making by AI for fairness, security, and transparency in the markets.
- Decentralized energy networks will be made more resilient, efficient, and consumer-driven as AI becomes the core of grid management, making the method to a sustainable energy future and fully intelligent.

7.2 *Blockchain for Peer-to-Peer Energy Trading*

- Peer-to-peer (P2P) energy trade is living being revolutionized by blockchain technology, establishing a accessible energy market, secure, and distributed [38]. The distribution of electricity has usually been controlled by governmental utilities, but producers—energy consumers and providers—can now distribute, purchase, and share additional energy with other individuals in their own community due to blockchain technology.
- Through blockchain automates transactions, smart contracts, thus providing tamper-proof record-keeping, real-time settlement, and reduced dependency on intermediaries.
- This decentralized approach fosters allowing households, energy democratization, businesses, and local energy cooperatives to participate in a resilient energy system and more flexible.
- Confidence and improved transparency and are two of block chain's important reimbursement for peer-to-peer trade in energy [39]. Since every operation has been recorded on a dispersed database, all network users can corroborate and adjust it.
- This promotes proportionate, held, and fraud-proof energy exchanges. Additionally, prosumers may collect and exchange digital tokens for extra power using blockchain-enabled tokenization, opening up fresh revenues while supporting the use of alternative sources of energy like solar and wind. Energy markets can function with real-time pricing mechanisms, where blockchain algorithms automatically match sellers and buyers based on demand and supply conditions.

- Future integrations of blockchain-based P2P energy trading are with smart grids, IoT, and AI-driven energy organization systems [40]. Advancements in DeFi will empower microgrid operators and Virtual Power Plants (VPPs) to use blockchain for automated settlements, dynamic pricing of energy, and cross-border energy transactions.

However, this calls for more grown-up narrow frameworks that support blockchain-based energy markets while ensuring defense for customers, data security, and grid stability. As governments and utilities move towards acceptance blockchain-powered decentralized energy models, P2P trading will be one of the major drivers in the structure of a consumer-centric energy future and more sustainable, cost-effective.

7.3 Integration of Electric Vehicles (EVs) and Vehicle-To-Grid (V2G) Systems

- The integration of electric vehicles (EVs) and Vehicle-to-Grid (V2G) systems is one of the innovatory innovations in decentralized energy systems that improve lattice elasticity, energy storage and demand-side management. As the acceptance of EVs increases, their batteries are converted into movable energy storage space units, competent of captivating power from the grid and also supplying power back to the network. V2G technology can store overload renewable energy throughout periods of low demand and discharge it back to the grid during peak demand, which improves the efficiency of the grid and stability [41]. This two-way energy flow reduces the stress on traditional power plants, increases renewable energy usage, and improves decentralized grid resilience.
- V2G systems are enabled through smart charging infrastructures, IoT-based energy management platforms as well as AI-driven predictive analytics. These technologies allow utilities and the grid operating companies control the discharging and charging cycles remotely to keep a balanced energy distribution. Moreover, go-ahead pricing models and demand answer mechanisms give confidence EV owners to contribute in V2G programs, earning plunder from causal stored energy throughout peak hours [42]. The addition of blockchain technology additional increase the declaration of protected contact of power, computerized settlements, as well as simplicity in the trading of energy stuck between utilities, EV owners, and microgrid operators.
- Looking forward, the success of EV and V2G integration will be determined by standardized regulations, advanced battery technologies, and increased charging infrastructure. Policymaking needs to establish grid-friendly EV charging policies and encourage public–private partnerships, as well as invest in renewable-powered charging networks [42]. V2G technology can successfully shape the vision of a sustainable, decentralized, intelligent energy future with the autonomous electric fleets and smart energy ecosystems growing at tremendous rates-the ultimate goal

for EVs being not only a mode of transportation but also contributor to the stability of the grid as well as energy independence.

8 Conclusion

Recent advancements in contemporary energy systems, such the smart grid and digital power plants, have enabled the move from central to decentralized energy management feasible. On the forefront of revolutionary developments like the big data analytics, artificial intelligence, Internet of Things and these strategies guarantee ecological responsibility, reliability, and energy efficiency. A common approach to think about smart grids is as an effective framework that offers uninterrupted integration of demand-side control, real-time monitoring and renewable energy sources. In the mean time, VPPs optimize energy distribution and ensure demand flexibility and through the integration of Distributed Energy Resources (DERs) through a single, intelligent entity. The potential benefits of VPPs and smart grids significantly outweigh the drawbacks, which include infrastructure costs, regulatory obstacles, and cybersecurity threats. They promote energy independence, contribute decarbonization goals and they generate new financial possibilities via creative business models. Future sustainable energy will be driven in substantial ways by the development and application of these kinds of technologies as the world advances regarding healthier and stronger energy systems. Future generations' resilience and sustainability can be guaranteed via smart grids and VPPs, that provide an opportunity for collaboration among fields investigators, and politicians.

References

1. Goia, B., Cioara, T., Anghel, I.: Virtual power plant optimization in smart grids: a narrative review. Future Internet **14**(5), 128 (2022)
2. Ghasempour, A.: Internet of things in smart grid: architecture, applications, services, key technologies, and challenges. Inventions **4**(1), 22 (2019)
3. Wang, X., Liu, Z., Zhang, H., Zhao, Y., Shi, J., Ding, H.: A review on virtual power plant concept, application and challenges. IEEE Innov. Smart Grid Technol. Asia (ISGT Asia) **2019**, 4328–4333 (2019)
4. Rouzbahani, H.M., Karimipour, H., Lei, L.: A review on virtual power plant for energy management. Sustain. Energy Technol. Assess **47**, 101370 (2021)
5. Luo, F., Dorri, A., Ranzi, G., Zhao, J., Jurdak, R.: Aggregating buildings as a virtual power plant: Architectural design, supporting technologies, and case studies. IET Energy Syst. Integr. **4**(4), 423–435 (2022)
6. Roozbehani, M.M., Heydarian-Forushani, E., Hasanzadeh, S., Elghali, S.B.: Virtual power plant operational strategies: Models, markets, optimization, challenges, and opportunities. Sustainability **14**(19), 12486 (2022)
7. Sikorski, T., Jasiński, M., Ropuszyńska-Surma, E., Węglarz, M., Kaczorowska, D., Kostyla, P., Leonowicz, Z., Lis, R., Rezmer, J., Rojewski, W., Sobierajski, M., Solak, B.: A case study on distributed energy resources and energy-storage systems in a virtual power plant concept: Tech. Aspects. Energies **13**(12), 3086 (2020)

8. Marinescu, B., Gomis-Bellmunt, O., Dörfler, F., Schulte, H., Sigrist, L.: Dynamic virtual power plant: a new concept for grid integration of renewable energy sources. IEEE Access **10**, 104980–104995 (2022)
9. Liu, R., Liu, Y., Jing, Z.: Impact of industrial virtual power plant on renewable energy integration. Glob. Energy Interconnect. **3**(6), 545–552 (2020)
10. Bhuiyan, E.A., Hossain, M.Z., Muyeen, S.M., Fahim, S.R., Sarker, S.K., Das, S.K.: Towards next generation virtual power plant: technology review and frameworks. Renew. Sustain. Energy Rev. **150**, 111358 (2021)
11. Liu, X., Li, Y., Wang, L., Tang, J., Qiu, H., Berizzi, A., Valentin, I., Gao, C. Dynamic aggregation strategy for a virtual power plant to improve flexible regulation ability. Energy **297**, 131261 (2024)
12. Chadokar, L., Kirar, M.K., Yadav, G.K., Salaria, U.A., Sajjad, M.: Aggregation and bidding strategy of virtual power plant. J. Electr. Eng. Technol. **20**(1), 199–216 (2025)
13. Wang, K., Cheng, B., Ren, Y., Wang, S., Ji, R., Kong, X.: Flexible resource dynamic aggregation regulation method of virtual power plant to ensure more renewable energy generation. Process. Saf. Environ. Prot. **180**, 339–350 (2023)
14. Nafkha-Tayari, W., Ben Elghali, S., Heydarian-Forushani, E., Benbouzid, M.: Virtual power plants optimization issue: a comprehensive review on methods, solutions, and prospects. Energies **15**(10), 3607 (2022)
15. Liu, X., Lin, X., Qiu, H., Li, Y., Huang, T.: Optimal aggregation and disaggregation for coordinated operation of virtual power plant with distribution network operator. Appl. Energy **376**, 124142 (2024)
16. Deng, J., Guo, Q.: Decentralized energy management system of distributed energy resources as virtual power plant: economic risk analysis via downside risk constraints technique. Comput. Ind. Eng. **183**, 109522 (2023)
17. Abdelkader, S., Amissah, J., Abdel-Rahim, O.: Virtual power plants: an in-depth analysis of their advancements and importance as crucial players in modern power systems. Energy Sustain. Soc. **14**(1), 52 (2024)
18. Cioara, T., Antal, M., Mihailescu, V.T., Antal, C.D., Anghel, I.M., Mitrea, D.: Blockchain-based decentralized virtual power plants of small prosumers. IEEE Access **9**, 29490–29504 (2021)
19. Xie, Y., Zhang, Y., Lee, W.J., Lin, Z., Shamash, Y.A.: Virtual power plants for grid resilience: a concise overview of research and applications. IEEE/CAA Journal of Automatica Sinica **11**(2), 329–343 (2024)
20. Islam, M.R., Vu, L., Dhar, N., Deng, B., Suo, K.: Building a resilient and sustainable grid: a study of challenges and opportunities in AI for smart virtual power plants. In: Proceedings of the 2024 ACM Southeast Conference, pp. 95–103 (2024)
21. Mishra, S., Bordin, C., Wu, Q., Manninen, H.: Resilient expansion planning of virtual power plant with an integrated energy system considering reliability criteria of lines and towers. Int. J. Energy Res. **46**(10), 13726–13751 (2022)
22. Venkatachary, S.K., Alagappan, A., Andrews, L.J.B.: Cybersecurity challenges in energy sector (virtual power plants)-can edge computing principles be applied to enhance security? Energy Inf. **4**(1), 5 (2021)
23. Alajlan, R., Rahman, M.H., Alnaeem, M., Almaiah, M.: A literature review on cybersecurity risks and challenges assessments in virtual power plants: current landscape and future research directions. IEEE Access (2024)
24. Rao, S.P., Tiruvalluru, R.S., Balaji, S.R.A., Tomomewo, O.S., Ranganathan, P.: Virtual power plants security challenges, solutions, and emerging trends: a review. In: 2024 Cyber Awareness and Research Symposium (CARS), pp. 1–11. IEEE (2024)
25. Ruan, G., Qiu, D., Sivaranjani, S., Awad, A.S., Strbac, G.: Data-driven energy management of virtual power plants: a review. Adv. Appl. Energy 100170 (2024)
26. Cavus, M.: Integration smart grids, distributed generation, and cybersecurity: strategies for securing and optimizing future energy systems (2024)

27. Sugunaraj, N., Balaji, S.R.A., Chandar, B.S., Rajagopalan, P., Kose, U., Loper, D.C., Mahfuz, T., Chakraborty, P., Ahmad, S., Kim, T., Apruzzese G, Ranganathan, P.: Distributed energy resource management system (DERMS) cybersecurity scenarios, trends, and potential technologies: a review. IEEE Commun. Surveys Tutorials
28. Vardakas, J.S., Zorba, N., Verikoukis, C.V.: A survey on demand response programs in smart grids: pricing methods and optimization algorithms. IEEE Commun. Surveys Tutorials **17**(1), 152–178 (2014)
29. Kong, X., Lu, W., Wu, J., Wang, C., Zhao, X., Hu, W., Shen, Y.: Real-time pricing method for VPP demand response based on PER-DDPG algorithm. Energy **271**, 127036 (2023)
30. Mahmud, A.S.M.: A real-time demand response pricing model for the smart grid (2018)
31. Roy, S., Das, D.C., Sinha, N.: Optimizing smart city virtual power plants with V2G integration for improved grid resilience. In: 2024 IEEE International Conference on Interdisciplinary Approaches in Technology and Management for Social Innovation (IATMSI), vol. 2, pp. 1–6. IEEE (2024
32. Lage, M., Castro, R.: A practical review of the public policies used to promote the implementation of pv technology in smart grids: the case of portugal. Energies **15**(10), 3567 (2022)
33. Shaukat, N., Islam, M.R., Rahman, M.M., Khan, B., Ullah, B., Ali, S.M., Fekih, A.: Decentralized, democratized, and decarbonized future electric power distribution grids: a survey on the paradigm shift from the conventional power system to micro grid structures. IEEE Access **11**, 60957–60987 (2023)
34. Garrity, T.F.: Innovation and trends for future electric power systems. In: 2009 Power Systems Conference, pp. 1–8. IEEE (2009)
35. Koshy, S., Rahul, S., Sunitha, R., Cheriyan, E.P.: Smart grid–based big data analytics using machine learning and artificial intelligence: a survey. Artif. Intell. Internet Things Renew. Energy Syst. **12**, 241 (2021)
36. Mostafa, N., Ramadan, H.S.M., Elfarouk, O.: Renewable energy management in smart grids by using big data analytics and machine learning. Mach. Learn. Appl. **9**, 100363 (2022)
37. Strielkowski, W., Vlasov, A., Selivanov, K., Muraviev, K., Shakhnov, V.: Prospects and challenges of the machine learning and data-driven methods for the predictive analysis of power systems: a review. Energies **16**(10), 4025 (2023)
38. Wongthongtham, P., Marrable, D., Abu-Salih, B., Liu, X., Morrison, G.: Blockchain-enabled peer-to-peer energy trading. Comput. Electr. Eng. **94**, 107299 (2021)
39. Abdella, J., Tari, Z., Anwar, A., Mahmood, A., Han, F.: An architecture and performance evaluation of blockchain-based peer-to-peer energy trading. IEEE Transactions on Smart Grid **12**(4), 3364–3378 (2021)
40. Jamil, F., Iqbal, N., Ahmad, S., Kim, D.: Peer-to-peer energy trading mechanism based on blockchain and machine learning for sustainable electrical power supply in smart grid. Ieee Access **9**, 39193–39217 (2021)
41. İnci, M., Savrun, M.M., Çelik, Ö.: Integrating electric vehicles as virtual power plants: a comprehensive review on vehicle-to-grid (V2G) concepts, interface topologies, marketing and future prospects. J. Energy Storage **55**, 105579 (2022)
42. Kumar, M., Vyas, S., Datta, A.: A review on integration of electric vehicles into a smart power grid and vehicle-to-grid impacts. In: 2019 8th International Conference on Power Systems (ICPS), pp. 1–5. IEEE (2021)

Foundation of Green Metaverse in Synchronization with Fuzzy Logic

Gunjan Mukherjee

Abstract The emergence of the Metaverse poses as an, interconnected virtual ecosystem which has raised the global interest across various disciplines. Metaverse offers unprecedented opportunities for communication, commerce, inputs, environmental variablesand collaboration, its rapid development brings significant concerns regarding sustainability, energy consumption, and environmental impacts causing great influences on the human life. The concept of the Green Metaverse addresses these challenges by integrating ecological concepts and sustainable technologies into Metaverse infrastructure and applications. In this context, fuzzy logic emerges as a powerful computational intelligence tool which is capable of handling the inherent and ample uncertainties and imprecise information associated with environmental systems, user behavior, and energy optimization. The association with different natural phenomena also caters a good amount of information to the users. This chapter provides an appropriate framework for the close understanding of the very new concept of Green Metaverse along with the fuzzy logic system towards the unfurling of the most eco-friendlily virtual world in the near future.

1 Introduction

The Metaverse is a new and exciting concept in technology. It is a shared virtual space where people can interact with each other and digital environments using advanced technologies like Virtual Reality (VR), Augmented Reality (AR), Artificial Intelligence (AI), and blockchain. In the Metaverse, users appear as avatars, explore 3D worlds, attend virtual events, play games, or even do business. It's like living in a digital universe that exists all the time, where anyone can join from anywhere. The technology based concept can extend beyond the common human periphery. Such technology based virtualization will be the main aspect of modern day research. However, building and running the Metaverse [1] takes a lot of computing power. Powerful servers, high-speed internet, and energy-hungry devices are needed to support such

G. Mukherjee (✉)
Department of Computational Sciences, Brainware University, Kolkata, India
e-mail: mukherjee.gunjan@gmail.com

S. Kautish et al. (eds.), *Green Metaverse: Fuzzy Logic Approaches to Sustainable Virtual Worlds*, Information Systems Engineering and Management 74,
https://doi.org/10.1007/978-3-032-08183-4_15

a system. Also, many blockchain-based activities, like cryptocurrency mining and NFTs, use a lot of electricity. All this leads to a high carbon footprint, which means it adds to pollution and climate change. As more people and companies use the Metaverse, the environmental impact becomes a serious concern.

To address this, the idea of a Green Metaverse [2] has been introduced. A Green Metaverse means designing and using virtual technologies in a way that is friendly to the environment. This includes using energy-efficient systems, powering servers with renewable energy (like solar or wind), and creating eco-friendly virtual designs. The goal is to reduce pollution and energy use while still offering advanced digital experiences. The main objective of such technological outgrowth is the judicious usages of energies towards the making of healthy living environment. The building of a Green Metaverse [3] is not easy. There are many things to consider, like unpredictable user behavior, changing data, and different types of digital devices. Making smart decisions in such complex situations requires a system that can handle uncertainty. The fuzzy logic system strongly take care of such metaverse system helping to reduce the uncertainties to the utmost degree . Unlike regular logic, which only allows "true or false" answers, fuzzy logic [4] can work with values in between, like "somewhat true" or "very high." It helps computers make better decisions when information is not clear or complete. Fuzzy logic [5] is very useful in systems that need to work in real-time and respond to changing situations. The versatile application potential of such Fuzzy logic [6] system can be very much fruitful towards the reframing of such virtual concept with optimum levels of uncertainties . In the Green Metaverse,

Fuzzy logic can help in many different ways towards the overall optimization such as

control energy use in data centers by adjusting systems based on demand.

Predict user activity and allocate resources efficiently.

Manage environmental conditions in virtual smart cities.

Make gaming and virtual experiences more sustainable.

By using fuzzy logic, the Green Metaverse can become smarter, more flexible, and more eco-friendly. It helps balance user needs with environmental goals.

In this chapter, we explore the foundation of the Green Metaverse and how fuzzy logic [7] can be a key tool in making it work. We look at their features, benefits, challenges, and how they can be combined to build a digital world that is both advanced and sustainable.

2 Literature Review

The concept of the Green Metaverse has gained much importance in regards of the environmental influences and energy consumption in the virtual ambiance. Recent research underscores the ceertain demands for meeting up with the sustainable practices such as renewable energy sources, edge and cloud computing optimization to be included. The further demands for energy-efficient network protocols to mitigate the carbon footprint of metaverse operations [8, 9] has been felt. Within this framework, fuzzy logic [10] appears to be a powerful tool for managing the fundamental and basic uncertainty and imprecision in user behavior, environmental conditions, and system states existing in metaverse platforms.

The dynamic resource allocation has been felt to reduce the computational powers , network bandwidth and other allied cooling systems. The fuzzy inference system has been effectively utilized to mitigate the over usages of such consumption.The achievement of energy savings with user experiences has been highlighted [11] (Chen and Gupta 2023). The practical exposition of fuzzy logic based [12] optimization has enhanced VR streaming efficiency by balancing quality and power usage. It also facilitated eco-friendly blockchain application in order to reduce energy-intensive validations, and controlled the cooling of server with fuzzy logic driven thermal management strategies, resulting in energy reductions of up to 20% without compromising performance [13] . In addition to these advancements, some challenges still remain in the scenario for construction of the scalable fuzzy models which can be capable of managing the complex, high-dimensional data which is typical of metaverse environments. The achievement of real-time, low-latency fuzzy inference [14] essential is very much essential as interactive applications. The overall procedure has become considerably promising on the integration of Fuzzy logic techniques for getting more improved kind of predictive accuracy and the adaptive control power. The criticality of the green computing environments posed sometimes the real challenge towards the adaption of metaverse (Zhang and Patel, 2024). Overall the Fuzzy logic represents a promising and valuable interdisciplinary approach to building sustainable, energy-efficient metaverse ecosystems.

3 The Metaverse: Definition and Structure

The formal definition of the Metaverse concept components can be written as

> **Virtual Environments (VEs)**: Digitally created 3D worlds accessible through VR/AR interfaces. **Avatars and Interactions**: Personalized representations of users engaging in real-time activities. **Digital Assets and Economy**: Blockchain-based tokens and NFTs representing ownership and economic value. **Decentralization**: Peer-to-peer networks and DAOs (Decentralized Autonomous Organizations) supporting governance.

The infrastructure involves high computational loads, bandwidth, and energy, contributing to ecological stress.

The Metaverse is envisioned as a persistent, decentralized, and immersive virtual environment that integrates technologies such as Virtual Reality (VR), Augmented Reality (AR), Artificial Intelligence (AI), and blockchain. While its applications span education, entertainment, healthcare, and commerce, the intensive computational demands of these technologies raise significant concerns about their environmental footprint. Thus, the concept of a Green Metaverse has emerged, aiming to align the growth of virtual environments with sustainability goals.

Principles of the Green Metaverse

The Green Metaverse adheres to a set of foundational principles designed to minimize environmental impact while supporting technological innovation.

Energy Efficiency: The architecture of Metaverse platforms must adopt energy-efficient computing infrastructures. This includes low-power data centers, optimized hardware (e.g., GPUs), and adaptive workload management to reduce idle resource usage. **Sustainable Hardware:** VR/AR devices and supporting infrastructure should utilize recyclable, biodegradable materials and be designed for durability and modularity to minimize e-waste. **Green Cloud Computing:** Hosting Metaverse platforms on green cloud services powered by renewable energy sources can significantly reduce the carbon footprint. **Eco-conscious Software Design:** Optimized rendering, reduced frame rates where feasible, and selective content streaming can reduce computational overhead. **Blockchain Sustainability:** Use of energy-efficient consensus algorithms such as Proof of Stake (PoS) and hybrid models over energy-intensive Proof of Work (PoW) for managing digital ownership and transactions. **Circular Economy Integration:** Encouraging virtual commerce, asset reuse (e.g., 3D models), and digital product life cycle management aligned with sustainability goals.

Challenges Facing the Green Metaverse

The green metaverse poses plenty number of challenges . The challenge points can be summarized as follows.

High Energy Consumption: Rendering 3D environments in real-time, maintaining persistent simulations, and enabling global user interaction demand enormous computational power. **Blockchain's Carbon Footprint:** Many Metaverse platforms depend on blockchain for digital ownership (NFTs), which traditionally involves energy-intensive PoW mechanisms. **Lack of Standardization:** Absence of global sustainability standards for virtual environments leads to fragmented and inefficient development practices. **Device Manufacturing Impact:** The production and disposal of AR/VR devices generate electronic waste and deplete finite resources. **User Awareness:** End-users often lack awareness of the environmental costs of their digital interactions, reducing the impact of green initiatives. **Scalability of Green Solutions:** Renewable energy integration, efficient cooling systems, and edge computing are still developing, making global deployment challenging.

4 Mathematical Interpretation of the Metaverse

The Metaverse can be mathematically represented as a dynamic, interactive virtual environment where users, objects, and services coexist and interact in real time. This interpretation uses concepts from graph theory, set theory, functions, and fuzzy logic.

4.1 *Metaverse as a Dynamic Graph*

We represent the Metaverse at time t as a dynamic labeled graph:

$$\mathcal{M}(t) = (V(t), E(t), A(t))$$

$V(t)$: Set of vertices representing users, objects, locations, and services. $E(t) \subseteq V(t) \times V(t)$: Set of edges denoting interactions or connections. $A(t)$: Set of attributes (metadata) associated with vertices and edges.

4.2 *User Modeling*

Each user $u_i \in V$ is defined by:

$$u_i = (\mathbf{x}_i, a_i, \pi_i(t))$$

$\mathbf{x}_i \in \mathbb{R}^d$: Feature vector (preferences, identity, demographics). a_i: Avatar or digital representation of the user. $\pi_i(t)$: Behavior or policy function guiding user interactions over time.

4.3 Virtual Environment State

The state of the virtual world is given by:

$$\mathcal{E} : \mathbb{R}^d \times \mathbb{T} \to \mathbb{R}^k$$

$\mathbb{R}^d$: Spatial coordinates. $\mathbb{T}$: Time domain. $\mathbb{R}^k$: Environmental parameters (e.g., light, temperature, bandwidth).

4.4 Interaction Function

Interactions among entities are modeled by:

$$I(u_i, u_j, t) = f(\mathbf{x}_i, \mathbf{x}_j, c_{ij}(t))$$

where f is the interaction function and $c_{ij}(t)$ is the context (e.g., location, mutual interests).

4.5 Digital Asset Transactions

Digital assets (e.g., NFTs) are defined as:

$$A_k = (\mathrm{id}_k, \omega_k(t), \delta_k)$$

id_k: Unique asset identifier. $\omega_k(t)$: Ownership function mapping asset to user at time t. δ_k: Metadata or value of the asset.

Transaction condition:

$$\omega_k(t + \Delta t) = u_j \quad \text{if } T(u_i, u_j, A_k) = \text{valid}$$

4.6 Energy and Sustainability Modeling

Let $P(u_i, t)$ denote power consumed by user u_i at time t. The total energy usage is:

$$E_{\text{total}} = \int_0^T \sum_{i=1}^{N} P(u_i, t)\, dt$$

where N is the number of active users. In a Green Metaverse, the goal is to minimize E_{total} while maintaining service quality.

4.7 Fuzzy Logic for Decision-Making

A fuzzy inference system processes uncertain data as:

$$y = F(\tilde{x}_1, \tilde{x}_2, \ldots, \tilde{x}_n)$$

where:

$\tilde{x}_i$: Fuzzy variables such as "high load", "low energy". F: Fuzzy inference rule set.

Example fuzzy rule:

IF load is high AND energy is low THEN reduce resolution

4.8 System Dynamics Equation

The state of the Metaverse evolves as:

$$\mathcal{M}(t+1) = \Phi(\mathcal{M}(t), \Pi, A, \mathcal{E}, I)$$

where:

Φ: Transition function modeling system dynamics. Π: Set of user policies or behaviors. A: Asset and economic activities. $\mathcal{E}$: Environmental updates. I: User-object and user-user interactions.

This formulation provides a foundation for modeling, simulation, and optimization of the Metaverse using formal methods.

Fuzzy Logic is an approach to computing based on "degrees of truth" rather than the usual true or false (1 or 0) Boolean logic. It was introduced by Lotfi A. Zadeh in 1965 as a means to model the uncertainty and vagueness inherent in human reasoning. Unlike classical logic where variables must be either completely true or false, fuzzy logic [15] variables may have a truth value that ranges between 0 and 1.

Basic Concepts of Fuzzy Sets

Fuzzy Sets: A fuzzy set is a class of objects with a continuum of grades of membership. Each element has a membership function value between 0 and 1. **Membership Function** ($\mu(x)$)**:** This function quantifies the degree to which an element belongs to a fuzzy set. For example, in the fuzzy set "tall people," the height 6'2" might have a membership value of 0.9. **Linguistic Variables:** These are variables whose values are words rather than numbers (e.g., *temperature* could be *low, medium, high*). **Fuzzy Rule Base:** Fuzzy logic uses a collection of IF-THEN rules. For example: `IF temperature is high THEN fan speed is high`. **Inference Mechanism:** This performs reasoning using the rule base and applies fuzzy logic [16] operators (AND, OR, NOT) to evaluate the rules. **Defuzzification:** Converts the fuzzy output of the inference engine into a crisp value. Common methods include *centroid*, *bisector*, and *mean of maxima*.

Mathematical Representation

Let X be a universe of discourse and A a fuzzy subset of X. Then:

$$A = \{(x, \mu_A(x)) \mid x \in X\}$$

where $\mu_A(x) : X \rightarrow [0, 1]$ is the membership function of fuzzy set A.

Centroid Method (Center of Gravity)

The most commonly used defuzzification method. It finds the center of area under the aggregated output membership function.

$$x^* = \frac{\int_a^b x \cdot \mu_A(x)\, dx}{\int_a^b \mu_A(x)\, dx}$$

where:

x^*: Defuzzified output. $\mu_A(x)$: Aggregated membership function. $[a, b]$: Domain of the output variable.

1. Bisector Method

This method finds the vertical line that divides the area under the aggregated membership function into two equal halves.

Find x^* such that:

$$\int_{a}^{x^*} \mu_A(x)\,dx = \int_{x^*}^{b} \mu_A(x)\,dx$$

2. Mean of Maximum (MoM)

This method takes the average of all values of x for which the membership function reaches its maximum value.

$$x^* = \frac{1}{n}\sum_{i=1}^{n} x_i \quad \text{where } \mu_A(x_i) = \mu_{\max}$$

3. Smallest of Maximum (SoM)

This method selects the smallest value of x where the membership function reaches its maximum.

$$x^* = \min\{x \mid \mu_A(x) = \mu_{\max}\}$$

4. Largest of Maximum (LoM)

This method selects the largest value of x where the membership function reaches its maximum.

$$x^* = \max\{x \mid \mu_A(x) = \mu_{\max}\}$$

5 Integration of Fuzzy Logic with the Green Metaverse

The integration of fuzzy logic into the Green Metaverse offers a new kind of approach which manages the transformative approach towards the complexity, uncertainty, and sustainability challenges that arises as virtual worlds. The overall concept becomes more intelligent, and resource-intensive. The Green Metaverse is an evolution of the conventional Metaverse structure. This focuses on reducing environmental impact by promoting energy-efficient technologies, sustainable hardware usage, and eco-conscious digital experiences. However, as Metaverse platforms grow in scale , the diversity enlarges towards the encompassing of real-time 3D rendering and decentralized economies. The consequence of AI-driven virtual agents and massive user interactions inherently involve imprecise data, ambiguous conditions, and nonlinear decision-making contexts. Such uncertainties is literally overcome by means of the fuzzy concept. a soft computing strategy. Unlike traditional binary logic systems that operate on absolute true or false values, fuzzy logic enables decision-making using degrees of truth, reflecting how humans naturally reason under uncertainty. The fuzzy logic can be applied to regulate energy consumption, optimize virtual resource allocation, and balance performance with environmental goals. The fuzzy inference systems can dynamically adjust rendering quality, network load, and compute resource distribution based on different linguistic inputs. Some instances of the application of the fuzzy logic concept are "moderate congestion," "high user engagement," or "low battery level." These inputs, when evaluated through fuzzy rules, can help reduce power usage in VR/AR headsets or mobile devices without compromising user experience. The datacenter and different power houses play very crucial roles in supplying powers to the metaverse. Differemt edge computing infrastructure can employ the fuzzy controllers towards the optimisation [17] of the HVAC operations. The renewable energy is the alternative replacement of traditional energy concept which minimises computational workloads. Variables such as ambient temperature, processing demand, and grid availability can be processed fuzzily to ensure sustainable operations. In virtual commerce and blockchain-based asset trading central to the Metaverse economy fuzzy decision-making models can assess environmental impact indicators such as energy cost per transaction or carbon intensity of digital ownership systems. These models can guide users and systems toward greener alternatives, such as favoring proof-of-stake protocols over more energy-intensive proof-of-work systems. Additionally, in immersive simulations and gamified virtual spaces, fuzzy logic supports adaptive system behavior. For example, it can help autonomous avatars and non-player characters react in more human-like, eco-conscious ways by interpreting fuzzy descriptors like "crowded environment" or "low interaction density" to modify their actions in energy-saving ways. Fuzzy logic also facilitates interoperability between heterogeneous systems within the Green Metaverse, such as integrating smart city data with AR layers or combining sensor inputs from IoT devices with AI-driven virtual experiences. These systems often operate with different standards and uncertain inputs, which fuzzy logic can harmonize using rule-based reasoning. Ultimately, the integration of fuzzy logic into the

Green Metaverse not only empowers intelligent, context-aware system adaptation but also ensures some virtual digital innovation aligns with sustainability and environmental ethics. By embedding fuzzy models at multiple layers from infrastructure to interface, This integration supports the dynamic, scalable, and environmentally responsible evolution of the Metaverse, making it not only a technological wonder but also a climate-conscious digital ecosystem for future generations.

Fuzzy Logic Integration Algorithm for Metaverse

Given: Crisp inputs: $x_1, x_2, \ldots, x_n$ (e.g., user inputs, environmental variables) Output decisions: y (e.g., NPC actions, environment responses)

Step 1: Define Linguistic Variables and Membership Functions

For each input x_i, define fuzzy sets A_{ij} with membership functions:

$$\mu_{A_{ij}} : X_i \to [0, 1]$$

where $\mu_{A_{ij}}(x_i)$ is the degree of membership of x_i in fuzzy set A_{ij}.

Similarly, define output fuzzy sets B_k with membership functions:

$$\mu_{B_k} : Y \to [0, 1]$$

Step 2: Fuzzification

Convert each crisp input x_i into fuzzy values:

$$\mu_{A_{ij}}(x_i) \quad \forall j$$

Step 3: Rule Base

Define a set of M fuzzy IF-THEN rules of the form:

$$\text{Rule } r: \quad \text{IF } x_1 \text{ is } A_{1j_1} \text{ AND } x_2 \text{ is } A_{2j_2} \text{ AND } \ldots x_n \text{ is } A_{nj_n} \text{ THEN } y \text{ is } B_k$$

Step 4: Inference (Rule Evaluation)

Compute the firing strength w_r of each rule r by combining input memberships, for example, using the minimum operator:

$$w_r = \min\left(\mu_{A_{1j_1}}(x_1), \mu_{A_{2j_2}}(x_2), \ldots, \mu_{A_{nj_n}}(x_n)\right)$$

or product operator:

$$w_r = \prod_{i=1}^{n} \mu_{A_{ij_i}}(x_i)$$

Step 5: Aggregation of Rule Outputs

The aggregated output fuzzy set $\mu_B(y)$ is computed by combining weighted consequents:

$$\mu_B(y) = \max_{r=1}^{M} \left(w_r \wedge \mu_{B_{k_r}}(y)\right)$$

where $\wedge$ denotes fuzzy AND (often minimum).

Step 6: Defuzzification

Convert the aggregated fuzzy output to a crisp value y^* via defuzzification methods, for example, the centroid method:

$$y^* = \frac{\int y\, \mu_B(y)\, dy}{\int \mu_B(y)\, dy}$$

Step 7: Execute Output

Use y^* to trigger the corresponding metaverse action or decision.

Optional Step 8: Adaptation

Update membership functions $\mu_{A_{ij}}$ and rules based on feedback for improved system behavior.

The fuzzy logic application towards the optimization in metaverse for finding the optimization in finding the uncertainties can be tracked by taking into account the user engagement levels along with the proximity values. The plotting of the graph in terms of Fuzzy membership function is shown in the Fig. 1.

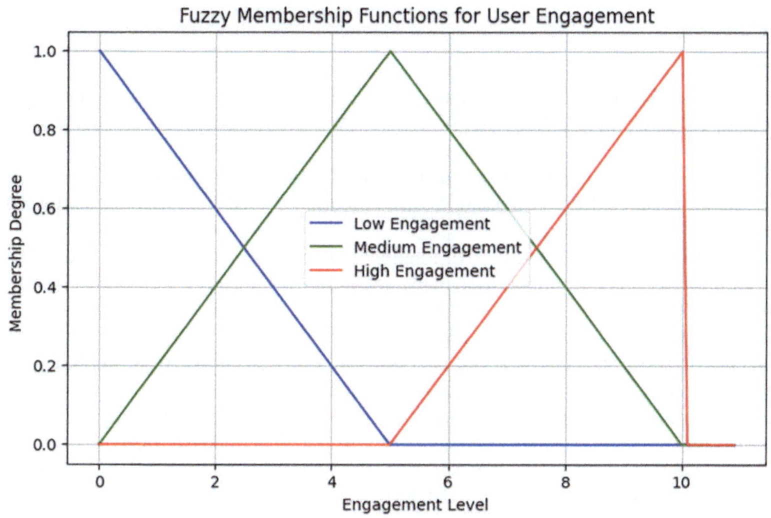

Fig. 1 Fuzzy Membership functions for member engagements

6 Future Directions and Research Opportunities

Hybrid Models: Integrating fuzzy logic with machine learning and deep learning for adaptive control. **Green Blockchain**: Fuzzy-driven dynamic adjustment of consensus mechanisms to balance speed and energy. **Edge-Fog Computing**: Distributed architectures optimized with fuzzy logic for reduced latency and energy cost.

7 Conclusion

The convergence of the Metaverse with sustainable computing principles marks a pivotal step toward a responsible digital future. Fuzzy logic, with its flexibility and robustness, can effectively address the complexities and uncertainties associated with green transitions in virtual environments. By embedding fuzzy logic into the core operations of the Green Metaverse, we move closer to realizing eco-conscious digital ecosystems that harmonize technological advancement with environmental stewardship. The future research can be directed towards the optimization of the environmental pollution. The reduction of the carbon footprint can be the striking outcome of the current research. The diminishing of the uncertainties by means of the Fuzzy logic concept can improve the overall virtual environment and such structure can bring in the overall sustainability of such system in some greenish manner.

References

1. Liu, F., Yuan, Y., Wu, J.: When the metaverse meets carbon neutrality: challenges and opportunities. IEEE Commun. Mag. **61**(5), 78–84 (2023). https://doi.org/10.1109/MCOM.2023.1234567
2. Deveci, M.: Accelerating the integration of the metaverse into urban transportation using fuzzy trigonometric-based decision making. Soft. Comput. **27**(3), 2101–2118 (2023). https://doi.org/10.1007/s00500-023-08013-1
3. Elfakki, A.O., Zhang, Y., Yu, H.: An intelligent tool based on fuzzy logic and a 3D virtual learning environment to support sustainable learning. Sustainability **15**(2), 889 (2023). https://doi.org/10.3390/su15020889
4. Novak, V., Perfilieva, I., Mockor, J.: Mathematical principles of fuzzy logic. Springer (1999)
5. Biacino, L., Gerla, G.: Fuzzy logic, continuity and effectiveness. Arch. Math. Logic **41**(7), 643–667 (2002). https://doi.org/10.1007/s001530100114
6. Arabacioglu, B.C.: Using fuzzy inference system for architectural space analysis. Adv. Eng. Inform. **24**(3), 386–395 (2010). https://doi.org/10.1016/j.aei.2009.12.004
7. Takagi, T., Sugeno, M.: Fuzzy identification of systems and its applications to modeling and control. IEEE Trans. Syst. Man Cybernet. SMC **15**(1):116–132 (1985). https://doi.org/10.1109/TSMC.1985.6313399
8. Wang, J., Tian, Y., Wang, Y., Yang, J., Wang, X., Wang, S., Kwan, O.: A framework and operational procedures for metaverses-based industrial foundation models. IEEE Trans. Syst. Man Cybern.: Syst. **53**(4), 2037–2046 (2022)
9. Zhao, Y., Huang, Y., Lu, P., Shi, W., Ma, B., Su, X., Qiao, X.: Green and Economical Interaction in Metaverse: Challenges, Architecture, and Future Directions. IEEE Network (2024)
10. Meana-Llorián, D., Pelayo, F., González, L.: IoFClime: fuzzy logic-based IoT solution for energy-efficient temperature control. Sensors **17**(12), 2930 (2017). https://doi.org/10.3390/s17122930
11. Kumar, N., Acharya, A., Gendelman, H.E., Byrareddy, S.N.: The 2022 outbreak and the pathobiology of the monkeypox virus. J. Autoimmun. **131**, 102855 (2022)
12. Bellocchi, G., Rivington, M., Matthews, K., Acutis, M.: Validation of biophysical models: issues and methodologies. Rev. Agronom. Sustain. Develop. **30**(1), 109–130 (2010). https://doi.org/10.1051/agro/2009003
13. Tran, H.T., Nguyen, H.V., Bui, V.P., Ha, T.N., Nguyen, V.T., Nguyen, D.H., Le, M.T.: Encoding reality with semantic interpretation in metaverse interactions. AEU-Int. J. Electron. Commun. **187**, 155512 (2024)
14. Ibrahim, D.: An overview of soft computing. Procedia Comput. Sci. **102**, 34–38 (2016). https://doi.org/10.1016/j.procs.2016.09.372
15. Bellocchi, G., Mazzoncini, M., Bonari, E.: A fuzzy logic-based strategy for evaluating agroecological models. Ecol. Model. **306**, 174–181 (2015). https://doi.org/10.1016/j.ecolmodel.2015.03.015
16. Zhang, Y., Li, H., Han, X.: Sustainable development of metaverse: role of green computing. IEEE Trans. Sustain. Comput. **8**(1), 123–135 (2023). https://doi.org/10.1109/TSUSC.2023.1234567
17. Chen, X., Chen, M.: Fuzzy rough sets for investment decision-making in the green metaverse economy. Appl. Soft Comput. **135**, 110001 (2023). https://doi.org/10.1016/j.asoc.2022.110001
18. Ebert, D., Feuermann, M., Gaudet, P., Harris, N.L., Hill, D.P., Lee, R., Mi, H., Moxon, S., Mungall, C.J., Muruganugan, A., Mushayahama, T.: The gene ontology knowledgebase in 2023. Genetics **224**(10.1093), 50 (2023)
19. Liu, H., Park, K.S.: Exploring the impact of metaverse tourism experiences on actual visit intentions: an integrated model of presence, the technology acceptance model, and the theory of planned behavior. Int. J. Tourism Res. **26**(1), e2616 (2024)
20. Jung, Y., Nam, J., Choi, H., Park, J.: Deciphering the determinants: what guides creators in choosing their metaverse platforms?. Technol. Forecast. Soc. Change **210**, 123843 (2025)

21. Patel, D., Manickam, R.: Enhancing maintenance learning in the industrial metaverse: integrating virtual reality and machine learning. In: International Conference on Industry 4.0 and Advanced Manufacturing, pp. 153–163. Springer Nature Singapore, Singapore (2024, January)
22. Ahmed, M., Aziz, K.: Fuzzy cognitive map-based evaluation of smart city components in the metaverse. Exp. Syst. Appl. **195**, 116538 (2022). https://doi.org/10.1016/j.eswa.2021.116538

Sustainable Practices in the Metaverse Ecometaverse: Virtual Carbon Footprint Tracker

Aswathy K. Cherian, B. Shyam Kaarthikeyen, M. K. Vidhyalakshmi, and Rajimol Raju

Abstract This chapter presents the EcoMetaverse: Virtual Carbon Footprint Tracker, a unique approach promoting sustainability in the metaverse. As virtual worlds and digital environments continue to evolve in our reality, it is very crucial to consider their impacts on the environment. This chapter outlined a software platform used to track, visualize, and reduce carbon footprints based on activity in virtual spaces. The system determined emissions related to actions such as virtual travels and online events alongside digital consumption, providing immediate feedback about how users were affecting the environment. The chapter also framed how gamification and reward systems facilitated eco-friendly behaviors amongst individuals by providing users with motivations to implement sustainable actions. The chapter addressed the educational opportunities associated with the platform. This exploration of eco-friendly living illustrates the broader implications of sustainability within digital ecosystems while highlighting the role metaverse may play in cultivating a responsible and sustainable virtual future. The EcoMetaverse platform provides users the opportunity to create sustainable lifestyles within modern technologies and as a consequence facilitate awareness of their impact within virtual and real domains on a large scale.

A. K. Cherian · M. K. Vidhyalakshmi
Department of Computing Technologies, School of Computing, SRM Institute of Science and Technology, Potheri, Kattankulathur, India
e-mail: aswathyc1@srmist.edu.in

M. K. Vidhyalakshmi
e-mail: vidhyalm1@srmist.edu.in

B. Shyam Kaarthikeyen (✉)
SRM Institute of Science and Technology, Potheri, Kattankulathur, India
e-mail: shyamkaarthikeyan@gmail.com

R. Raju
Department of Information Technology, Amal Jyothi College of Engineering, Kottayam Dt., Kerala, India
e-mail: rajimolraju@amaljyothi.ac.in

S. Kautish et al. (eds.), *Green Metaverse: Fuzzy Logic Approaches to Sustainable Virtual Worlds*, Information Systems Engineering and Management 74,
https://doi.org/10.1007/978-3-032-08183-4_16

Keywords Metaverse · Carbon footprint · Sustainability · Virtual environments · EcoMetaverse · Digital consumption · Green computing · Environmental impact · Gamification · Virtual reality · Sustainable behavior · Emission tracking

1 The Digital Carbon Dilemma: What the Metaverse Means for the Environment

The digital revolution has given us more connectivity and creativity than ever before. But this shift has not been without a secret environmental cost. Although the metaverse may be perceived as a tidy, ethereal substitute for real-world activity, it relies on concrete infrastructure that guzzles energy.

1.1 Energy Requirements of Virtual Worlds

Metaverse platforms are driven by huge data centers that run continuously 24/7, burning electricity for servers, air conditioning, and network hardware. These centers account for a significant part of the world's internet usage in energy. Every avatar interaction, digital asset exchange, or virtual conference has a carbon price—though unseen by the end user.

High-definition graphics, real-time rendering, VR headsets, and blockchain economies (i.e., NFTs, virtual land ownership) exacerbate the load. The immersive quality of these experiences tends to need more powerful hardware and more frequent software updates, which means greater power usage on both client and server ends.

1.2 Carbon Footprint of Digital Assets

The production, exchange, and storage of digital assets—particularly those attached to decentralized technologies—can all leave substantial environmental impacts.

Blockchain transactions, and indeed transactions in proof-of-work protocols in particular, are notoriously power-hungry. Even outside of blockchain, the storage and copying of high-volume digital commodities (e.g., 3D models, skins, and media) necessitate enormous backend infrastructure.

1.3 Behavioral Factors and Digital Overconsumption

Overindulgence in the digital, like prolonged periods spent within virtual spaces, constant commuting between metaverse areas, and overload accumulation of virtual resources, mimics the unsustainable activities in the physical realm. Such behaviors are not monitored much and hence are obscure to users so that they cannot comprehend the actual effect of their behavior.

1.4 The Need for Transparency and Tools

Even with the increasing concern, most users are still unaware of the environmental cost of their online existence. There is a need for platforms that do not only measure this impact, but also present it in an accessible and motivational form. The EcoMetaverse seeks to address this with an open, user-focused instrument that gives users the power to participate responsibly in virtual environments.

2 Introducing EcoMetaverse: Architecture of Virtual Carbon Footprint Tracker

The EcoMetaverse platform is conceptualized as an intelligent, user-centric software system that allows for the measurement, visualization, and reduction of carbon emissions originating from digital activities in the metaverse. Built on the principles of sustainability, data science, and fuzzy logic, it bridges the gap between abstract environmental concerns and actionable user insights.

2.1 Conceptual Framework

At its essence, the EcoMetaverse runs on a modular structure made up of primary layers:

- Data Collection Layer: Records instant user engagements, virtual travel frequency, avatar movement, event attendance, and digital asset usage.
- Processing and Intelligence Layer: Utilizes fuzzy logic algorithms to decode the ambiguity and uncertainty of user behaviors and compute estimated carbon emissions.
- User Experience Layer: It gives intuitive dashboards, gamified feedback systems, rewards, and educational modules that in turn influence behaviour and/or make people more aware of their data.

The user is not flooded with technical data but has meaningful, understandable outputs to engage with three.

2.2 *Fuzzy Logic for Behavioral Interpretation*

Why fuzzy logic? User behavior in digital spaces is heterogeneous and hard to quantify exactly. To illustrate, how do we decide what constitutes "excessive" use of virtual transport? Or what kind of "energy-efficient" attendance at a virtual conference? Binary logic would be unable to capture such grays. Fuzzy logic admits degrees of membership and uncertainty and is therefore well-suited to analyzing behaviors like:

- Teleportation or zone-switching frequency
- Time spent in completely immersive VR environments
- Virtual resources energy-intensive versus energy-light.

The system relies on a collection of fuzzy rules, including:

- IF travel_frequency is HIGH AND avatar_rendering is HEAVY THEN emission_score is VERY HIGH
- IF usage_time is MEDIUM AND event_type is LOW-CARBON THEN emission_score is MODERATE.

The rules are the core of the emission estimation engine.

2.3 *Real-Time Feedback and Visualization*

One of the distinguishing characteristics of EcoMetaverse is its real-time carbon footprint tracker. Users are immediately shown graphical representations of their carbon footprint from their current actions. These are presented through:

- Dynamic Dashboards: Provide emissions over time, by activity (travel, event, asset use)
- Eco-Heatmaps: Display areas or actions of high emissions in the virtual environment
- Progress Trackers: Engage users by showing decreases in footprint and revealing new eco-levels.

By making data transparent, the platform can foster personal responsibility and awareness.

2.4 Integration Capabilities

EcoMetaverse is interoperable. It can connect to popular metaverse engines such as Unity, Unreal Engine, and WebXR platforms through light APIs. It has the ability to accommodate plugins for decentralized metaverse platforms (e.g., Decentraland, The Sandbox), enabling developers to integrate tracking functionalities without sacrificing performance.

The platform can also draw data from:

- Estimates of local device energy usage (through user opt-in)
- Cloud provider server logs
- Virtual event calendars and attendance tracking.

2.5 Ethics and Privacy

Monitoringdigital behavior raises ethical concerns. EcoMetaverse applies strict privacy-by-design principles:

- No personally identifiable information (PII) is retained without opting in.
- Emission data is anonymized and encrypted.
- Users can opt in or out of tracking, and dictate data retention policies.
- Transparency reports and data dashboards help build trust with users and stakeholders.

3 Gamification and Behavioral Nudging Towards Green Engagement

Virtual sustainability cannot be based on awareness alone—it needs to be motivated, engaged, and behaviorally changed. The EcoMetaverse uses gamification and behavioral nudging aspects to make environmental stewardship a rewarding and fun experience. These are intended to transition user behavior from consumption passivity to active sustainability in the metaverse.

3.1 Psychology of Gamified Behavior

Gamification takes game design features—points, badges, levels, leaderboards— and uses them in non-game settings. EcoMetaverse uses these features in a deliberate way to promote users to lower their virtual carbon footprint. Behavioral science research indicates that users are more inclined to embrace sustainable actions when:

- They get immediate feedback and visual indicators
- Their actions are part of shared accomplishments
- They get rewarded or publicly acknowledged for green behavior

EcoMetaverse leverages these triggers through a multi-level incentive mechanism and goal-oriented interactions.

3.2 EcoPoints and Reward Tiers

All eco-friendly activities in the metaverse are rewarded in the form of EcoPoints—a carbon-reduction-based virtual currency. For instance:

- Taking low-carbon virtual travel earns + 50 EcoPoints
- Participating in green-certified events earns + 100 EcoPoints
- Converting to low-impact digital assets yields + 30 EcoPoints.

As people gain points, they become awarded with EcoLevels (such as "Digital Recycler", "Green Voyager", "Virtual Carbon Neutral") and digital badges that they can display on their profile. These may be created in NFT or collectible asset form to enhance personalization and value.

3.3 Leaderboards and Social Impact

To encourage a feeling of competition and social interaction, EcoMetaverse features:

- Global and local leaderboards showcasing leading eco-performers
- Team challenges where teams work together to meet emission targets
- Seasonal activities that reinforce regular, long-term sustainable behavior

Achievements may be shared by users within their social networks, gaining publicity and leading others to adopt eco-friendly habits.

3.4 Subtle Behavior Change Through Nudging Techniques

Gamification is augmented by nudging behaviors—unobtrusive prompts that lead users to make improved decisions without constricting choice freedom. In EcoMetaverse, nudging occurs in the following forms:

- Color-coded labels on travel modes (green for low-carbon, red for high)
- Pop-up reminders when entering high-emission areas ("Would you like to switch to EcoMode?")

- Carbon Impact Previews displaying estimated emissions prior to visiting virtual events.

These nudges are intended to educate and guide behavior without guilt-tripping or coercion.

3.5 Emotional Engagement with Avatars

In order to enhance the emotional connection between behavior and impact, users' avatars change according to their eco-behavior. For example:

- Green users can unlock nature-inspired skins or accessories
- High carbon footprint can manifest in avatar visual indicators (e.g., smog effect, energy depletion)
- Avatars also get "eco-boosts" that increase performance in low-footprint areas.

This affective reinforcement connects the metaphorical idea of carbon emissions with individual identity and expression in the metaverse.

4 Educational Modules and Virtual Sustainability Literacy

Although real-time monitoring and gamification engage people, long-term digital sustainability in virtual worlds relies on educated users. The EcoMetaverse platform features a strong educational component that constructs sustainability literacy—environmental concept knowledge, digital footprint, and sustainable decision-making in virtual worlds.

4.1 The Role of Education in Digital Sustainability

Metaverse users are heterogeneous, and the levels of environmental consciousness vary. The relationship between virtual behavior and ecological impact may be abstract without guided learning. EcoMetaverse education is not offered as rigid content but as adaptive, scenario-based, and interactive learning congruent with the metaverse's immersive environment.

Through the infusion of environmental education into everyday virtual activity, users acquire:

- Contextual comprehension of digital emissions
- Skills to identify high-impact actions
- Tools to lower and offset their carbon footprint.

4.2 Interactive Learning Modules

EcoMetaverse has modular lessons that may be visited by users at their convenience. Some topics are:

- Digital Carbon 101: Learning how online actions produce emissions
- Green Design for Developers: Creating low-impact virtual worlds
- Virtual Mobility and Emissions: Picking green travel in digital realms
- Sustainable Virtual Events: Best practices for organizing green conferences
- Offsetting Virtual Carbon Footprints: What it means and how to do it.

Each module is gamified with quizzes, simulations, and instant feedback. Completion of modules earns EcoPoints and may unlock avatar customizations or digital certifications.

4.3 Virtual Simulations and Labs

Users can also engage in sandbox environments where they simulate the environmental impact of building and interacting in the metaverse. These include:

- EcoSim: A scenario simulator in which users create and sustain a virtual eco-community
- Carbon Lab: An environment for experimentation with emissions consequences of multiple user actions (e.g., scheduling servers, reducing graphics, virtual travel reduction)
- Energy Explorer: An analysis tool to understand effects of different server sources (renewable energy vs. fossil fuels) on digital carbon emissions.

These simulations help internalize cause-and-effect relationships through experimental exploration.

4.4 Certifications and Knowledge Sharing

After completing learning paths, users are issued electronic certifications (e.g., "Certified Virtual Environmentalist") that can be embedded in their profiles or attached to professional credentials. EcoMetaverse also supports peer learning through:

- Peer-generated tutorials and tips
- Sustainability community hubs
- Green-building team challenges.

This peer-driven model converts passive users into active teachers, reinforcing a shared responsibility model.

4.5 Educational Partnerships

EcoMetaverse is specifically designed for the purpose to connect with educational institutions, NGOs, and environmental organizations. This collaborative partnership that can add richness in the form of current research, real-world applicability, and authenticity to the content. The platform provides opportunities for:

- Institutions to present virtual lectures or seminars
- Organizations to sponsor learning modules against SDGs
- Students to undertake projects and research in EcoMetaverse for academic credits.

This brings the metaverse from being a mere entertainment space into a very serious, impactful space for learning.

5 From Data to Action: Analytics, Real World Impact Insight

Action without data is guesswork. EcoMetaverse is not just a tracker but an engine that delivers decisions. The platform translates raw behavior data into insights that lead to sustainable actions both within the virtual world and in real life. This section discusses how EcoMetaverse closes the loop between digital behavior, analytics, and action-oriented sustainability.

5.1 Aggregated Emission Analytics

In addition to personal tracking, EcoMetaverse provides aggregate-level information for platform managers, developers, and policy makers. The analytics aid in the identification of:

- High-emissions zones or areas in the virtual space
- Patterns of usage causing environmental strain
- Low-carbon behaviors popularly practiced and worth encouraging
- Bottlenecks in the system where energy inefficiency takes place.

For example, if a virtual city in the metaverse is regularly indicated as a carbon hotspot, the administrators can restructure transportation systems, refresh graphics settings, or reward usage of renewable assets.

5.2 Personalized Sustainability Reports

Individual users get monthly and yearly summaries of:

- Total emissions saved or created
- Most and least sustainable actions
- Progress towards personal sustainability objectives
- Suggestions for improvement based on trends in behavior.

These reports utilize clear-to-read graphics such as pie charts, bar graphs, and eco- scores (A–E grades) to be easy to understand. Users are able to compare against community norms and share findings socially to encourage peer accountability.

5.3 AI-Driven Suggestions

EcoMetaverse features an AI helper trained on sustainability data and user behaviors. The helper:

- Recommends best times to join events with light server loads
- Suggests low-emission virtual products or regions
- Reminds users about upcoming green challenges or special opportunities to earn bonus EcoPoints
- Provides low footprint substitutes for high impact activities.

By making eco choices easy and proactive, the platform increases engagement and sustainability at the same time.

5.4 Bridging Virtual Actions with Real-World Effects

Perhaps the strongest innovation of EcoMetaverse is its hybrid impact mechanism of a connectivity between digital sustainability and world effects. Examples are:

- Carbon offset integration: People can redeem EcoPoints as donations to plant trees.
- Green tokens: Redeemable as credits toward low prices on green products in partner ecosystems.
- Collaborative drives: Metaverse communities can pool resources to sponsor causes, unlocking virtual accomplishments when real targets are achieved. This two benefit system supports the concept that eco friendly behavior in the virtual environment is not symbolic and it has real global implications.

5.5 *Developer and Platform-Level Feedback*

The platform provides metaverse developers and administrators with emission dashboards to help them with optimization:

- Optimize server provisioning and performance
- Audit the carbon footprint of the new features
- Create eco friendly infrastructure and experiences
- Achieve sustainability certifications for very low impact platforms.

By providing developers with capabilities to act, EcoMetaverse becomes an ecosystem scale solution, rather than simply a user utility.

6 Conclusion

The expanded metaverse is an exciting possibility, but also an urgent obligation. As we increase our online spaces, we must develop our ethical consciousness in virtual environments [7]. The EcoMetaverse space exemplifies this vision by addressing sustainability within digital ecosystems. The EcoMetaverse platform tracks digital carbon emissions and fosters a culture of sustainability through education, engagement, and actionable insights [2].

Using fuzzy logic, EcoMetaverse accounts for user behaviors that other platforms often overlook. It contextualizes ambiguous actions in relation to environmental impact and delivers insights in an accessible manner [12]. EcoMetaverse features personalized dashboards, real-time feedback, interactive educational experiences, and gamified incentives that motivate users to understand and reduce their digital carbon footprint [9, 11].

Furthermore, EcoMetaverse bridges physical and digital realms by linking virtual actions with real-world environmental consequences [3, 10]. It embodies a holistic vision for a sustainable future, integrating carbon offset mechanisms such as reforestation and promoting environmental awareness among digital citizens [1, 4–6]. Thus, EcoMetaverse represents a step toward aligning technological advancement with the principles of sustainable development and responsible innovation.

References

1. Andre, A.S.G., Elder, T.: On worldwide electricity usage of communication is under. Trends in technology to 2030. Six Probl 117–157 (2015)
2. Boavida, A., Oliveira, L., Fernandes, J.: Carbon footprint of digital services: a literature review and guidelines for future work. Sustainability **13**(11) (2021)
3. Bourzac, K.: Greening the metaverse: the effort to reduce environmental cost of virtual worlds. IEEE Spectrum (2022). https://spectrum.ieee.org/greening-the

4. European Commission: Policy: The Environmental Effect of the Metaverse Briefings. European Union Publications (2022)
5. Facebook Bets on the Metaverse Come with Considerable Risks. The New York Times
6. Raghavan, B., Ma, J.: The energy and emergy of the internet. In: Proceedings of the ACM SIGCOMM Conference on Internet Measurement (2011)
7. Kapoor, A., Dwivedi, Y.K.: Metaverse and sustainable development objectives: a chance for accountable innovation. J. Bus. Res. **150**, 433–442 (2022). https://doi.org/10.1016/j.jbusres.2022.04.063
8. Nebel, B., Mittelbach, M.: Carbon accounting for ICT: from device to system level (2021)
9. Morgado, E., Serrano, J.: Gamification and sustainability: the promise of playful participation. Sustainability **12**(23), 9985 (2020)
10. UNEP: Digital transformation and climate change: effects and opportunities. United Nations Environment Programme (2020)
11. Xie, Q., Wang, S., Tan, X.: Metaverse: The Part of Gamified Eco-Feedback Systems. Intell. Virtual Reality Hardw. **3**(4), 320–336 (2021)
12. Zhou, Y., Zhang, H., Li, J.: Fuzzy logic systems for sustainable smart environments: survey. J. Clean. Prod. **384**, 135493 (2023)

Technological Innovations for Sustainability in Manufacturing Industrial Technology Education

Abdulmajeed Bidemi Abdulsalam

Abstract Sustainable value attainment entails delivering benefit to all stakeholders by balancing economic, environmental, and social goals. In manufacturing, this means rethinking design strategies at the product, process, and systems levels to fit into closed-loop material flows across different life stages. It also requires a deep comprehension of complex interrelations between socio-technical systems and the natural environment for sustainable value production in an ongoing and successful way. Prevailing models of education, as contained in conventional academic silos, lack the interdisciplinary foundation necessary to effectively address sustainability issues. Integration of knowledge across a number of disciplines has, therefore, become a necessity for addressing the complexity of sustainability problems. The prevailing model of industrial technology education has to be revamped to create sustainable value in manufacturing. This transformation must be prospective and innovative, based on a paradigm shift in education that prepares future engineers, technologists, scientists, and managers with the required expertise, hands-on experience, and necessary competencies. In this report, current trends in developing modern technical education programs are discussed.

Keywords Technological innovation · Industrial and technology education · Sustainability · Sustainable manufacturing · Curriculum development

1 Introduction

The conjunction of technology and sustainability has become an important issue in several areas, especially in industrial technology education. Industrial technology education is essential in preparing students for prosperous careers in an ever-changing

---

A. B. Abdulsalam (✉)
Department of Electrical Electronics Technology Education, Federal College of Education (Technical) Akoka, Lagos, Nigeria
e-mail: abdulmajeed.pg202319304@st.futminna.edu.ng

S. Kautish et al. (eds.), *Green Metaverse: Fuzzy Logic Approaches to Sustainable Virtual Worlds*, Information Systems Engineering and Management 74,
https://doi.org/10.1007/978-3-032-08183-4_17

technological environment. Industrial Technology Education (ITE) is a crucial discipline that provides persons with the knowledge, skills, and techniques required for employment in various industries. It is the combination of academic concepts with applications, aiming to equip students with essential skills needed to thrive in an environment driven by technology. Because of the imperative nature of sustainability, educational institutions have to incorporate technological advances into their curriculum so that the professionals of tomorrow will be conversant with and are capable of executing sustainable practices.

Industrial and Technological Education for sustainability is very much crucial in fostering sustainability by providing students with the necessary knowledge and skills to solve the environmental problems faced today. This view of the education system highlights the integration of technology with industrial practices for sustainable development. Industrial and Technology Education for sustainability refers to the infusion of sustainability principles into industrial and technological instruction. It puts the emphasis on environmentally sustainable behaviors, efficient resource management, and the advancement of technologies that are aligned with sustainable development goals. One focus of ITE for sustainability is on Renewable Energy Technology. In this regard, students learn about solar, wind, and bioenergy systems that reduce dependency on fossil fuels. By comprehending the application and management of such technologies, the future experts are able to foster a sustainable energy environment.

The ITE may also cover concepts of sustainable production and green design. This would involve instructing students about the best use of resources, reduction in wastes, and the use of sustainable materials [2]. Methods like Life Cycle Assessment (LCA) assist students in ascertaining the environmental consequences of items from manufacturing to the disposal stage, thereby encouraging them to think towards solutions that mitigate adverse impacts on the world. A significant aspect is the fostering of innovation and entrepreneurial activity in sustainable development. Students learn to promote these new technologies while critically analyzing their environmental implications. This entrepreneurial framework could well lay the foundation for enterprises applying sustainability and the circular economy, wherein materials are reused and repurposed. Besides that, ITE underlines that there is a necessity for multi-disciplinary collaboration occasioned by sustainability issues. Therefore, by integrating multi-disciplinary knowledge from engineering, environmental science, and social studies, students could resolve sustainability problems in a more holistic way, covering the technical, economic, and social factors. Industrial and Technological Education for sustainability is meant to train students for the eco-friendly industries, while simultaneously developing the members of society capable of becoming responsible citizens of the world. ITE nurtures sustainable knowledge and technology, with the implication that the upcoming generations are to promote innovations and instill positive change into their immediate society and beyond.

The invention signifies that the introduction of new technologies or new models to an existing one offers better solutions and provides increased efficiency or, rather, addresses issues across a few areas. This takes in improvements in engineering, information technology, biology, electronics, et cetera in the name of sustainability.

Sustainability is estheticized as a final goal predicated on continual economic growth with minimum or negligible capacity of pollution or environmental hazards or sustainable economic growth with concerted social considerations, giving the present generation's guarantee of natural capital for their basic survival to future generations. Sustainability encompasses the philosophy of sustainable development, which is the capacity to progress at the current level of civilization without compromising the capability of future generations in order to satisfy the needs of the current generation [13, 15]. The nature of technical innovation is the fact that it has the capacity to spearhead progress and transformation, affecting industries, economies, and society in general. These include the creation of the internet, improvements in renewable energy and the creation of artificial intelligence. Innovation fuels rapid industrial growth.

Product and process technological innovations are generally known to drive technical advancement, providing a competitive advantage that subsequently stimulates industrial expansion. Sustainable products and processes are innovative by nature and provide social and environmental advantages. Industrial manufacture is a key backbone to the generation of wealth and social wealth; a nation's economy to a great extent depends on its manufacturing capability and diversity of products and services it generates for its citizens and exports. Both developed and developed nations have emphasized the important function performed by manufacturing in generating employment, enhancing social welfare, and propelling national economies. Economic development and social well-being are also directly related to education and training quality and availability; quality and availability of good education and labour force training are essential for the economic and social development of any nation. Sustainability-oriented activities are also crucial to governments, communities, and individuals alike. With that, promoting innovation is imperative to boosting sustainable manufacturing, which in turn results in more sustainable development, while education and training are pivotal to enabling such a prospect.

Sustainable manufacturing has three primary elements namely products, processes, and systems. To be able to produce sustainably, all these parameters must have the following attributes: (a) a decrease in adverse environmental impacts, (b) maximum use of energy and resources, (c) the least generation of wastes, (d) safety-oriented operations, and (e) better health of the individual, but not compromising or enhancing the quality of the product and production process. Sustainable production represents a contemporary path to the manufacturing of advanced functional products with the help of sustainable production processes and technology that utilize capabilities across the supply base. Accordingly, an integrated strategy of sustainable production, focusing on the product, process, and system levels, should ultimately result in sustainable value creation for all stakeholders.

This requires a participatory system to create value through sustainable production, which is developed from a systems thinking view that identifies the interdependencies between manufacturers, consumers, the product or service produced, and the greater social and natural world, rather than concentrating on producers bringing value into being in isolation without consideration for the interests of other parties or the environmental effects of those interactions. This study proposal addresses the impact of technological progress on industrial technical training sustainability, e.g., diversified methodology, setbacks, and consequences for instructors and business.

2 Operational Terms

Technological Advances in Education Technological advances in education entail the use of digital technologies, online learning environments, simulation software, and collaborative technologies that increase learning efficiency and accessibility. Utilization of computer-based simulations facilitates experiential learning in a simulated environment where students can think out and solve sophisticated industrial processes [1].

Sustainability in Industrial Technology Industrial technology covers a range of areas such as manufacturing, engineering, and computer technology, each of which has various effects on the environment. Sustainability is defined as activities that address present need without impairing the ability of future generations to satisfy their own needs. The inclusion of sustainability in teaching industrial technology makes the students aware of the environmental implications of their activities and the need for the application of sustainable practices.

3 Key Innovations Promoting Sustainability

1. Hybrid Learning Environments Blended learning integrates conventional face-to-face teaching with elements of online learning, which allows greater flexibility and access to resources. Blended learning can enhance individualized learning experiences, research suggests, because students can learn about sustainability at their own pace and get instant feedback.
2. Digital Fabrication Technologies Digital fabrication technologies like 3D printing and computer-aided design (CAD) are transforming industrial technology education. The technologies permit the students to produce prototypes and models with minimum waste of material and maximum efficiency. The use of digital fabrication stresses resource optimization in sustainable operations [7].
3. Virtual Reality (VR) and Augmented Reality (AR) Virtual Reality (VR) and Augmented Reality (AR) technology provides lifelike learning experiences that have the ability to reproduce actual industrial settings.

Virtual Reality (VR) and Augmented Reality (AR) are used to provide effective three-dimensional interactive visual experiences for various applications. Virtual Reality (VR) has been viewed as an exciting technology for the hospitality and tourism sector with the mind-boggling capacity to deceive the senses into feeling presence in a simulated 3D world developed by a computer. Augmented Reality (AR) marries the physical and digital realms to make real-world experiences more meaningful by overlaying virtual information. A consumer can view a painting or device in the physical world, stand their tablet or smartphone in front of it, and on the screen have the artwork or device with more informative details, thereby enriching reality. The devices can enhance comprehension of intricate technological ideas since they show the effect of sustainable practices [14]. Students can visualize the product life cycle, comprehending how decisions related to design influence sustainability.

4. Data Analytics and Internet of Things (IoT) Merging data analytics and IoT for industries enables students to learn how real-time data influences decision-making. The Internet of Things (IoT) that enables various objects to be interconnected over the internet for easy convenience has existed for over a decade.

The newest IoT development trend is not just targeted at systems and devices but also at data analysis. This is primarily due to the fact that data obtained from sensors or systems typically contains important information that could significantly enhance system performance or customer service, should an effective data analysis solution be enacted. Students may investigate how efficiency and sustainability can be enhanced through projects such as analyzing energy use or waste management [11]. Challenges in Implementing Innovations Even with potential for technological innovations to improve sustainability education, there are still some challenges:

1. Limitation of Resources Most schools lack the financial and infrastructural capability when adopting new technology. The high initial price tag and recurring maintenance charges of sophisticated technology hinder schools from incorporating them into their curriculum.
2. Faculty Training and Development The effective adoption of technology necessitates that the faculty be adequately experienced with contemporary tools and techniques. Constant professional development is needed for the faculty to effectively educate and implement innovation in sustainable practice.
3. Curriculum Alignment Incorporating technology innovation into the curriculum requires reviewing learning goals and outcomes. Staff must align the application of technology with learning outcomes in sustainability so that students acquire the requisite skills.

4 Sustainable Benefits Derived from Innovative Products, Processes, and Systems

1. Resource constraints Most learning institutions are subject to economic and infrastructural constraints in the use of new technology. The expense of the initial investment and maintenance of technology that is updated can deter schools from incorporating them in their curricula.

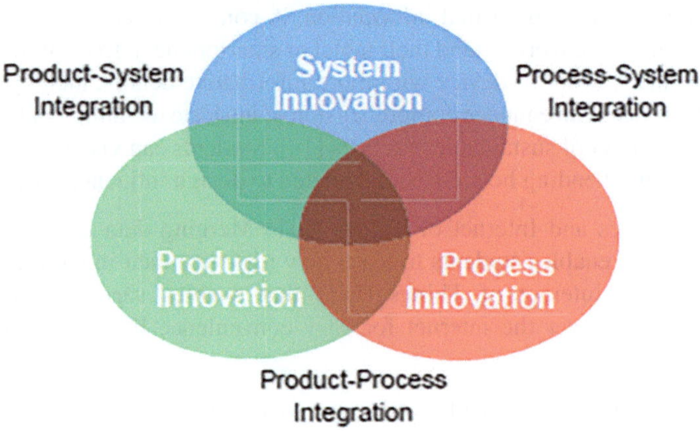

Fig. 1 Multilevel innovation in sustainable manufacturing: product, process, and system perspectives. Adapted from Jawahir et al. [8]

2. Faculty Development and Training The proper integration of technology requires that teachers be skilled in modern tools and approaches. Ongoing professional development is essential to enable faculty members to teach and apply developments in sustainable practices competently.
3. Curriculum Alignment Integrating technology advancements into current curriculum requires revisiting goals and objectives of learning. Faculty must align application of technology with learning objectives in terms of sustainability so that students achieve the required competencies (Fig. 1).

An integrated approach to sustainable manufacturing can establish a strong foundation for creating eco-friendly products through environmentally responsible processes and structured systems [8, 16, 17]. Figure 2 illustrates the new elements associated with sustainable manufacturing for each aspect of innovation. This approach seeks to build a comprehensive framework for sustainable value in manufacturing by unifying diverse value-generating factors. From this unified model, multiple forms of value—technological, economic, political, and environmental—can be derived and enhanced.

5 Cutting-Edge Training Initiatives in Sustainable Manufacturing

Instructional and developmental initiatives aimed at emerging scientists and engineers, featuring distinctive material on sustainable manufacturing, are highly advanced and are expected to inspire fresh motivation and accomplishments in these fields. The Global Educational Action Plan, originally proposed during early dialogues at the 1992 Rio Earth Summit and reinforced by later global meetings,

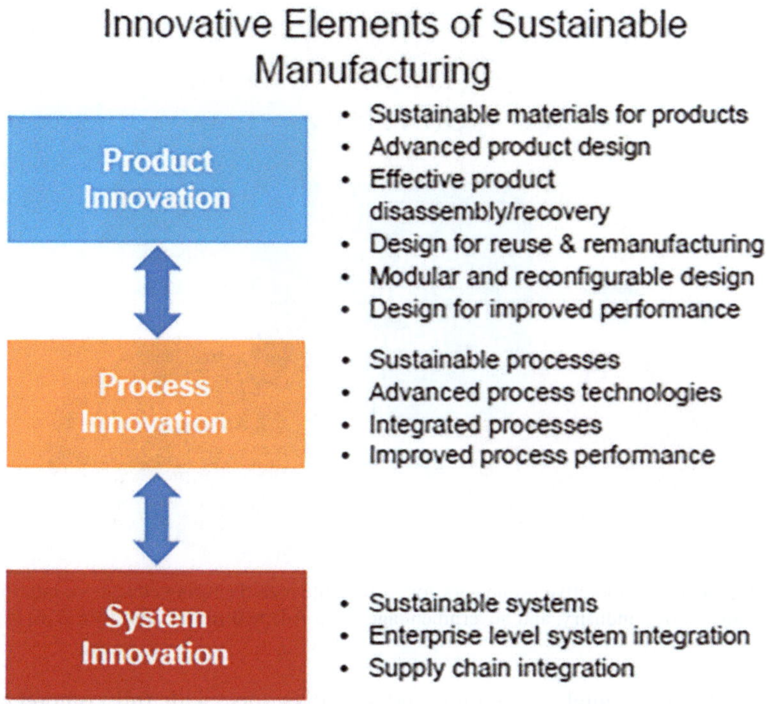

Fig. 2 Conceptual representation of key innovative features in sustainable manufacturing across product, process, and system levels. Adapted from Jawahir et al. [8]

now reflects a broader understanding: education should no longer be viewed merely as an endpoint, but rather as a driving force for advancing sustainable development. The need for gaining competences, knowledge, and skills towards sustainable manufacturing and greener consumption patterns is realized as a fundamental need for fulfilling education needs, both in the developed and developing regions of the world. A new curriculum for sustainable manufacturing must be established on a tripartite platform among the three principal stakeholders: academia, industry, and state and federal organization's (Fig. 3).

This strategic partnership puts training and education in the forefront of the three parties, as shown in Fig. 3, in order to facilitate achieving needed social and environmental effects accompanied by matching economic returns through enhanced interactions. Traditional higher education courses concentrate primarily on the training of engineers and scientists in physical and natural sciences, materials engineering, design engineering for products, and manufacturing science. These fields are currently taught in isolation without any particular relevance to practical applications such as social and human sciences. The new curriculum will focus on highlighting multidisciplinary and interrelated subjects of relevance to environmental and socioeconomic concerns, providing foundations for "sustainable manufacturing." A key emphasis will be given to creating innovative teaching and learning modules

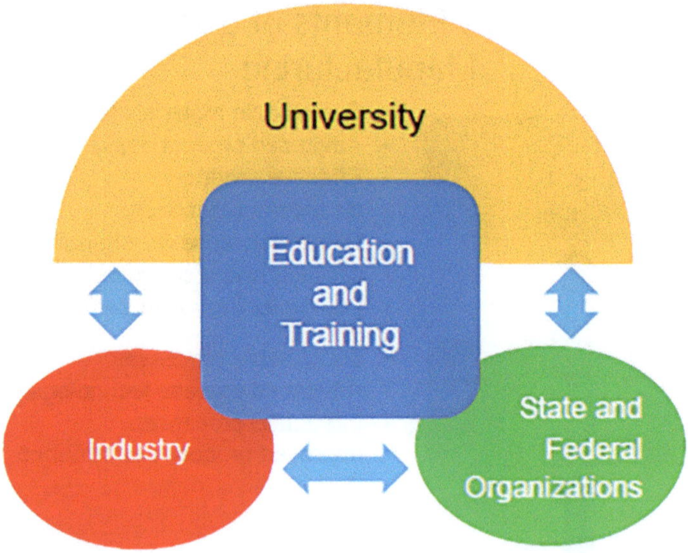

Fig. 3 Figure 3: Collaborative framework for sustainable manufacturing education: interlinkages between universities, industry, and government agencies. Adapted from Badurdeen et al. [4]

involving environmental, economic, and social factors, with full comprehension of natural cyclical processes characteristic of bio-complexity and material recyclability. Principles of sustainability design will be incorporated, encompassing the most important areas of applied sustainability, i.e., the 6Rs (reduce, reuse, recycle, recover, redesign, and remanufacture), and creating a perpetually sustainable material circulation in a closed-loop process encompassing four phases of the product life cycle: pre-manufacturing, manufacturing, use, and post-use. The suggested curriculum would include marketing, innovation, management, ethics, legislation, and policies, thereby being capable of giving future scientists and engineers a comprehensive base of knowledge to understand and implement science-based principles of sustainability in manufacturing.

Manufacturing engineering science courses will include discussion on optimizing the process performance, guaranteeing quality, enhancing health and safety, and studying sustainable manufacturing practice. Our earlier research has shown the development of cumulative learning between undergraduate and graduate levels and in Ph.D. programs [9].

The proposed educational and training model for sustainable manufacturing encompasses a broader and more integrative approach than traditional degree programs. It comprises five key elements designed to address both academic and industrial needs. These include: (1) a newly structured formal degree program incorporating classroom instruction, laboratory-based practical's, and project-based learning at both undergraduate and postgraduate levels; (2) an interdisciplinary certification program aimed at equipping industry professionals with a holistic understanding of sustainability across engineering and production domains; and (3) specialized short-term courses tailored for practical industrial applications, enabling continuous skill development and on-the-job learning.

6 Conclusion

Industrial technology education holds significant potential in advancing sustainability through the strategic application of technological innovations. By leveraging modern pedagogical tools such as blended learning platforms, digital fabrication technologies, smart learning environments (including virtual and augmented reality), and the Internet of Things (IoT), learners can be better prepared to tackle sustainability challenges in professional settings. Sustainable manufacturing, when approached creatively, can act as a catalyst for broader sustainable development goals.

For true sustainability to be achieved, economic advancement, social well-being, and environmental stewardship must be pursued concurrently. Innovation in product design, process optimization, and system-level efficiency—particularly through lifecycle extension—will be essential in driving this transformation. However, navigating the complex trade-offs that come with sustainable manufacturing demands a new generation of professionals equipped with interdisciplinary education and specialized competencies.

While commendable progress has been made in reforming curricula to reflect evolving sustainability imperatives, further efforts are necessary to fully embed multidisciplinary and interdisciplinary frameworks. Such an evolution is critical not only to dismantle traditional academic silos but also to foster a more agile and sustainability-driven workforce capable of generating long-term value in an increasingly complex global industrial landscape.

The proposed curriculum is based on technological development to serve the purpose of sustainability as well as environmental and social repercussions. One of the criteria for assessing the teaching experience of the suggested sustainable development curriculum is stated. Application of the set standards significantly enhances the capacity of technologists to perform sustainable design. Sustainability instruction must depend on education as well as active participation and collaboration. We have to direct our new models for sustainable production education into modifying the attitude of future generations from the age of early childhood and institutionalized through formal instruction at the undergraduate, graduate, and professional levels. Fixing the accompanying challenges is paramount to successful implementation. Ongoing research and partnership between academics and practitioners in industry will enhance the efficacy of these technologies and herald a sustainable future.

References

1. Alam, A.: Leveraging the power of 'modeling and computer simulation'for education: An exploration of its potential for improved learning outcomes and enhanced student engagement. In: 2023 International Conference on Device Intelligence, Computing and Communication Technologies (DICCT), pp. 445–450. IEEE (2023)
2. Appolloni, A., Jabbour, C.J.C., D'Adamo, I., Gastaldi, M., Settembre-Blundo, D.: Green recovery in the mature manufacturing industry: the role of the green-circular premium and sustainability certification in innovative efforts. Ecol. Econ. **193**, 107311 (2022)
3. Badurdeen, F., Brown, A., Gregory, R., Fu, H., Schroeder, M., Sekulic, D., Vincent, L., Luhan, G.A.: Reframing Interdisciplinary Approaches to Systems Thinking for Sustainability, Proceedings of the International Symposium on Sustainable Systems & Technologies, May 15–17, 2013, Cincinnati, OH (2013)
4. Badurdeen, F., Jayal, A.D., Jawahir, I.S.: Towards a System Approach for Developing Sustainable Products from Sustainable Manufacturing, Bangalore Workshop, Bangalore India, (2009)
5. Elsawy, M., Youssef, M.: Economic sustainability: Meeting needs without compromising future generations. Int. J. Econ. Financ. **15**(10), 1–23 (2023)
6. Hendler, S.: Exploring coordination practices in digital–physical product development. J. Manuf. Technol. Manag. **32**(3), 742–771 (2021)
7. Immadisetty, A.: Sustainable innovation in digital technologies: a systematic review of energy-efficient computing and circular design practices. Int. J. Comput. Eng. Technol. **15**(06), 1056–1066 (2024)
8. Jawahir, I.S., Badurdeen, F., & Rouch, K.E.: Innovation in sustainable manufacturing education (2013). https://doi.org/10.14279/depositonce-3753
9. Jawahir, I.S., Rouch, K.E., Dillon Jr., O.W., Holloway, L., Hall, A. Design for sustainability (DFS): new challenges in developing and implementing a curriculum for next generation design and manufacturing engineers. Int. J. Engng Ed. 0/0: 1–12 (2007)
10. Mahesh, K.M., Aithal, P.S., & Sharma, K.R.S.: Green HRM and teaching sustainability in higher education institutions: for promoting sustainable education and sustainable development goals. Int. J. Case Stud. Bus. IT Educ. (IJCSBE) **8**(1), 260–270 (2024)
11. Mahmood, S., Sun, H., Ali Alhussan, A., Iqbal, A., El-Kenawy, E.S.M.: Active learning-based machine learning approach for enhancing environmental sustainability in green building energy consumption. Sci. Rep. **14**(1), 19894 (2024)
12. Milliken, B., Traini, H., Stewart, J.: Attempts toward blended teaching and personalized learning in school-based agricultural education. J. Agric. Educ. **65**(3), 304–321 (2024)
13. Nagatsu, M., Davis, T., DesRoches, C.T., Koskinen, I., MacLeod, M., Stojanovic, M., Thorén, H.: Philosophy of science for sustainability science. Sustain. Sci. **15**, 1807–1817 (2020)
14. Nayyar, A., Mahapatra, B., Le, D., Suseendran, G.: Virtual Reality (VR) & Augmented Reality (AR) technologies for tourism and hospitality industry. Int. J. Eng. Technol. **7**(2.21), 156–160 (2018)
15. Shahadu, H.: Towards an umbrella science of sustainability. Sustain. Sci. **11**, 777–788 (2016)
16. Skalli, D., Charkaoui, A., Cherrafi, A., Garza-Reyes, J.A., Antony, J., & Shokri, A.: Analyzing the integrated effect of circular economy, Lean Six Sigma, and Industry 4.0 on sustainable manufacturing performance from a practice-based view perspective. Bus. Strategy Environ. **33**(2), 1208–1226 (2024)
17. Syed, S.: Sustainable manufacturing practices for zero-emission vehicles: analyzing the role of predictive analytics in achieving carbon neutrality (2024). Available at SSRN 5024423.

Impact of Metaverse and Financial Technologies on the Future of Digital Finance

Bhakti Parashar, Sandeep Kautish, and Vaibhav Parashar

Abstract Metaverse is increasingly emerging as a transformative digital environment that takes advantage of immersive technologies to enable virtual engagement and innovative forms of interaction. Its expansion effects are being felt in various industries, the financial sector is undergoing particularly notable changes. Technologies such as blockchain, digital assets, cryptocurrencies and decentralized finance (DEFI) are re-defining traditional financial structures, creating new structures for price exchange in virtual places. This chapter delays integration of intelligent systems with financial innovations in terms of metaverse, checks their ability to redefine financial services and promote a creative economy that combines virtual and real -world assets. A major challenge identified is the prevalent lack of a progressive mindset within the financial industry, which hinders the development of advanced capabilities and digitally operated economy, obstructing the distribution of customer-focused experiences. The study emphasizes strategic innovation and immediate need for flexible approach to meet the transfer demands of digital consumers. As financial ecosystems develop, metaverse are ready to become an important platform to bridge digital and physical finance. This chapter also states that hugging metaverse in financial services is not only a technological advancement, but a strategic requirement for continuous economic relevance and development.

Keywords Metaverse · Financial services · Innovation · Technology · Digital · Virtual · Economy

B. Parashar (✉)
VIT-Bhopal University, Kothri Kalan, India
e-mail: sharma.bhakti@gmail.com

S. Kautish
Physics Wallah Limited, Noida, India

V. Parashar
SGSU, Bhopal, India

S. Kautish et al. (eds.), *Green Metaverse: Fuzzy Logic Approaches to Sustainable Virtual Worlds*, Information Systems Engineering and Management 74,
https://doi.org/10.1007/978-3-032-08183-4_18

1 Introduction

The digital finance scenario includes a wide range of participants, including traditional institutions such as banks and insurance firms, with unconventional entrances such as Fintech Start-up and technology companies and mobile network operators. These stakeholders engage in dynamic relations—through partnership, form the acquisition, or directly compete in emerging financial locations. Innovations such as Peer-to-Pier Lending and Crowd funding give examples of how digital finance is re-shaping traditional financial services by introducing more decentralized and accessible models. Basic technologies such as blockchain, with cloud computing and other digital infrastructure, carry forward progress and change within the digital banking industry [1].

According to the coin market cap, Metaverse- or Meta Universe-represents a digitally manufactured space that reflects various dimensions of the physical world. Within this immersive environment, users participate in both social and economic activities, reflecting the real -world interaction. As a result, it has been designed to significantly affect the integration of metaverse how individuals navigate and many aspects of life adapt to fast digital experiences. Throughout history, individuals have imagined Utopian societies as a means of avoiding real-life difficulties and inequalities. Metaverse stands as a contemporary development of this ideal, which offers an immersive virtual environment, where people can re -add reality and expand their creative abilities. By crossing the physical world, it begins an intense change of how we conceptualize the conversation, economy and social organization—completely new ecosystems, business models and moral structures [2].

Despite being at the stages of its development, the metaverse offer tangible opportunities in various domains including entertainment, education, commerce and governance. It is not just a digital migration, but a platform promoting innovation and economic development. According to city group estimates, Metaverse may potentially serve five billion users by 2030, with an estimated market evaluation between $ 8 trillion and $ 13 trillion [3]. These forecasts have gained strong interest from corporate leaders, counselling agencies and financial institutions. Emerging technologies-like cryptocurrency, non-fungible tokens (NFTS), smart contracts and other fintech innovation-interactive environments are playing an important role in shaping financial systems within the innovative environment [4]

Metaverse- A shared, emergent virtual space, which is born from digitally increased physical reality, virtual environment and internet integration—is becoming an important driver in re -shaping digital banking and finance landscape. This developed domain is not only a platform for social interaction or entertainment, but a basis for transformative financial innovation. As blockchain technology and cryptocurrency are embedded within these wide virtual ecosystems, a new type of financial infrastructure is emerging-one that supports decentralized, user-centric economic activity. These development marks a significant departure from traditional financial systems, which enable more autonomy, transparency and flexibility. The rise of metaverse provides a mixture of promising opportunities and complex challenges for both individuals and businesses. It enables innovative forms of communication, commerce and creativity, while in the virtual environment also introduces significant concerns around privacy, regulation and digital rights. Navigating this rapidly developed place will demand a fine understanding of its implications, especially global interest and investment continue to accelerate [5].

An example of this innings is an example of Green Metaverse tokens (GMT), a purpose-manufactured digital asset, sewn to the price exchange within transactions and metaverse platforms. Such special currencies indicate how financial devices are being reconsidered to meet unique demands of virtual environment, in this process the major principles of digital banking, investment strategies and asset management have been potentially defined.

1.1 Objective of the Study

This chapter examines the impact of metaverse and fintech developed on digital finance. This checks how the immersive environment can increase customer engagement through virtual branches, personal advice and gamified education. The study has examined new financial assets such as NFTs, virtual real estate, and tokens, in which the role of blockchain and DEFI is safe, and enable cross-border payments and digital identity in transparent systems. It also addresses regulatory challenges including jurisdiction, AML, KYC and cyber security in virtual economies. Finally, it evaluates strategic adaptation by financial institutions and fintech firms, focusing on digital changes, technology investment, talent acquisition and cooperation within metaverse.

1.2 Significance of the Metaverse and Financial Technologies in Shaping the Future of Digital Finance

The rapid convergence of metaverse and Financial Technologies (Fintech) marks a transforming phase for digital finance, which takes intensive implications equally to individuals and global financial systems. significance of metaverse and financial technologies in shaping the future of digital finance is given in Fig. 1.

1.2.1 Charting a New Financial Landscape

Metaverse is developing beyond a gaming environment in a complex digital economy with unique assets, transactions and communities. Understanding this change is necessary to guess how financial services will be favorable within these virtual realms, which enable new products, services and professional models that cross

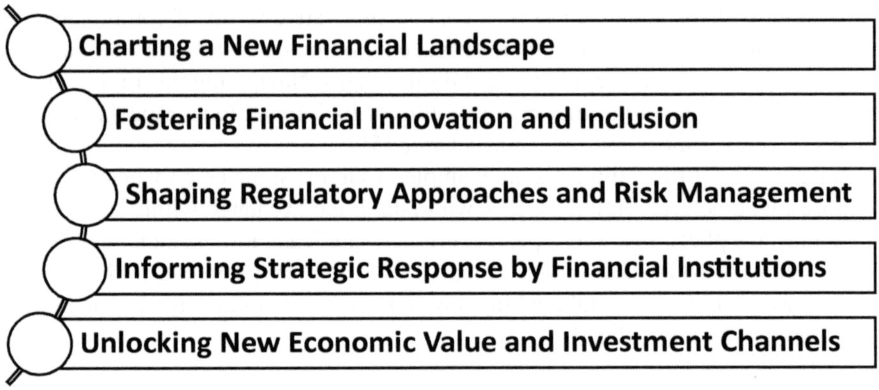

Fig. 1 Significance of the metaverse and financial technologies in shaping the future of digital finance. *Source* Author's own

traditional digital banking. This includes new forms of digital collateral, immersive marketplace and virtual asset management.

1.2.2 Fostering Financial Innovation and Inclusion

Fintech has expanded financial access worldwide, and has the ability to carry forward its integration—its integration with metaverse and digital currencies supported by digital currencies. This research helps to explain how the virtual environment can offer attractive, attractive financial services, especially the cost -effective, can benefit them without traditional banking access through alternative paths for lending, investment and payment.

1.2.3 Shaping Regulatory Approaches and Risk Management

The speed of innovation within metaverse and fintech challenges the current regulatory structure. Searching for this sector is important for emerging risks such as identification fraud, asset protection, money laundering in virtual economies and identifying interoperability issues. These insights support the development of adaptive regulatory policies that maintain financial stability, protect consumers, and maintain market integrity in digital finance.

1.2.4 Informing Strategic Response by Financial Institutions

Established financial players will have to navigate these disruptive trends to remain competitive. Research under the influence of Metaverse on finance provides valuable guidance to banks, insurers and asset managers to move consumer behavior for banks, insurers and asset managers, evaluate virtual property opportunities and integrate metaverse technologies in their existing systems.

1.2.5 Unlocking New Economic Value and Investment Channels

Metaverse makes ways for economic development through digital asset construction and exchange. This study highlights how virtual economies can attract real-world capital, open new investment opportunities in digital real estate, NFTs, and other assets, and bridge physical and virtual markets—ultimately expanding wealth creation and financial innovation.

2 An Overview of Financial Technologies (FINTECH)

The financial sector holds a vital role in both societal development and individuals' everyday activities across the globe. Historically, this industry has experienced significant shifts influenced by political, geographical, and legislative transformations. However, as noted by several scholars [6–8], the emergence of FinTech signals the beginning of a transformative new chapter for banking institutions. Despite its growing influence, this evolution remains relatively underexplored in academic discourse and presents substantial strategic and operational challenges for financial sector leaders [9–13].

2.1 Fintech Evolution and Development

According to Arner et al. [14], the development of financial technology (fintech) can be divided into many different ages, each of which is marked by significant technical, institutional and behavioral changes in the financial services industry. These ERA -Fintech 1.0, 2.0, 3.0, and emerging Fintech 3.5-Progressive changes in interacting with consumers and institutions are run by the convergence of progressive changes, digital infrastructure, innovation and regulatory optimization.

Each stage indicates a unique change in market structure and consumer engagement, including founder from infrastructure to decentralized, inclusive and intelligent digital finance solutions [14], the development of financial technologies from the its foundation till the future of fintech is discussed in Fig. 2

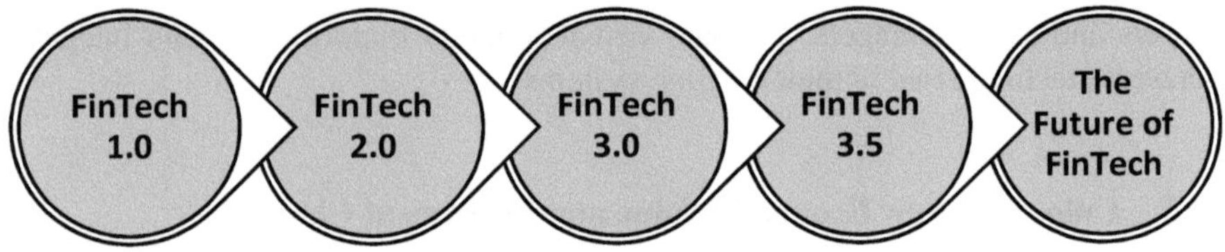

Fig. 2 Development of financial technologies. *Source* Author's own

2.1.1 FinTech 1.0 (1866–1967): The Foundation of Financial Infrastructure

- **1866**: Completion of the first transatlantic telegraph cable, enabling near-instantaneous communication between continents.
- **1918**: Introduction of Fedwire in the U.S., the first real-time gross settlement system.

These innovations laid the groundwork for electronic fund transfers and the modernization of financial services.

2.1.2 FinTech 2.0 (1967–2008): The Digital Transformation of Finance

- **1967**: Deployment of the first Automated Teller Machine (ATM) by Barclays Bank in London.
- **1971**: Launch of NASDAQ, the world's first electronic stock exchange.
- **1973**: Establishment of Swift, standardization of interbank financial communication
- **1998**: Establishment of Payal, Revolution in online payment.

This era saw the digitization of banking services, which ended in the global financial crisis of 2008, which exposed the systemic weaknesses and set the phase for the next phase of fintech development.

2.1.3 FinTech 3.0 (2008–Present): The Rise of FinTech Startups and Decentralization

- **2009**: Introduction of Bitcoin, the first decentralized cryptocurrency, utilizing blockchain technology.
- Proliferation of mobile banking and payment platforms, enhancing financial accessibility.
- Adoption of Open Banking initiatives, allowing third-party developers to build applications and services around financial institutions.

This period is marked by increased competition, customer-centric services, and regulatory adaptations to accommodate new financial models.

2.1.4 FinTech 3.5: Global Expansion and Financial Inclusion

- Emergence of mobile money services like M-Pesa in Kenya, facilitating financial inclusion.
- Significant FinTech growth in countries like China and India, driven by high mobile penetration and supportive regulatory environments.

FinTech 3.5 underscores the shift towards inclusive financial systems, bridging gaps in underserved regions.

2.1.5 The Future of FinTech: Personalization and Integration

- **Artificial Intelligence (AI)**: Enhancing personalized financial services and risk assessment.
- **Embedded Finance**: Integrating financial services into non-financial platforms, offering seamless user experiences.
- **Decentralized Finance (DeFi)**: Expanding peer-to-peer financial services without traditional intermediaries.

The purpose of these progress is to create a more efficient, accessible and user-centric financial ecosystem.

3 Concept of Metaverse

The concept of metaverse continues to develop, it is usually understood as a post-reality environment-a running, mutual, multi-user digital space that merges the elements of physical reality with virtual dimensions [15]. As an incarnation of extended reality (XR), metaverse include virtual reality (VR), enhanced reality (AR), and various levels of mixed reality (MR), which creates hybrid experiences that blur between real and digital [16]. This convergence is made possible through advanced technologies and immersive devices, enabled users to interact with virtual environment and institutions through individual avatars and multisensory inputs. Additionally, metaverse contains a network of interconnected immersive social locations through multi-use platforms. Thus, it introduces important challenges and vast opportunities for professionals, seeking to take advantage of its abilities [17] and to detect its versatile dimensions [18] for scholars.

Although the metaverse continues, it often serves more as a buzzword designed to catch the attention of users, corporations and investors [19]. Despite this, many of its underlying concepts are not completely novels. Scholars have long detected related dynamics in the virtual environment, including the avatar-deception connection (e.g., Zhang et al. [20]), collaborative methods (e.g., Pinkvert and Olivier [21]), competitive or cooperative virtual settings include interaction patterns (e.g., Weiss and Shiele [22]).

Recent studies, especially user onboarding [23], emotional effects on digital interactions [24], and Metavors gaming references [25] feel expenses. In addition, the researchers have begun to consider the implications of metaverse in areas such as new research agenda, practical applications and marketing [26], information system [27], and broad multi -spoke domains [18]. Nevertheless, educational discourses are fragmented, with the ongoing debate about the basic understanding of definitions, vocabulary and technology.

3.1 Evolution of Metaverse

The concept of metaverse has made a significant change since its establishment, which is developing into the digital ecosystem that is developed rapidly from the initial speculation. The term "Metaverse" was first introduced by Neel Stephenson in his 1992 Science Fiction novel Snow Crash, where he imagined a virtual reality place by avatar and digital economies [28].

In the early 2000s, platforms such as second life presented the first glimpse in an immersive virtual environment, where the users could talk, trade, and even drive businesses. These platforms prepared the basis for understanding the user's behavior in virtual places and highlighted the social and economic capacity of the digital avatar [29].

Development accelerated with the rise of online gaming and social virtual world, which introduced millions for real -time interaction in the simulated 3D space. Studies of this era examined avatar-user relationships, collaborative dynamics and virtual economy development [22, 23, 30]. Although these platforms were yet to be labeled as part of "metaverse", they introduced many of its main features.

There was a significant leap with integration of emerging technologies such as blockchain, NFT, and extended reality (XR), which enabled decentralized ownership and enhanced immersion. In 2009, the release of bitcoin and later blockchain innovations allowed digital assets to achieve real-world value and verification qualifications-a working metaverse economy components [26, 31].

In recent years, The Metaverse have expanded beyond entertainment to include education, finance, retail and workspaces. It is now being investigated from various disciplinary perspectives, including marketing [26], information system [27], and interdisciplinary research [18]. Scholars have also highlighted their emerging challenges, such as the dynamics of emotional interaction [24], entry-level adoption barriers [23], and gaming and social platforms [25].

3.2 *Emerging Technologies Shaping the Future of the Metaverse*

Immersive realities include virtual reality (VR), enhanced reality (AR), and convergence of mixed reality (MR), which enables a strong sense of appearance within the intensive level of user immersion and the presence within the virtual space, these emerging metaverse technologies are explained in Fig. 3.

Markets in Metaverse depicts the complex digital economies, where virtual goods, services, and assets are exchanged, online commercial activity and a transformative jump in user interaction models. In contrast, the gamification in metaverse takes advantage of sports-based mechanics and motivational structures to enhance the user's participation and deepen the engagement, which creates complex and rewarding experience within the virtual domain. These main technical columns-Immersive Realities, Metaverse Markets, and Games-Digital realities, enabling novel forms of expression, commerce, and socio-cultural engagement, forms a fundamental architecture of expansion and multidimensional metaverse.

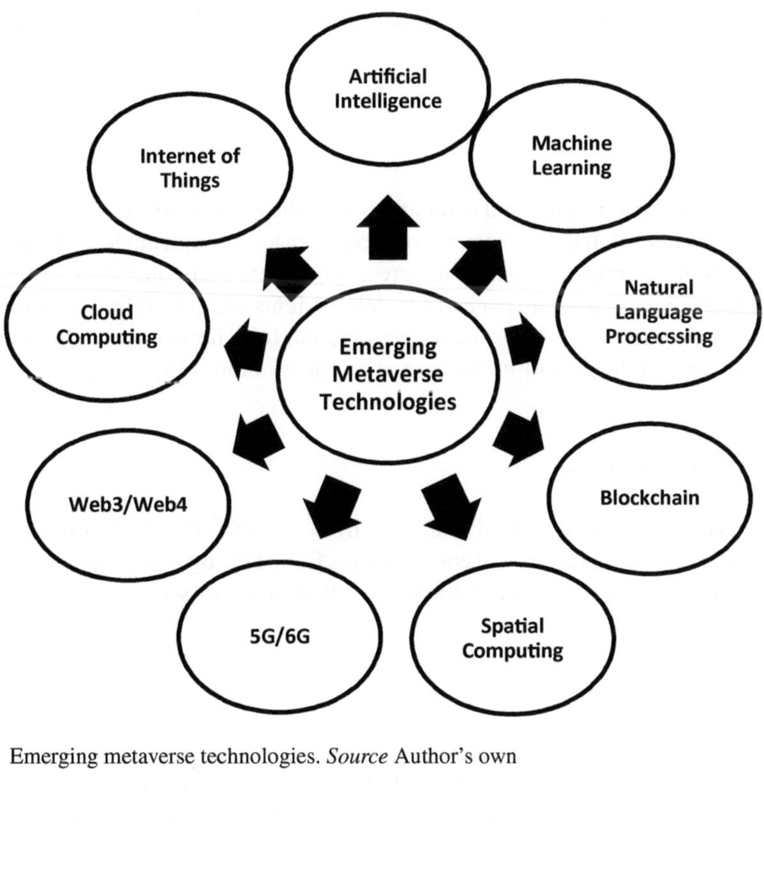

Fig. 3 Emerging metaverse technologies. *Source* Author's own

To understand the technical scenario of metaverse well, it is necessary to fly into separate components of immersive realities, markets in markets, and matters in metaverse, which are detailed in the following subdivisions.

3.2.1 Immersive Realities

Immersive realities are used by brands to introduce interactive services and products that enrich user experiences in domains such as retail, entertainment and social contact [32, 33]. Various definitions occupy the essence of immersive virtual spaces. According to [34], Immersive Virtual Reality forms a 3D computer-generated environment that significantly affects the sense of reality of users. Srikanth [35] has described immersive VR as a sensation of being physically located in a non-physical place, created through visual and hearing stimuli. At its core, immersive VR strives to envelop users in a digitally fabricated realm, providing a heightened sense of presence within synthetic surroundings ("Immersive Virtual Reality," 2008).

Immersive Realities encompass technologies such as Virtual Reality (VR), Augmented Reality (AR), and Mixed Reality (MR), all of which fall under the broader umbrella of Extended Reality (XR). Each of these technologies can be defined as follows:

Virtual Reality (VR)

VR involves computer-generated simulations that allow users to interact with a digital 3D environment [36]. It delivers immersive experiences through devices like headsets and haptic tools, creating a sense of "telepresence" where users can manipulate virtual objects and navigate digital spaces [37]. VR systems are generally categorized as non-immersive (desktop VR), semi-immersive, and fully immersive, with each level providing varying degrees of interaction and sensory input [38].

Augmented Reality (AR)

AR acts as a bridge between the digital and physical worlds by overlaying computer-generated content onto real-world environments [39]. Through AR, users can interact with virtual elements in real time, thereby turning ordinary settings into interactive spaces [40].

Mixed Reality (MR)

The MR mixes the real and virtual world to create interactive environment where digital and physical elements are co -existence and affect each other [41]. It units the immersive abilities of VR and AR, which provides users the ability to manipulate virtual items that react to the real world [42]. Mixed reality (MR) devices use advanced sensors and camera systems to interpret the user's physical environment, allow for real-time, reference-sensitive digital overlays. This integration enables users to engage in a more comfortable and comfortable manner with virtual elements in their real environment, allowing their feelings of immersion and interaction within the digital environment [43]. Within metaverse, spatial computing serves as a fundamental component of MR, which supports the distribution of adaptable and environment-environmental virtual materials.

Extended Reality (XR)

Across a collective term, all immersive technologies including virtual reality (VR), enhanced reality (AR), and mixed reality (MR) have been included, leading to the boundaries of human interaction and sensory connections in digital realms [44]. By merging these techniques, XR distributes a diverse spectrum of immersive experiences—completely from virtual simulation to real -world promotional enhancement [45]. This integration effectively dissolves the difference between physical and virtual places, establishing a harmonious and adaptable digital landscape [46]. Through the XR, users associate with digital materials that parallel and enhance the real -world interaction, it changes how they look at the virtual environment [47] and navigate.

3.2.2 Markets in the Metaverse

The underlying technical fabrics of Metaverse Powers change the traditional model of sophisticated digital marketplace, commerce and price exchange [47]. This virtual economy, built on the convergence of AR, VR, blockchain and decentralized systems, represents a significant change in economic structures. Major technologies running this change include:

Decentralized Ledger Technologies (DLT) and Blockchain

The blockchain acts as the backbone of the digital economy of the metaverse by offering transparency, safety and irreversibleness. It addresses the necessary concerns around digital ownership and authenticity [48]. Through blockchain, decentralized marketplaces emerge, which enable users to trade digital assets, virtual goods and NFTs with the cryptographic assurance of Provence [49].

Smart Contracts

These programmable agreements once fulfilled predetermined conditions facilitate automatic execution of conditions, eliminating the need for mediators [50]. In Metaverse Commerce, smart contracts enable secure and efficient transactions, reduce the opposition risk, and maintain confidence in colleagues to colleague's digital exchanges [46].

Tokenization of Assets

Competent by blockchain, Tokenization converts both digital and real-world assets into traditional tokens-such as virtual land, digital collectable or financial equipment [51]. Tokenization asset increases market access by improving liquidity, allows partial ownership, and wide user's participation in the economy of metaverse [52].

Non-fungible Tokens (NFTs) and the Digitalization of Value

NFTs provide a mechanism for establishing ownership of unique digital assets. NFT reappears the understanding of digital assets in domains such as NFT virtual real estate, art and tech-based artifacts [50], by anchoring value in personality. While working within a decentralized structure, NFTs empower both the creators and consumers, leading to re -defined economic participation in virtual markets [53].

Virtual Economies and Digital Exchange Platforms

Within metaverse, dynamic digital economies thrive, which is lined by native digital currencies corresponding to virtual ecosystem. Digital exchange platforms act as important intermediaries in these economies, enabling smooth conversion and transfer of diverse digital currencies and assets [51]. These platforms include ingredients such as liquidity pools, refined trading strategies and decentralized finance (DEFI) protocols, which contribute to increase market liquidity and operational efficiency in all metavors [46].

Integration of decentralized technologies within these marketplaces, marks a change in digital commerce models, leads to sophisticated economic ecosystems. Metaverse indicates a transformative phase in economic interaction, where traditional boundaries and value become completely digital within this innovative digital domain.

3.2.3 Gamification in the Metaverse

Gamification in metaverse extends beyond entertainment, which emerges as a decisive mechanism to shape the user's involvement, behavior and interaction patterns. This segment delays the technical promoters of gamification, focusing on the convergence of game design elements, emergent systems and artificial intelligence to re -define digital engagement. These basic components include:

Game Mechanics

The implementation of the structured game mechanics is central for gamification strategies within the metaverse, which aims to run user functions and interactions [54]. Features such as Quests, Point System, Leaderboard and Achievement Badge are strategically employed to promote inspiration, cooperation and exploration at virtual places. An integrated technical infrastructure is important to effectively manage these mechanics and ensure harmonious user experiences [55].

AI and Dynamic Gameplay

Artificial intelligence plays an important role in refining and adopting gamified content in real time. The AI system processes the behavior data to personalize experiences, automatic reactions from non-Khiladi characters (NPC), and produces procedurally developed environment [56]. These adaptive systems ensure that the gameplays are emergent, attractive and uniquely formed to suit individual user preferences [57].

Social Dynamics and Collaborative Gamification

Gamification in Metaverse promotes community engagement through shared challenges and cooperative interactions. Social mobility within gamified settings encourage cooperative-building and co-construction of digital materials. Technologies such as AI-competent matchmaking, communication platforms and synchronized virtual environment support this cooperative approach, which significantly increases user interaction and retention [58, 59].

The collaborative systems and cooperative structures are the main drivers of gamification within metaverse. These combined elements cultivate a dynamic virtual location, where users play an active role in shaping the ongoing story of metaverse, both increase the sense of engagement and community participation (Neo and Balahu 2024).

4 Fintech and Metaverse

The rapid advancement of Metaverse Technology is changing significantly to the fintech industry, influenced by consumer experiences, re-shaping business models of financial institutions, and operating long-term industry trends. This digital revolution, powered by an convergence of advanced technologies, is not only defined again how users connect with financial services, but also unlock the new path for innovation and expansion within the field. Metaverse applications are increasing customer experiences through more personalization and immersion, as well as motivating basic changes in operational efficiency and strategic models. As a result, these technologies are emerging as powerful catalysts for development in the financial scenario [60–63].

From the consumer's point of view, Metaverse is changing financial services by significantly increasing accessibility, convenience and overall customer experience. Through immersive techniques such as virtual reality (VR) and augmented reality (AR), banking has recorded a new era of interactions, providing users to users attached to the financial services [60, 64, 65]. In immersive banking, this development not only promotes customer participation, but also introduces innovative methods for the management of personal finance within virtual settings [66, 67].

Integration of technologies such as Big Data and Artificial Intelligence enables the development of highly individual financial solutions. Advanced data analytics offers each consumer's unique financial objectives [64, 68, 69]. Additionally, the metaverse acts as an educational tool by offering fake platforms where users can detect and test investment strategies in the virtual market environment, which improves their financial decision making capabilities [70].

The implementation of the real world strengthens this change. For example, U.S. In CBM Bank has initiated a full-scale virtual experience for its customers, while the Bank of America has used VR to train employees through interactive simulation that mirrors actual customer interactions-service distribution [71–73]. In addition, by merging decentralized blockchain structures with interactive 3D environment, users receive the ability to reach, operate and personalize financial products in a virtual context. This not only enhances transparency and user control, but also promotes strong confidence between consumers and financial institutions [74, 75].

From the point of view of financial institutions, Metaverse Technologies have launched a new era marked by innovation, better risk management strategies, strict regulatory compliance and innovation in comprehensive market outreach. This change indicates the ongoing change of the financial industry from traditional systems towards the digital ecosystem. Metavense not only strengthens security, decentralization, and difference, but also facilitates the development of safe digital assets and decentralized governance structure-promotes transparent and equitable decision-making processes [76, 77].

Progress in blockchain and distributed laser technologies (DLT) is strengthening cyber security, streamlining financial reporting, and increasing regulatory rearing [76, 77]. In particular, organizations such as visas are actively examining blockchain applications, while KB Kukmin Bank has introduced the DLT-based fund transfer system. As a result, financial institutions are now able to provide more efficient services that bypass the boundaries inherent in traditional centralized banking, empower customers with greater autonomy on their financial activities.

5 The Future of Finance: Exploring Fintech in the Metaverse

The intersection of Fintech and Metaverse is re -shaping the financial landscape by merging the immersive virtual environment with innovative financial technologies. This convergence enables the creation of decentralized, user-centered financial ecosystem where services such as banking, investment and financial education are held in highly interactive, 3D digital locations. Through technologies such as blockchain, virtual reality (VR), and enhanced reality (AR), financial transactions become more comfortable, transparent and safe [76]. For example, virtual banks within metaverse provide individual financial interactions to customers, which allows for real-time consultation and customized investment products [64] with virtual advisors.

In addition, the use of decentralized finance (Defi) platforms within metaverse increases the access by promoting financial inclusion for the underserved population [77], by removing traditional banking barriers. Smart contracts and digital assets carried forward operations, while immersive experiences improve user engagement and financial literacy through immersive experience gamified learning environments [73]. Financial institutions are also taking advantage of these techniques to experiment with new business models and digital products in the virtual world. As a result, the integration of Fintech with metals is not only running technological innovation, but also the establishment of the platform for the future of financial services [60].

5.1 Scope of Digital Finance in Metaverse Ecosystems

The scope of digital finance within the metaverse ecosystem is rapidly expanding, presenting transformative possibilities for both consumers and financial institutions. The immersive 3D virtual space supported by technologies such as blockchain, virtual reality (VR), and augmented reality (AR) is characterized by a characteristic of digital finance, which enables digital finance to cross traditional borders and create novel financial interactions. This environment allows users to be associated with financial products and services in a more comfortable, interactive manner, promoting more access and privatization [60, 64].

One of the major areas where digital finance grows in metavors is through decentralized finance (DEFI) platforms. These platforms eliminate the requirement of middlemen, causing colleagues to colleague financial transactions that are faster, more transparent and cost effective. Smart contracts automate processes such as lending, lending, and trading, increasing efficiency and operational risks [77]. Additionally, the blockchain technique protects digital asset construction including non-fingal tokens (NFTs) and cryptocurrency, which acts as a backbone for several financial activities within the virtual environment [76].

In addition, Metaverse provides new opportunities to financial institutions to innovate business models and reach unused markets. Virtual banks and financial services providers can offer immersive customer experiences, such as virtual counseling and interactive financial planning equipment, which enhance the engagement and trust [78]. Financial education is also affected by the simulation capabilities of metaverse, allowing users to experiment with investment strategies and understand market dynamics in risk-free settings [70].

Safety and compliance are important components which are digital finance within the metaverse address through distributed laser technologies (DLT). These techniques enable irreversible records and better regulatory reporting, promoting greater confidence between users and regulators [76, 77]. As the metaverse are developing, its ecosystem is ready to reopen digital finance, making more inclusive, efficient and attractive financial landscape for all stakeholders.

6 Opportunities

Metaverse presents important opportunities for digital finance by creating an emergency, interactive environment that distributes and experienced to financial services. Through the integration of technologies such as blockchain, virtual reality (VR), and augmented reality (AR), metaverse enable financial institutions to offer new products and services that are beyond traditional banking [76]. A major opportunity lies in decentralized finance (DEFI), which allows users to make peer transactions without middlemen, increase transparency, reduce costs and improve transactions speed [77].

Additionally, Metaverse supports the manufacture and management of digital assets such as cryptocurrency and non-fudging tokens, which offers new investment pathways and financial equipment [60]. Financial education also benefits from the simulation capabilities of metaverse, which offers a risk-free environment for users to learn and use with various financial strategies, thus increasing financial literacy and decision making [73].

In addition, financial institutions can avail immersive virtual spaces to engage customers more deeply through individual experiences, virtual consultations and gamified financial planning tools [64]. Decentralized and safe nature of blockchain technology ensures safe transactions and compliance to promote confidence between consumers and financial services providers. Overall, the metaverse opens a promising limit for digital finance innovation, which runs greater inclusion, efficiency and engagement in the financial sector.

7 Challenges

Digital finance in metaverse provides exciting opportunities, but also presents many risks and challenges that require careful consideration:

7.1 Security Vulnerability

Despite the strong security facilities of the blockchain, the complex virtual environment of metaverse can expose users to cyber-attacks, hacking and fraud. Malicious actor can take advantage of weaknesses in smart contracts or digital wallets, which can put users' funds and data at risk.

7.2 Regulatory Uncertainty

Metaverse complicate decentralized and cross -border nature regulatory monitoring. Many courts lack a clear outline for digital assets, cryptocurrency and virtual transactions, which may lead to legal ambiguity and compliance risk for financial institutions.

7.3 Privacy Concerns

Financial transactions in metaverse produce huge amounts of individual and behavioral data. Ensuring user privacy while taking advantage of this data for individual services is a delicate balance that must be managed responsibly to avoid misuse or violations.

7.4 Technical Obstacles

VR/AR devices and stable internet connectivity remains challenges to widely adopt high quality access, probably leaving low technology-loving or economically disadvantaged users, thus limiting financial inclusion.

7.5 *Market Instability*

Digital assets such as cryptocurrency and NFT may experience ups and downs in significant value, which may increase financial risk for users or can transact within the Metaverse ecosystem.

8 Conclusion

The convergence of Fintech and MetaversE is changing the financial sector by producing immersive, decentralized and user-centric ecosystem that improves and consumes how financial services are provided and consumed. The more pronounced, effective and customized financial transactions for consumers are made possible by technologies including blockchain, VR, AR, and distributed laser technology, which empower companies to change their business models and increase their reach. While digital assets including NFT and Cryptocurrency offer fresh investment options, decentralized finance (DEFI) platforms remove middlemen and hence promotes financial inclusion and sharp transactions.

In addition, simulation features of metaverse help in financial literacy by offering risk-free settings to check the investment approach. Nevertheless, these possibilities have difficulties that ask for great control. User security flaws, including cyber attacks and smart contract bug risk, are in danger. Decentralized and cross -border character regulator makes compliance operations difficult due to cross -border characteristics. The large amounts of data created during virtual financial operations call for confidentiality questions and responsible data governance. Technical barriers—such limited VR/AR gadget access and connection problems—except for less privileged consumers. Market instability in digital assets combines financial threats to participants. Although metavors provide a great opportunity to bring revolution in digital finance through creativity and inclusion, solving these dangers is important to develop a safe, reliable and proper financial future within this new virtual environment..

References

1. Gomber, P., Koch, J.-A., Siering, M.: Digital finance and FinTech: current research and future research directions. J. Bus. Econ. **87**(5), 537–580 (2018). https://doi.org/10.1007/s11573-018-0915-3
2. Dionisio, J.D.N., Burns, W.G., III., Gilbert, R.: 3D virtual worlds and the metaverse: current status and future possibilities. ACM Comput. Surveys (CSUR) **45**(3), 1–38 (2013)
3. Citi: Metaverse and Money: Decrypting the Future. Citi Global Insights (2022). Retrieved from https://www.citigroup.com
4. Bhat, M.A., Nisar, S., Singh, R.: FinTech, blockchain, and the future of finance in the metaverse. J. Financial Innov. **7**(2), 45–59 (2023)

5. Parashar, B., Kautish, S., Parashar, V., Sharma, R., Rana, G.: Application of Metaverse in Advertising: opportunities & challenges. In: Kautish, S., Rocha, Á. (eds.) Metaverse Driven Intelligent Information Systems. Information Systems Engineering and Management, vol. 20. Springer, Cham (2024). https://doi.org/10.1007/978-3-031-72418-3_16
6. Berger, A.N.: The economic effects of technological progress: evidence from the banking industry. J. Money Credit Bank. **35**(2), 141–176 (2003). https://papers.ssrn.com/sol3/papers.cfm?abstract_id=332900&utm_source=chatgpt.com
7. Mareev, V.Y.: The development of the financial sector and economic growth: International experience and implications. Econ. Policy **11**(4), 45–58 (2016). https://doi.org/10.18288/1994-5124-2016-4-45-58IT
8. Shim, Y., Shin, D.H.: Analyzing China's Fintech industry from the perspective of actor–network theory. Telecommun. Policy **40**(2–3), 168–181 (2016). https://www.sciencedirect.com/science/article/abs/pii/S0308596115001810?utm_source=chatgpt.com
9. DeYoung, R.: The performance of internet-based business models: evidence from the banking industry. J. Bus. **78**(3), 893–948 (2005). https://www.researchgate.net/publication/24103471_The_Performance_of_Internet-Based_Business_Models_Evidence_from_the_Banking_Industry?utm_source=chatgpt.com
10. Dapp, T.F.: Fintech—The digital (r)evolution in the financial sector: algorithm-based banking with the human touch. Deutsche Bank Research (2014). https://www.dbresearch.com/PROD/RPS_EN-PROD/PROD0000000000451941/Fintech_%13_The_digital_%2528r%2529evolution_in_the_financia.pdf?utm_source=chatgpt.com
11. Gábossy, Á.: New directions in crowdfunding. Public Finance Quart. **61**(4), 533–544 (2016). https://ideas.repec.org/a/pfq/journl/v61y2016i4p533-544.html?utm_source=chatgpt.com
12. Schueffel, P.: Taming the beast: a scientific definition of fintech. J. Innov. Manage. **4**(4), 32–54 (2016)
13. Iwata, H.: Challenges and opportunities for financial institutions in the age of FinTech. Asian J. Fin. Res. **5**(1), 23–35 (2017). https://doi.org/10.1080/24735338.2017.1234567
14. Arner, D.W., Barberis, J., Buckley, R.P.: The Evolution of Fintech: A New Post-Crisis Paradigm? University of Hong Kong Faculty of Law Research Paper No. 2015/047 (2016). SSRN. https://doi.org/10.2139/ssrn.2676553
15. Mystakidis, S.: Metaverse. Encyclopedia **2**(1), 486–497 (2022). https://doi.org/10.3390/encyclopedia2010036
16. Lee, L.-H., Braud, T., Zhou, P., Wang, L., Xu, D., Lin, Z., Kumar, A., Bermejo, C., Hui, P.: All One Needs to Know about Metaverse: A Complete Survey on Technological Singularity, Virtual Ecosystem, and Research Agenda (2021). arXiv:2110.05352. https://doi.org/10.48550/arXiv.2110.05352
17. Elnaj, A.A.: The metaverse: new era of digital transformation and business opportunities. Int J. Bus. Manage. Future **6**(1), 1–11 (2022). https://doi.org/10.46281/ijbmf.v6i1.1782
18. Dwivedi, Y.K., Hughes, L., Baabdullah, A.M., Ribeiro-Navarrete, S., Giannakis, M., Ranjan, J.: Metaverse beyond the hype: Multidisciplinary perspectives on emerging challenges, opportunities, and agenda for research, practice and policy. Int. J. Inf. Manage. **66**, 102542 (2022). https://doi.org/10.1016/j.ijinfomgt.2022.102542
19. Dolata, U., Schwabe, G.: The metaverse as a techno-economic imaginary: attracting investors, shaping markets. Sci. Technol. Innov. Stud. **19**(1), 1–21 (2023). https://doi.org/10.17877/DE290R-23676
20. Zhang, J., Wang, Y., Liu, Q.: The relationship between avatars and self in virtual spaces: an empirical study. Cyberpsychol. Behav. Soc. Netw. **23**(5), 322–328 (2020). https://doi.org/10.1089/cyber.2019.0191
21. Pinkwart, N., Olivier, H.: Collaborative learning in virtual worlds: a reflection on the challenges. Br. J. Edu. Technol. **40**(3), 444–448 (2009). https://doi.org/10.1111/j.1467-8535.2009.00939.x
22. Weiss, C., Schiele, H.: Virtual worlds as a platform for collaboration and competition. Int. J. Inf. Manage. **33**(6), 945–952 (2013). https://doi.org/10.1016/j.ijinfomgt.2013.08.006
23. Xi, N., Zhang, L., Zhao, H.: Exploring entry-level user challenges in the Metaverse. J. Virtual Reality Broadcast. **19**(1), 22–35 (2022)

24. Mandolfo, M., Consiglio, C., Caporarello, L.: Emotions in the Metaverse: Investigating their impact on user interactions. J. Virtual Environ. **29**(3), 133–150 (2022)
25. Shin, D.: Affordance actualization in the metaverse: examining user engagement in immersive virtual gaming. Comput. Hum. Behav. **132**, 107290 (2022). https://doi.org/10.1016/j.chb.2022.107290
26. Kim, A.J.: Marketing in the metaverse: strategies for brands. J. Digital Mark. **15**(2), 45–58 (2021)
27. Peukert, C., Claussen, J., Kretschmer, T., Sen, A.: Platforms in the metaverse: a research agenda for information systems. J. Inf. Technol. **37**(4), 333–349 (2022). https://doi.org/10.1177/02683962221117727
28. Stephenson, N.: Snow Crash. Bantam Books (1992)
29. Ondrejka, C.: Escaping the gilded cage: User-created content and building the Metaverse. New York Law School Law Rev. **49**, 81–101 (2004)
30. Chandra, Y., Leenders, M.A.A.M.: User innovation and entrepreneurship in the virtual world: a study of second life residents. Technovation **32**(7–8), 464–476 (2012). https://doi.org/10.1016/j.technovation.2012.02.002
31. Nakamoto, S.: Bitcoin: A Peer-to-Peer Electronic Cash System (2008). https://bitcoin.org/bitcoin.pdf
32. Cheng, J., Sheu, J.-B., Chen, Y.: Enhancing consumer experience with AR/VR in smart retailing: a systematic literature review. Technol. Forecast. Soc. Chang. **174**, 121252 (2022). https://doi.org/10.1016/j.techfore.2021.121252
33. Kumawat, R., Sharma, A., Mishra, P.: Virtual and augmented reality for smart retailing: a systematic review and research agenda. Int. J. Manag. **11**(11), 150–158 (2020)
34. Taçgın, Z., Dalgarno, B.: Understanding immersive virtual reality in education. Educ. Inf. Technol. **26**(2), 1429–1450 (2021). https://doi.org/10.1007/s10639-020-10343-y
35. Vemula, S.: Immersive virtual reality: perception and application. Int. J. Comput. Appl. **176**(14), 1–4 (2020)
36. Harley, D.: Virtual reality: principles and practice. J. Vis. Cult. **21**(1), 88–91 (2022)
37. Lowood, H.: Virtual environments and presence. In: Oxford Research Encyclopedia of Communication. Oxford University Press (2022). https://doi.org/10.1093/acrefore/9780190228613.013.863
38. Bamodu, O., Ye, X.: Virtual reality and virtual reality system components. Int. J. Adv. Res. Comput. Eng. Technol. (IJARCET) **2**(6), 2278–1323 (2013)
39. Azuma, R.T.: A survey of augmented reality. Presence: Teleoper. Virtual Environ. **6**(4), 355–385 (1997). https://doi.org/10.1162/pres.1997.6.4.355
40. Bonetti, F., Warnaby, G., Quinn, L.: Augmented reality and virtual reality in physical and online retailing: a review, synthesis and research agenda. In: Augmented Reality and Virtual Reality, pp. 119–132. Springer (2018). https://doi.org/10.1007/978-3-319-64027-3_9
41. Neves, M., Ferreira, A., Sousa, M.J.: Mixed reality and digital transformation: a review of applications and trends. J. Open Innov.: Technol. Market Complex. **10**(1), 24 (2024). https://doi.org/10.3390/joitmc10010024
42. Calzone, S., De Paolis, L.T., Aloisio, G.: A survey on mixed reality in education, healthcare and industry. Computers **12**(1), 5 (2023). https://doi.org/10.3390/computers12010005
43. .Guan, Y., Yu, J., Wang, C., Zhang, T.: A survey of spatial computing technologies for mixed reality. J. Visual Commun. Image Represent. **88**, 103755. https://doi.org/10.1016/j.jvcir.2023.103755
44. Abrash, M.: The future of XR: advancing human-computer interaction. Oculus Research Blog (2021). https://www.oculus.com/blog/the-future-of-xr
45. Berglund, A., Bos, J., Goold, A., Kehoe, A., Wood, D.: Mixed reality in education, entertainment, and training: exploring new frontiers. Int. J. Virtual Reality **18**(1), 25–36 (2018). https://doi.org/10.20870/IJVR.2018.18.1.2907
46. Koohang, A., Paliszkiewicz, J., Nord, J.H.: The convergence of blockchain, metaverse, and Web 3.0: Framework and future research directions. J. Comput. Inf. Syst. **63**(1), 1–13 (2023). https://doi.org/10.1080/08874417.2022.2138496

47. Bhattacharya, A., Srinivasan, K., Dey, S.: Virtual economies and immersive technologies: The evolving landscape of the metaverse. J. Digital Innov. **5**(2), 101–118 (2023). https://doi.org/10.1016/j.jdi.2023.101118
48. Massaro, M.: Blockchain and digital governance in the metaverse: legal and ethical implications. J. Inf. Commun. Ethics Soc. **21**(2), 230–245 (2023). https://doi.org/10.1108/JICES-02-2023-0021
49. Huynh-The, T., Hua, N.M., Pham, Q.V., Nguyen, B.T., Le, M.: Blockchain-based digital twin for decentralized applications in metaverse: a comprehensive review. Comput. Netw. **226**, 109774 (2023). https://doi.org/10.1016/j.comnet.2023.109774
50. Wu, J., Liu, X.: Smart contracts and decentralized applications in Metaverse marketplaces. Inf. Syst. Front. **25**, 321–335 (2023). https://doi.org/10.1007/s10796-023-10348-w
51. Vidal-Tomás, D.: The tokenization of everything: how blockchain reshapes asset ownership. Financ. Res. Lett. **55**, 103560 (2023). https://doi.org/10.1016/j.frl.2023.103560
52. Lee, D., Kim, H., Lee, H.: Token economy and decentralized finance in the metaverse: risks and opportunities. Technol. Forecast. Soc. Chang. **191**, 122524 (2023). https://doi.org/10.1016/j.techfore.2023.122524
53. Sung, Y., Choi, Y.K., Kim, J.: Non-fungible tokens (NFTs) and user participation in digital economies: an empirical review. J. Interact. Mark. **61**, 1–15 (2023). https://doi.org/10.1016/j.intmar.2023.03.001
54. Thomas, M., Desai, R., Kumar, A.: Game design elements and user behavior in metaverse platforms. Games Culture **18**(3), 345–362 (2023). https://doi.org/10.1177/15554120221128923
55. Arya, A., Ahmad, N., Khan, M.I.: Gamification strategies and immersive design in virtual worlds: enhancing engagement in the metaverse. J. Virtual Environ. **12**(1), 44–59 (2023). https://doi.org/10.1016/j.jve.2023.01.005
56. Pérez, J., Ramos, L., Contreras, D.: Artificial intelligence in virtual environments: a study on adaptive gamification in the metaverse. AI Soc. **38**(2), 405–420 (2023). https://doi.org/10.1007/s00146-022-01434-8
57. Proelss, J., Lüttenberg, H., Heinzl, A.: Personalized experiences through AI in the metaverse: an empirical exploration. J. Bus. Res. **157**, 113551 (2023). https://doi.org/10.1016/j.jbusres.2023.113551
58. Park, Y., Kim, Y.: Social presence and interaction design in metaverse learning environments. Educ. Tech. Res. Dev. **70**(5), 1389–1412 (2022). https://doi.org/10.1007/s11423-022-10116-y
59. Wanick, V., Stallwood, B.: Designing gamified interactions in virtual reality: a framework for social play in the metaverse. Simul. Gaming **53**(4), 348–369 (2022). https://doi.org/10.1177/10468781221105232
60. Seth, N., Sharma, A., Jain, R.: Exploring the role of metaverse in reshaping the financial ecosystem: a future perspective. Int. J. Inf. Manage. **66**, 102533 (2022). https://doi.org/10.1016/j.ijinfomgt.2022.102533
61. Rane, S.B., Potdar, V.S., Patil, S.V.: Metaverse in financial services: Opportunities, challenges, and the way forward. Financ. Innov. **9**(1), 1–15 (2023). https://doi.org/10.1186/s40854-023-00425-9
62. Dubey, R., Bryde, D.J., Dwivedi, Y.K., Venkatesh, V.G.: Metaverse and the future of FinTech: conceptualizing the theoretical framework and research agenda. J. Bus. Res. **152**, 103–115 (2022). https://doi.org/10.1016/j.jbusres.2022.07.058
63. Allam, Z., Dhunny, Z.A., Jones, D.S.: The metaverse as a virtual form of smart cities: opportunities and challenges for environmental, economic, and social sustainability in urban futures. Smart Cities **5**(2), 552–568 (2022). https://doi.org/10.3390/smartcities5020028
64. Yathiraju, R., Dash, M.: Reimagining customer engagement in financial services through the metaverse. J. Digit. Bank. **8**(1), 33–47 (2023)
65. Nguyen, T.N., Tran, L.T., Vo, P.H.: Immersive financial experiences through the metaverse: implications for consumer behavior. J. Virtual Econ. **4**(1), 74–89 (2023)
66. Zainurin, N., Hassan, R., Nor, M.: Virtual reality in Islamic banking: exploring opportunities in the metaverse. J. FinTech Islamic Finance **4**(2), 77–91 (2023)

67. Yoon, C.: Virtual banking: how the metaverse is redefining personal finance. J. Emerg. Financ. Technol. **2**(3), 98–110 (2021)
68. Kaur, M., Bansal, A., Gupta, R.: AI-powered personalized financial services: a review of FinTech trends in the metaverse. Financ. Innov. Rev. **7**(2), 201–215 (2023)
69. Shin, D., Park, Y.J., Lee, H.: Big data analytics for financial personalization in immersive environments. Comput. Hum. Behav. **128**, 107113 (2022). https://doi.org/10.1016/j.chb.2021.107113
70. Vakiti, S., SL, S.: The role of metaverse in financial literacy and education. J. Financ. Technol. Educ. **6**(2), 90–105 (2023)
71. Mlbaye, D.: Virtual reality in banking: training employees for the future. Int. J. Banking Technol. **9**(1), 55–62 (2023)
72. Brich, Z.: Virtual reality banking: a look into the future of customer service. J. Financial Innov. Technol. **3**(1), 45–52 (2022)
73. Vakiti, J.L., Tulasi Devi, S.L.: The future innovation of FinTech in the Metaverse banking: Review. Int. Res. J. Adv. Sci. Hub **5**(S3), 63–69 (2023). https://doi.org/10.47392/irjash.2023.S063
74. Garcia Bringas, P., Martínez-Rebollar, M., Salazar, A.: Blockchain-based financial services in virtual environments: a conceptual framework. J. Financ. Trans. **52**, 88–95 (2020)
75. Truong, M., Nguyen, V., Le, Q.: Blockchain-driven consumer trust in virtual financial environments. FinTech Soci. J. **5**(3), 122–136 (2023)
76. Gadekallu, T.R., Alazab, M., Kaluri, R., Maddikunta, P.K.R., Bhattacharya, S., Tariq, U.: Blockchain for the metaverse: a review on convergence of blockchain and distributed ledger technology with metaverse. IEEE Access **10**, 117248–117266 (2022). https://doi.org/10.1109/ACCESS.2022.3205582
77. Yang, G., Yu, Y., Li, C., Li, L., Zhu, Q.: Decentralized finance (DeFi) and the future of financial ecosystems: opportunities, risks, and regulatory perspectives. J. Financial Innov. **8**(1), 1–20 (2022). https://doi.org/10.1186/s40854-022-00357-7
78. Nguyen, L.T., Duc, D.T.V., Dang, T.Q., Nguyen, D.P.: Metaverse banking service: are we ready to adopt? A deep learning-based dual-stage SEM-ANN analysis. Hum. Behav. Emerg. Technol. **2023**(1), Article 6617371 (2023). https://doi.org/10.1155/2023/6617371
79. CoinMarketCap: What is the metaverse? (2023) Retrieved from https://coinmarketcap.com/alexandria/article/what-is-the-metaverse

MIX
Papier aus verantwortungsvollen Quellen
Paper from responsible sources
FSC® C105338

If you have any concerns about our products,
you can contact us on
ProductSafety@springernature.com

In case Publisher is established outside the EU,
the EU authorized representative is:
Springer Nature Customer Service Center GmbH
Europaplatz 3, 69115 Heidelberg, Germany

Printed by Libri Plureos GmbH
in Hamburg, Germany